Fracture Mechanics of Cementitious Materials

Fracture Mechanics
of Cementitious Materials

BRIAN COTTERELL
Professor of Mechanical & Production Engineering
Nanyang Technological University
Singapore

and

YIU-WING MAI
Professor of Mechanical Engineering
University of Sydney
Australia

CRC Press
Taylor & Francis Group
Boca Raton London New York

CRC Press is an imprint of the
Taylor & Francis Group, an informa business

A CHAPMAN & HALL BOOK

CRC Press
Taylor & Francis Group
6000 Broken Sound Parkway NW, Suite 300
Boca Raton, FL 33487-2742

First issued in paperback 2019

ISBN-13: 978-0-7514-0036-6 (hbk)
ISBN-13: 978-0-367-86607-5 (pbk)

Typeset in 10/12pt Times by Academic & Technical Typesetting, Bristol

A catalogue record for this book is available from the British Library
Library of Congress Catalog Card Number: 95-81444

Visit the Taylor & Francis Web site at
http://www.taylorandfrancis.com

and the CRC Press Web site at
http://www.crcpress.com

Preface

Since the early works of Griffith (1921), Orowan (1948) and Irwin (1957) linear elastic fracture mechanics (LEFM) has been successfully used by many researchers and engineers not only to obtain methodologies for the safe design of engineering materials and structures but also to develop new advanced materials by identifying the critical parameters in the fracture energy–microstructure relationship. Many high strength metals and ideally brittle ceramics and polymers have been adequately characterised by a single linear elastic fracture mechanics parameter such as the critical potential energy release rate G_c or the critical stress intensity factor K_c. However, the application of classical LEFM to cementitious materials like pastes, mortars and concretes (Higgins and Bailey, 1976; Kaplan, 1961; Kesler *et al.*, 1972) has not been as successful and critical failure cannot be defined by G_c or K_c alone. As it is understood now this is because of the large localised tension-softening or damage zone at the notch tip. Cementitious materials are therefore 'quasi-brittle'. Provided the damage zone can be modelled appropriately LEFM can still be used to characterise the fracture process and predict the failure loads of structrues. However, this requires a fundamental understanding of the stress-displacement relationship of the tension-softening zone in cementitious materials.

Cementitious materials are often assumed to have negligible strength in tension. As will be shown in this book this assumption has prevented the efficient use of these materials for many years. Fibres are therefore usually added to improve the tensile properties and the capacity for impact energy absorption. These fibres can be long (continuous) or short (discontinuous) and may be asbestos, steel, carbon, glass, polypropylene, polyethylene, nylon, Kevlar and many other natural fibres like cellulose, cotton, sisal, jute, bamboo etc. In fibre reinforced cementitious materials there is a large fibre bridging zone (FBZ) in addition to the matrix fracture process zone (FPZ) intimately associated with the tension-softening characteristic of the cement matrix. Fracture mechanics description of the failure of fibre reinforced cementitious materials has to include both the FPZ and the FBZ. Naturally, the constitutive relationship of the FBZ will depend on the geometric dimensions of the fibres and the nature and physico-mechanical properties of the fibre–matrix interphase.

Whilst several books have been written on the application of fracture mechanics to polymers (Williams, 1984), ceramics (Lawn, 1993), metals (Knott, 1973) and materials in general (Atkins and Mai, 1985) there is no single book that is devoted to the fracture mechanics of *both* cementitious materials and their fibre reinforced composites. The necessity for a consistent fracture mechanics approach to the failure of cementitious materials and its implications to design codes have been quite succinctly covered in the two Reports (ACI 446.1R–91 and 446.2R–92) by the ACI Committee 446 on Fracture Mechanics. Over the last 15 to 20 years there have been many research papers and reports on the fracture mechanics and fracture mechanisms of cementitious materials; and we have made our own contributions. Many of these papers however, appear in different mechanics and materials journals and various conference proceedings. This makes it very difficult for the postgraduate students, the practising engineers and the beginners in the field to seek the required information. We believe the time is ripe for a systematic introduction to the application of fracture mechanics for the failure of cementitious materials including their fibre composites. However, in writing this book we are very much guided by our own research experience; and in the choice of subjects and presentation we seek a balance of theories, experiments and applications. The mechanics and materials aspects are very much emphasised. The book should be of interest, as a reference text, to professional engineers, research scientists and concrete technologists who have little knowledge of fracture mechanics and who want to enter this fast developing field. It will be of particular interest to civil, structural and materials engineering postgraduate research students and their supervisors in universities and institutions.

The book contains eight chapters. Chapter 1 presents the fundamentals of both linear elastic and non-linear fracture mechanics theories to the readers and their potential application and limitation to characterise the fracture propagation of cementitious materials. In Chapter 2 the development of the FPZ in the cementitious matrices due to the phenomenon of tensile-softening and the FBZ due to fibre bridging across the crack faces is introduced. The associated fracture mechanisms are also discussed and the role of the interphase between the paste and aggregates and that between the fibre and matrix material are explained. Particularly, the strengthening and toughening mechanisms for the new inorganic material—*macro-defect-free cement*—invented by ICI (UK) Ltd in the 1980s have been clearly identified as due to the presence of the polymer. Chapter 3 gives an account of the many experimental techniques appropriate for the measurement of the FPZ and FBZ. It is shown that the two most important parameters in these two zones are the fracture strength f_t and the fracture energy G_{If}. The precise stress–displacement relationships are unimportant if these two parameters are correct.

Theoretical models for fracture in cementitious materials are developed in Chapter 4. Such models must consider the effect of the FPZ and be able to reproduce the load–displacement curves as in experiments. These include the equivalent crack model, the crack band model and the fictitious crack model. It is suggested that the most convenient model is the fictitious crack model with the LEFM K-superposition principle. Size and notch effects as well as the 'energy brittleness number' are also discussed here within the framework of fracture mechanics. In Chapter 5 fracture mechanics models are presented for crack growth in fibre reinforced cementitious composites. It is not necessary to model the matrix FPZ provided the fracture toughness of the reinforced matrix is used. The most important fracture parameters are K_{Ic}, f_t, E (Young's modulus), and G_{If}. Once determined they enable the load-displacement and crack-resistance curves for any geometry and size of specimens and structures to be predicted. Because of the large FBZ the crack-resistance curve is both geometry and size dependent. A unique crack-resistance curve can be obtained only if the FBZ is fully developed. This means prohibitively large specimens or structures.

Statistics is seldom considered in the fracture mechanics of cementitious materials although it is fully recognised that they are quasi-brittle and not very homogeneous. In Chapter 6 statistical fracture mechanics theories of ideally brittle solids and two-phase materials are given. It will be shown that the Weibull distribution cannot predict the strength dependence on size because of the non-proportional scaling of the localised tension-softening zone. Fibre reinforcement reduces the scatter of the strength distribution and increases the structural reliability. Even low stiffness fibres can improve the tensile strength of cementitious matrices. Chapter 7 presents the strength characteristics of cementitious materials under time-dependent loading from the viewpoints of both the deterministic single crack theory and the probabilistic statistical multiple crack approach. Cracks in these materials are susceptible to environment-assisted slow crack growth under static, dynamic and cyclic stresses. However, the mechanical fatigue due to cyclic stresses is still an unresolved problem. Under certain circumstances it is demonstrated that the statistical time-dependent fracture theories developed for single phase materials can be empirically used for two-phase materials with reasonable success.

Finally, in Chapter 8, the application of fracture mechanics to the design of concrete structures is demonstrated with known case studies in the literature. This shows how far fracture mechanics has now been adopted by the concrete community. It is expected that over the next few years the practical design rules will be modified to take into account the research results on the fracture mechanics of cementitious materials and their fibre composites.

We began work on fibre cements in the late 1970s for James Hardie & Coy. Pty. Ltd. looking for replacement fibres for asbestos and we quickly realised that fracture mechanics was scarcely used then to characterise the failure of fibre reinforced cements. There were many opportunities for making significant contributions on the fracture mechanics of these quasi-brittle materials. So when this project was finished our fundamental fracture mechanics work for cementitious materials was continued with funding from the Australian Research Council. Many of the experimental results and theoretical models of the FPZ and FBZ contained in this book are taken from past students and colleagues who worked on these projects during those times. In this regard, we thank both James Hardie & Coy. Pty. Ltd. and the Australian Research Council for their financial support. To R. Andonian, B.G. Barakat, R.M.L. Foote, Y.C. Gao, M.I. Hakeem, X.Z. Hu, J.K. Kim, M.V. Swain and L.M. Zhou we acknowledge their many contributions. Also, T.J. Chuang and B.R. Lawn of NIST and K.Y. Lam of National Singapore University have made numerous other contributions which are incorporated in one form or another in this book.

The concept of writing this book first came from Professor Narayan Swamy of Sheffield University in the late 1980s and to him we express our sincerest gratitude. We must apologise that it has taken so long to see the book in print and the fault is all ours. Finally, we thank our wives, Maureen and Louisa, for the patience and understanding when the book was written over several years in Sydney, Singapore and Hong Kong.

Brian Cotterell
Yiu-Wing Mai

Singapore and Sydney
September 1995

Foreword

Linear elastic fracture mechanics has been developed since the early twenties of this century but essentially for brittle materials. This theoretical concept proved to be a powerful tool to predict failure of materials such as glass, ceramics or high strength metals in a realistic way and under various given loading conditions. No wonder that this approach is now widely used for many years to solve practical problems. New and sophisticated experimental techniques and theoretical concepts have been elaborated continuously and most standards in this field are based on fracture mechanics. A number of excellent textbooks are available.

Already the first attempts to apply linear fracture mechanics to describe failure of concrete and concrete structures have shown clearly that crack growth in this type of composite material is not well represented by linear elastic fracture mechanics. Even modifications introduced in order to overcome obvious discrepancies have had little or no success. Therefore design rules for concrete structures are most often still based on predominantly empirical rules rather than being developed systematically within a wider and generally accepted theoretical concept.

It took a considerable time before a rational basis for a non-linear fracture mechanics approach was formulated for application to concrete; a material with pronounced strain softening. For the time being we live in a split situation. On one side we are able to predict failure of concrete and concrete-like materials by means of advanced numerical methods in great detail and with astonishing precision. In addition relevant material parameters are available for many practical applications. On the other side traditional standards still use the unrealistic strength criterion with a whole series of necessary correction factors. The fact that there is no comprehensive textbook on the theoretical basis and potential applications of non-linear fracture mechanics is certainly one major reason for this obvious gap.

Even at university level the relevant concepts of non-linear fracture mechanics are hardly taught and research activities are still widely diverging. A closer look into the relevant technical literature tells us immediately that some research groups take a lead and further improve the underlying theoretical concepts as well as the experimental and numerical methods while others continue to apply simplifying approaches which had long ago proved to be of little or no real meaning. Sometime it is very difficult to understand this unfortunate situation which may be characterized by 'le dialogue des sourds'.

At last, a book on fracture mechanics of cementitious materials written by two widely acknowledged expers Yiu-Wing Mai and Brian Cotterell has appeared. This book is written in a rigorous way and access to the corresponding original literature is made easy by an extensive list of references. It can be used as a textbook for students and at the same time as an appropriate introduction for engineers already active in practice. After the publication of this volume there is no more excuse for engineers not being informed on most recent developments in fracture mechanics of concrete and its consequences for design of concrete structures as well as for the analysis of concrete failures.

It is hoped that in the future research will be streamlined and progress will be accelerated. Implementation of relevant results into codes and practice of civil engineering will now be facilitated. It is rare that a textbook is so definitely needed as in the present case.

Folker H. Wittmann
Zürich, December 1995

Contents

Preface   *v*
Foreword   *ix*
Nomenclature   *xv*

1   Fundamentals of fracture mechanics   **1**

   1.1   Introduction   1
   1.2   Griffith's theory of fracture   1
   1.3   The compliance method of calculating the elastic potential energy release rate   5
   1.4   Energetic stability   8
   1.5   Linear elastic fracture mechanics (LEFM)   9
          1.5.1   Weight functions of Bueckner and Rice   13
          1.5.2   Measurement of the fracture toughness   13
   1.6   Crack opening displacement (COD)   16
          1.6.1   The use of Castigliano's theorem to calculate the COD   19
          1.6.2   Measurement of the CTOD   20
   1.7   The T-stress and higher order stress terms   20
          1.7.1   Mixed-mode fracture and crack paths   21
   1.8   Crack growth resistance   29
   1.9   Non-linear fracture mechanics (NLFM)   32
          1.9.1   The J-integral   33
          1.9.2   Measurement of J_{Ic} 35
          1.9.3   Crack growth resistance J_R 37
  1.10   Summary   38

2   Fracture mechanisms in cementitious materials   **39**

   2.1   Introduction   39
   2.2   Cementitious materials   40
          2.2.1   Fracture process zone (FPZ) and strain-softening characteristics   41
   2.3   Cementitious fibre reinforced composites   44
          2.3.1   The crack tip fibre bridging zone (FBZ)   50
   2.4   Fracture mechanisms in cementitious materials and fibre composites   51
          2.4.1   The micromechanisms of failure in concretes and the new cement pastes   51
          2.4.2   The micro-fracture mechanisms in fibre reinforced cements   63
          2.4.3   The role of interfaces in controlling strength and toughness   66
   2.5   Summary   69

3   Fracture parameters for cementitious materials   **71**

   3.1   Introduction   71
   3.2   Experimental techniques for measurement of the FPZ and FBZ   71
          3.2.1   Measurement of fracture process zone in cementitious materials   71
          3.2.2   The multi-cutting technique to measure the FPZ in cementitious
                   materials   73
   3.3   Measurement of the mode I strain-softening relationship for cementitious
          materials   77
          3.3.1   The direct tension method   78
          3.3.2   The J-integral method   78

　　　　3.3.3   Indirect method using a notch bend specimen   79
　　　　3.3.4   Compliance methods   80
　　　　3.3.5   Comment on the methods of determining the stress-displacement
　　　　　　　　 relationship   84
　　3.4   Modelling the mode I strain-softening relationship for cementitious materials   84
　　3.5   Measurement of mode I fracture energy   87
　　3.6   Modelling the mixed mode strain-softening relationship for cementitious
　　　　　materials   93
　　　　3.6.1   The mixed mode stress-strain relationship   93
　　　　3.6.2   The mixed mode stress-displacement relationship   96
　　3.7   Experimental techniques for the measurement of the FPZ and FBZ in fibre
　　　　　reinforced cementitious materials   97
　　3.8   Measurement of the mode I strain-softening relationship for fibre reinforced
　　　　　cementitious materials   101
　　3.9   The stress-strain relationship for fibre reinforced cementitious materials   102
　　　　3.9.1   Uncracked composites   102
　　　　3.9.2   Matrix cracking stress and the critical volume fraction   105
　　　　3.9.3   Fibre pull-out   106
　　　　3.9.4   Stress-strain curve for Type I composites   110
　　　　3.9.5   Stress-strain curve for Type II composites   111
　　3.10 The toughness of fibre reinforced composites   113
　　3.11 Summary   114

4   Theoretical models for fracture in cementitious materials   116

　　4.1   Introduction   116
　　4.2   The flexure of strain-softening materials   117
　　4.3   Equivalent crack models   118
　　4.4   The crack band model   124
　　4.5   The fictitious crack model   128
　　　　4.5.1   The finite element method   128
　　　　4.5.2   The boundary element method   131
　　　　4.5.3   The K-superposition method   131
　　4.6   Size and notch effects   139
　　4.7   Asymmetric fracture   148
　　　　4.7.1   The smeared crack model for mixed mode fracture   150
　　　　4.7.2   The fictitious crack model for mixed mode fracture   151
　　4.8   Summary   152

5   Theoretical models for fracture in fibre reinforced cementitious
　　materials   153

　　5.1   Introduction   153
　　5.2   Engineers' theory of bending analysis of Type II composites   153
　　5.3   Fracture behaviour of short fibre Type II reinforced cementitious
　　　　　composites   160
　　5.4   Crack growth models for fibre reinforced Type II cementitious composites   160
　　　　5.4.1   The K-superposition method applied to fibre reinforced Type II
　　　　　　　　 composites using a critical stress intensity factor   162
　　　　5.4.2   Crack growth resistance curves   167
　　　　5.4.3   The K-superposition method applied to fibre reinforced Type II
　　　　　　　　 composites modelling the FPZ as a fictitious crack   173
　　5.5   Size effect and the R-curve   177
　　5.6   Summary   179

6 The statistical nature of fracture in cementitious materials   181

6.1	Introduction	181
6.2	The strength distribution for ideal brittle solids	181
	6.2.1 The fracture mechanics approach to strength distribution	184
	6.2.2 Strength distribution materials that exhibit R-curves	188
6.3	The statistics of heterogeneous brittle materials	189
	6.3.1 The fracture of two-phase brittle materials	190
	6.3.2 Computer simulation of fracture in concrete modelled as a two-phase material	200
6.4	Statistics and size effect in cementitious materials	206
6.5	The statistics of fibre reinforced cementitious materials	210
	6.5.1 The strength of bundles	211
	6.5.2 Type I continuous fibre reinforced cementitious materials	212
	6.5.3 Type II discontinuous fibre reinforced cementitious materials	216
6.6	Summary	223

7 Time-dependent fracture behaviour of cementitious materials   224

7.1	Introduction	224
	7.1.1 Slow crack growth in cementitious materials	224
7.2	Modelling time-dependent crack growth in brittle materials	226
	7.2.1 Conventional single crack theory	226
	7.2.2 Statistical theory of time-dependent fracture	229
	7.2.3 Comparison of single crack and statistical fracture theories	231
	7.2.4 Time-dependent creep strain	233
	7.2.5 Application to cementitious materials	234
	7.2.6 Statistical time-dependent fracture in two-phase materials	235
	7.2.7 Comparison of the time-dependent statistics of homogeneous and heterogeneous materials	239
7.3	Summary	248

8 Application of fracture mechanics to the design of structures   249

8.1	Introduction	249
8.2	Application to monolithic structures	251
	8.2.1 The analysis of cracks in dams	252
	8.2.2 Case study of a fracture in a concrete bridge plinth	257
	8.2.3 *In situ* measurement of fracture properties	260
8.3	Punching shear failure of slabs	265
8.4	Reinforcement bonding and anchorage to concrete	266
	8.4.1 Reinforcement bond slip and tensile stiffening	267
	8.4.2 Anchorage to concrete	273
8.5	Concrete pipes	279
8.6	Summary	280

References   282

Index   293

Nomenclature

a	Crack length, or total length of true plus fictitious crack lengths.
a_0	Reference crack size.
a_e	Equivalent crack length.
a_i	Initial crack length.
A	Crack area or fibre area.
b	Ligament length.
B	Thickness. Risk of rupture.
c	Pore size.
c_f	Same as Δa_e.
C	Compliance.
C_m	Machine compliance.
C_p	The compliance of a specimen after saw-cutting through part of the FPZ.
C^*	Unloading compliance containing an FPZ.
d_a	Aggregate size.
d_f	Diameter of fibre or length of the FBZ.
d_p	Length of the FPZ.
$D = \begin{bmatrix} D_{nn} & D_{nt} \\ D_{tn} & D_{tt} \end{bmatrix}$	Instantaneous moduli matrix.
E	Young's modulus.
E'	Young's modulus of damaged material in the FPZ.
$\begin{aligned} E^* &= E \\ &= E/(1-\nu^2) \end{aligned}$	Young's modulus for plane stress for plane strain.
E_c	Young's modulus of composite.
E_f	Young's modulus of fibre.
E_m	Young's modulus of matrix.
E_t	Tangent softening modulus.
$f(a)$	Flaw size distribution function.
f_c	Compressive strength.
f_t	Tensile strength.
F, F_f	Load carried by fibre bundle. Fibre bridging force.
F_f	Fibre force.
g_{If}	Specific fracture energy.
G	Shear modulus. Crack extension force.
G_m	Shear modulus of matrix.

G_I, G_{II}, G_{III}	Rate of release of potential energy, or crack extension force, for the three modes of fracture.
G_{IIb}	The interfacial fracture energy between the fibre and matrix.
G_{If}, G_{IIf}	Mode I and II fracture energies.
G_i	Initiation fracture energy.
G_{Im}	The mode I fracture energy of the matrix.
$G_R(\Delta a)$	Crack growth resistance in terms of work of fracture.
G_u	Upperbound crack extension force.
h	Width of FPZ or depth of beam.
h_c, h_t	Depths of tensile and compressive zones.
h_y	Distance of first cracking strain from neutral axis.
H	Beam height.
J, J_R	Line integral. Crack growth resistance in terms of J.
k_I	Stress intensity factor for unit force.
K_a	Applied stress intensity factor.
K_e	Effective stress intensity factor.
K_F	Stress intensity factor due to fictitious force.
K_I, K_{II}, K_{III}	Mode I, II, III stress intensity factors.
K_i	Initiation fracture toughness.
K_{If}	Fracture toughness of cementitious material. If there is crack growth resistance, the plateau value of the mode I fracture toughness.
K_{Ic}	The plane strain fracture toughness or fracture toughness of reinforced matrix.
K_{Ic}^*	Reference fracture toughness of reinforced matrix.
K_{Ic}^e	Equivalent stress intensity factor.
K_m	Stress intensity factor due to stress in FPZ.
K_p	Fracture toughness of particle.
K_r	Stress intensity factor due to stress in the FPZ.
K_{ref}	Reference stress intensity factor.
$K_R(\Delta a)$	The crack growth resistance in terms of fracture toughness.
K_t	Crack tip total stress intensity factor.
l	Total length of real plus fictitious crack. Fibre length.
l_c	Critical length for fibre pull-out.
$l_{ch} = EG_{If}/f_t^2$	Characteristic length of the first kind.
$l_{ch}^* = l_{ch}(f_t/E)^2$ $= G_{If}/E$	Characteristic length of the second kind.
L	Length of specimen.
L_m	Spacing of multiple cracks.
m	Weibull modulus.
$m(a, x)$	Weight function.

M_c, M_u	Critical and ultimate bending moment.
n	Power law exponent, or number of fibres.
N	Number of cycles to failure, or number of intact fibres in a bundle.
p	Hydrostatic stress.
P	Load. Probability of failure.
P_m	Maximum load.
q_0	Initial clamping stress at fibre-matrix interface.
q_{th}	The threshold interfacial pressure at which the interfacial contact pressure between a fibre and the matrix just falls to zero before debonding.
r_p	Distance from the tip of the equivalent crack to the edge of the FPZ.
R	Stress ratio or radius of curvature.
s_E	Stress brittleness number.
$S(\sigma)$	Probability of survival at stress σ.
S	Shear strength or loading span.
t	Thickness of FPZ or time.
t_f	Lifetime.
T_i	Traction vector.
T_n, T_s	Normal and shear tractions at crack faces.
v	Volume fraction of fibres.
v_c	Critical volume fraction of fibres. For volume fractions greater than v_c the fibres alone can withstand a higher load than the composite before the matrix cracks completely.
v_m	Volume fraction of matrix.
v_p	Volume fraction of pore or second phase particle.
V	Volume.
V_p	Volume of PFZ.
w	Half-width of FPZ.
w_f	Specific fracture work.
w_s	Work of fracture per unit thickness of plate.
w_p	Specific fracture work of fibre pull-out.
W	Specimen width or a characteristic specimen dimension. Strain energy density function.
$\bar{W}$	Non-dimension specimen size of the first kind.
$\bar{\bar{W}}$	Non-dimension specimen size of the second kind.
W_f	Work of fracture.
β	Shear retention factor, $(\pi/2)(\sigma/f_t)$, or an angle.
ϵ^{co}	Strain due to elasticity of the uncracked material.
ϵ^{cr}	Strain due to opening of microcracks.
ϵ_{nn}, ϵ_{tt}	The strain referred to local coordinates, n and t.

ϵ_f	The mode I strain at complete tensile failure.
ϵ_t	The strain at the onset of mode I strain-softening.
ϵ_u	The ultimate strain in the matrix.
$\delta(x)$	The COD at a general position on a crack face.
δ_f	The CTOD at the tip of the true crack at the initiation of real crack growth in a cementitious material or final fibre pull-out in a fibre reinforced composite.
δ_m	The CTOD at the tip of the true crack at the moment of the formation of a continuous *cmatrix* crack in a composite.
δ_{ref}	Reference COD.
δ_t	The CTOD at the tip of the true crack.
Δ	Deflection.
Δ_i	Indicated deflection.
Δ_m	Machine deflection.
Δ_r	Residual displacement.
γ	Shear strain.
γ_u	Ultimate shear strain.
γ_s	Specific surface energy.
Γ	Surface energy.
λ	Ratio of mode II to mode I stress intensity factor.
Λ	Strain energy.
$\eta, \eta_l, \eta_\theta$	A parameter for determining J from Λ. Efficiency factors for fibre reinforcement.
θ	Angle.
ν	Poisson's ratio.
μ	Frictional coefficient, or mean fibre bundle stress.
ν_m, ν_f	Poisson's ratio of the matrix and fibre respectively.
Π	Potential energy of system.
Π_l	Potential energy of external loads.
ρ	Flaw density.
ρ_c	Density of concrete.
ρ_f	Fibre density.
ρ_p	Particle density.
ρ_w	Density of water.
σ	Normal stress.
σ_0	Reference stress.
σ_c	Cyclic stress amplitude, or critical fracture stress.
σ_{cmc}	The composite stress at which the matrix cracks.
σ_f	Fibre stress in a cementitious composite.
σ_{fb}	Fibre stress at debonding.
σ_{fbi}	Initial fibre stress at debonding.
σ_{fc}	Fibre fracture stress.
σ_{fp}	Fibre pull-out stress.

σ_{nn}, σ_{nt}	Normal stress and shear stress referred to local coordinates n, t.
σ_m	Matrix stress in a cementitious composite.
σ_{mc}	Critical matrix cracking stress for a cementitious composite.
σ_N	Nominal fracture strength.
σ_Y	Yield strength of a metal.
τ	Shear stress.
τ_b	Bond strength between fibre and matrix.
τ_u	Ultimate shear stress.
τ_f	Interfacial frictional shear stress.

1 Fundamentals of fracture mechanics

1.1 Introduction

Much of the design of structures is about scaling. In conventional strength of materials, prior to the development of fracture mechanics, the maximum stress level alone was thought to determine fracture and there were various stress-based criteria of failure. In very many design cases, such as small ductile metal components, the strength of materials approach is sufficient and designs can be based on stress alone. The concept of stress enabled the formulation of rational scaling laws. Galileo (1638) devoted a considerable portion of his book, *Dialogues concerning two new sciences,* to a discussion on the strength of materials and had a very clear understanding of the concept of stress and gave an early discussion on scaling.[1] Galileo deduced that the maximum bending moments sustainable by geometrically similar beams are proportional to the cube of a characteristic dimension whereas the bending moment due to the self-weight of a beam is proportional to the fourth power of a characteristic dimension. This size effect places a limit on the size of structures, both man-made and natural, which makes it impossible to build 'ships, palaces, or temples of enormous size in such a way that all their oars, yards, beams, iron bolts ... will hold together; nor can nature produce trees of extraordinary size because their branches would break down under their own weight' (Galileo, 1638). Until Griffith (1921) wrote his famous classic paper, *The phenomenon of rupture and flow in solids,* the strength scaling laws remained basically the same as Galileo's.

1.2 Griffith's theory of fracture

Griffith (1921) showed that the tensile strength of aged glass fibres was not constant, as is predicted by conventional strength of materials, but that fine glass fibres are stronger than thicker ones. Griffith's theory accounted for the size effect he observed.

There are two basic aspects to Griffith's innovative approach to fracture. First he realized that the relatively low strength of bulk glass was due to the presence of crack-like flaws and secondly he saw that a static crack in

---

[1] The Alexandrian Greeks, in the third century BC devised a very much earlier scaling law. They gave the correct equation to scale the size of torsion catapults to give the same range (Marsden, 1969; Cotterell and Kamminga, 1992).

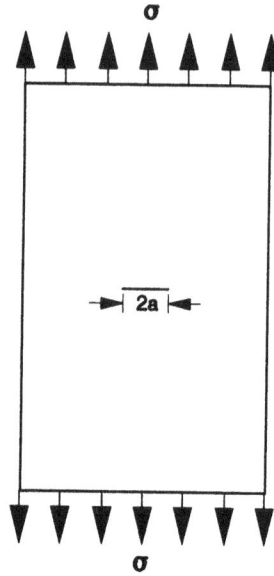

Figure 1.1 The classic Griffith crack.

an ideal elastic-brittle material could be modelled as a reversible thermo-dynamic system. The energies associated with a cracked ideal elastic-brittle solid, under static conditions, are the potential energy of the external loads, Π_1, the strain energy stored, Λ, and the surface energy, Γ. It can easily be shown that the potential energy of the system, $\Pi = \Pi_1 + \Lambda$, decreases as a crack extends. If this decrease in potential energy is equal to or greater than the necessary increase in surface energy $d\Gamma$, then the crack can grow. Griffith (1921, 1925) used the solution of Inglis (1913) to show that the rate of release of potential energy, or crack extension force as it is often called,[2] G_I, with respect to the crack area, $2aB$, in an infinite plate loaded by a normal stress, σ, is given by[3] (see Figure 1.1)

$$G_I = -\frac{1}{B}\frac{d\Pi}{d(2a)} = \frac{\sigma^2 \pi a}{E^*} \tag{1.1}$$

where B is the thickness of the plate. For plane stress, $E^* = E$, the Young's modulus, and for plane strain, $E^* = E/(1 - \nu^2)$, where ν is Poisson's ratio. If the specific surface energy is γ_s the necessary work to produce a unit area of fracture surface is $2\gamma_s B$. If the crack extension force is equal to or greater

---

[2] The unit of the rate of release of potential energy is J/m^2 which is the same as N/m. Hence G_I can be conceived as a force per unit length of the crack front in a similar fashion to how the work required to move a dislocation line of unit length through a unit distance is conceived as a force per unit length of a dislocation.
[3] In his original paper, Griffith (1921) made an error in proceeding to the limit which he corrected in his second paper (Griffith, 1925).

than this rate of fracture work, then the crack can propagate. So that the necessary condition gives the critical fracture stress, σ_c, as

$$\sigma_c = \sqrt{\frac{2\gamma_s E^*}{\pi a}} \qquad (1.2)$$

Most materials do not even approximate to an ideal elastic-brittle material and much more work than the surface energy is required to produce fracture. In general there will be a zone at the crack tip where the material is behaving non-elastically. Dissipative work has to be performed in this zone in addition to the surface energy. Orowan (1948) and Irwin (1948) saw that a plastic zone forms at the tip of a crack in high strength metals and, provided this zone is small, the plastic work required to create a unit area of fracture surface is a material constant that can be added to the surface energy. Since the surface energy is in most cases orders of magnitude less than the plastic work, it can be neglected. It has become normal practice to relate the plastic work to the fracture area of one surface, and if the specific fracture energy is G_{If} then the critical stress is given by

$$\sigma_c = \sqrt{\frac{E^* G_{If}}{\pi a}} \qquad (1.3)$$

In the case of high strength metals, the fracture energy associated with the plastic particle is a material constant. It is unnecessary to separate the continuum plastic work from the total work performed within the plastic zone, to obtain the essential non-continuum work of fracture, because each work component separately is a constant. The concept of including all the work performed within the plastic zone in the fracture energy has been generalized to other materials. In this generalization, the plastic zone is called the fracture process zone (FPZ) which can be defined as the smallest zone where the specific work performed, or fracture energy, is reasonably constant in a large specimen. In cementitious materials the FPZ is the strain-softened zone at the tip of the notch or crack where microcracking and debonding between the cement and the aggregate causes the stress to decrease with further straining (see Figure 1.2). However, whereas in high strength metals the fully developed plastic zone or FPZ can be much less than a millimetre in size, in cementitious materials the fully developed FPZ is usually large. Although the FPZ for cement paste is quite small, of the order of a millimetre (Higgins and Bailey, 1976), the FPZ in mortar is around 30 mm (Hu and Wittmann, 1989), and for concrete up to 500 mm (Chhuy et al., 1981). Outside the FPZ the cementitious material is essentially elastic, but the large FPZ makes the application of classic elastic fracture mechanics impossible in most laboratory size specimens.

In classic elastic fracture mechanics, the nominal fracture strength, σ_N, is, for a notched geometrically similar specimen, a function of the fracture energy, G_{If}, Young's modulus, E, and a representative length of the

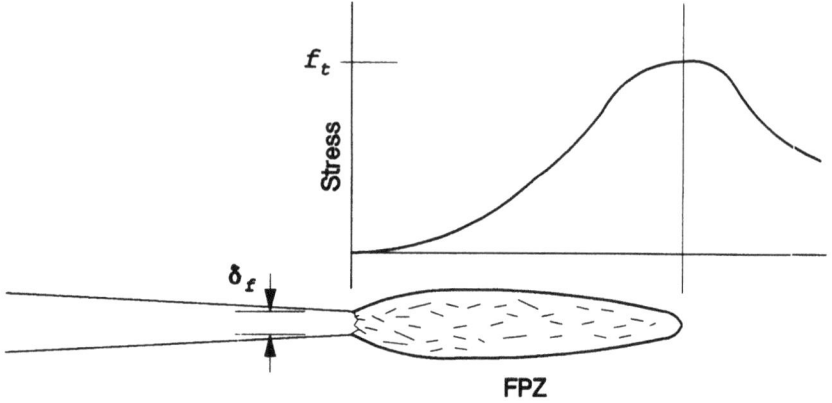

Figure 1.2 The fracture process zone and strain-softening.

specimen, W. Dimensional arguments show that the nominal stress is given by

$$\sigma_N \propto \left(\frac{G_{If} E}{W}\right)^{1/2} \qquad (1.4)$$

Hence, whereas the strength according to conventional strength of materials is constant, if there is a notch or crack in an essentially elastic material that is large compared with the size of the FPZ, then the strength scale is inversely proportional to the square root of a characteristic dimension. If the FPZ is not small then there is an extra parameter to be considered, the tensile strength f_t of the material which is the maximum stress in a strain-softening FPZ. Introducing this extra term into a dimensional argument it is found that the strength also depends upon a characteristic length, l_{ch}, defined by Hillerborg *et al.* (1976) as

$$l_{ch} = \frac{E G_{If}}{f_t^2} \qquad (1.5)$$

(a typical value of l_{ch} for concrete is 200 mm). The nominal fracture strength for a material with a finite FPZ is therefore, by dimensional arguments given by,

$$\sigma_N = F_1(\bar{W}) \left(\frac{G_{If} E}{W}\right)^{1/2} \qquad (1.6)$$

where $\bar{W} = W/l_{ch}$ is the non-dimensional size of the specimen, and $F_1(\bar{W})$ is a function of $\bar{W}$. Equation 1.6 is an appropriate form of scaling if $\bar{W} > 1$. If $\bar{W} \ll 1$, then the FPZ only partially develops before the nominal fracture strength is attained, and a more appropriate characteristic length is given by

$$l_{ch}^* = \frac{G_{If}}{E} = \left(\frac{f_t}{E}\right)^2 l_{ch} \qquad (1.7)$$

which we will call the characteristic length of the second kind to distinguish it from the more familiar characteristic length defined by Hillerborg *et al.* (1976). This second characteristic length is small and for concrete is of the order of $10^{-8} l_{ch}$. If $\bar{\bar{W}} \ll 1$ then a more appropriate scaling equation is based upon the non-dimensional length of the second kind $\bar{\bar{W}} = W/l_{ch}^{*}$ and is

$$\sigma_N = f_t F_2(\bar{\bar{W}}) \tag{1.8}$$

where $F_2(\bar{\bar{W}})$ is a function of $\bar{\bar{W}}$. As the non-dimensional size of the second kind of a specimen, $\bar{\bar{W}}$, tends to zero so the nominal strength will tend to a constant value that is dependent on the geometry. For small plain tensile specimens, the FPZ can spread quickly across the specimen before complete fracture. The fracture strength in this case is close to f_t, the maximum sustainable tensile strength. For plain bend specimens, the limiting nominal bending stress is much larger than f_t. The nominal strength of small specimens where $\bar{\bar{W}} \ll 1$ can be scaled according to conventional strength of materials. It is, of course, not necessary to use two different characteristic lengths, and the characteristic length of the second kind is only introduced here to emphasize the nature of the scaling of small specimens and throughout the rest of the book only the more usual definition of characteristic length, given by eqn 1.5, is used.

The size effect law (SEL), or relationship between the nominal fracture strength and the non-dimensional size, $\bar{W}$, of the specimen was first discussed for cementitious materials by Bažant (1984). The logarithmic variation of σ_N/f_t with $\bar{W}$ for a geometrically similar notched specimen is schematically illustrated in Figure 1.3a.

The SEL, as illustrated in Figure 1.3a, only applies to geometrically similar notched specimens or specimens where there is a high stress concentration. In very large specimens of this type, the FPZ is fully developed at final fracture. Smooth bend specimens, with no large stress concentration, do not develop a full FPZ before fracture except in the largest specimens. The SEL for smooth geometrically similar specimens is illustrated in Figure 1.3b. Only in the limit as the size tends to infinity, does fracture occur when the maximum stress in the FPZ reaches the tensile strength, f_t.

1.3 The compliance method of calculating the elastic potential energy release rate

This method enables the crack extension force, G, to be calculated from the compliance, whether obtained numerically, for example from a finite element solution, or experimentally. The compliance, C, of a cracked body increases with the crack size. In a perfectly elastic body there is no residual displacement on unloading (see Figure 1.4a) and under an equilibrium increase in fracture

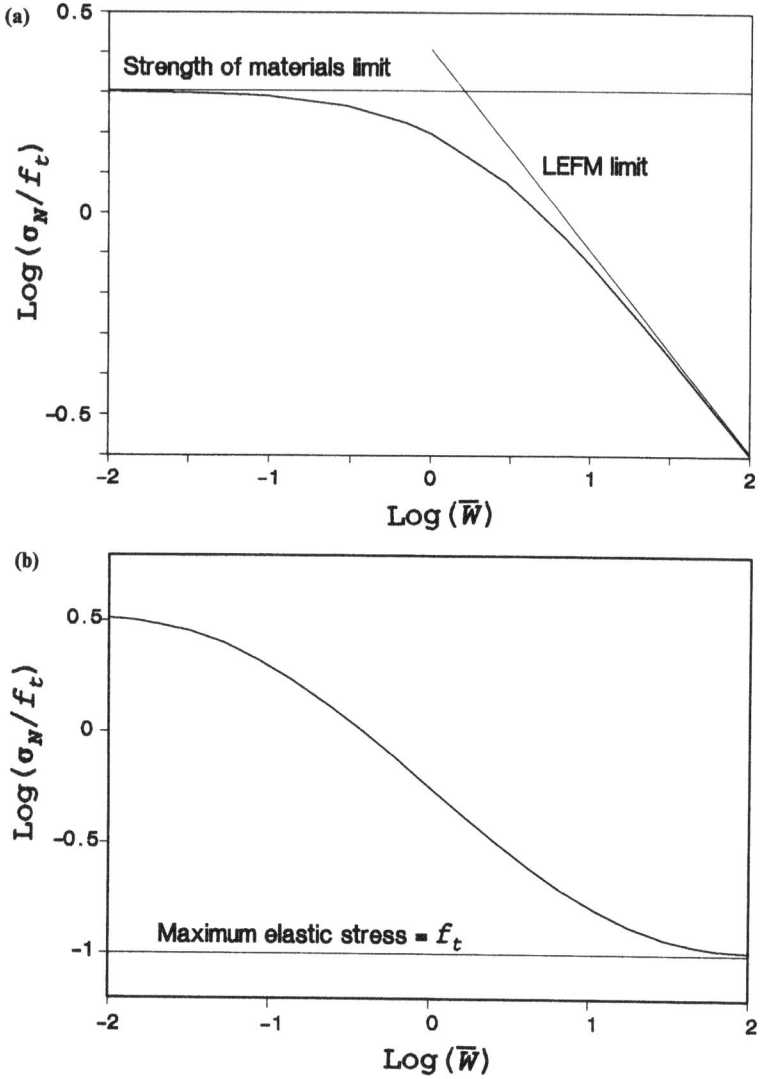

Figure 1.3 The size effect law: (a) geometrically similar notched specimens, (b) geometrically similar smooth specimens.

area, dA, the release in energy that is available for fracture is given by

$$\frac{d\Pi}{dA} = -\left[\frac{d\Pi_1}{dA} + \frac{d\Lambda}{dA}\right] \tag{1.9}$$

which is the area OAB. The strain energy stored, Λ, is given by

$$\Lambda = \frac{P\Delta}{2} = \tfrac{1}{2}P^2C \tag{1.10}$$

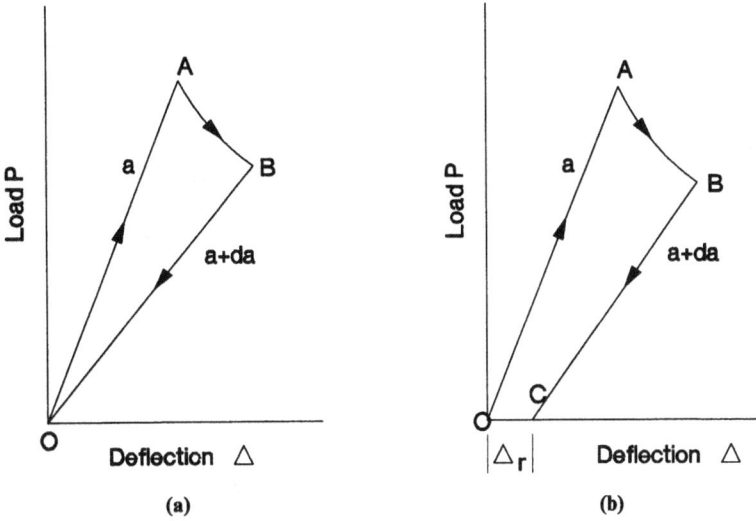

Figure 1.4 The compliance method: (a) ideal perfect elastic material, (b) real material.

and the rate of increase in strain energy with fracture area, dA, is given by

$$\frac{d\Lambda}{dA} = PC\frac{dP}{dA} + \frac{P^2}{2}\frac{dC}{dA} \tag{1.11}$$

At the same time the rate of change in potential energy of the load is

$$\frac{d\Pi_l}{dA} = -P\frac{d\Delta}{dA} = -\left[P^2\frac{dC}{dA} + PC\frac{dC}{dA}\right] \tag{1.12}$$

Hence the crack extension force, G, is given by

$$G = -\frac{d\Pi}{dA} = \frac{P^2}{2}\frac{dC}{dA} \tag{1.13}$$

which is independent of the load-deflection path.

In a real material there will be some non-elastic deformation at the crack tip, and on unloading some residual strain energy will remain locked in the specimen (see Figure 1.4b). The work done in extending the fracture area by dA will be less than the area OABC, but greater than the area OAB. Therefore, in the presence of residual displacements an upper bound to the crack extension force, G_u, is given by

$$G_u = \frac{P^2}{2}\frac{dC}{dA} + P\frac{d\Delta_r}{dA} \tag{1.14}$$

The lower bound to G is then given by eqn 1.13. Similar expressions for the crack extension force have been given by Wecharatana and Shah (1982) and Mai and Hakeem (1984).

In cementitious materials the behaviour is essentially elastic. The main contribution to residual displacements comes from two sources. One source is the residual dilatation of the FPZ, caused by microcracks failing to close perfectly and aggregate pull-out not being reversed. The other source is wedging open of the crack by debris. There may be some energy dissipated on unloading through friction between the aggregate and the matrix. However, for cementitious materials, crack extension force is probably better given by the lower than the upper bound.

1.4 Energetic stability

Fracture in elastic-brittle materials is often unstable and catastrophic, but inherent stability depends upon geometry (Atkins and Mai, 1985). Obviously all specimens are more likely to be stable under fixed grip than constant load conditions, because the compliance increases with crack growth. For stable fracture of a material whose fracture energy is constant, the crack extension force must decrease with crack growth. Thus from eqn 1.13 the condition for stable crack growth is given by

$$\frac{dG}{dA} = P\frac{dC}{dA} + \frac{P^2}{2}\frac{d^2C}{dA^2} < 0 \qquad (1.15)$$

Therefore for stable crack growth under constant load $(dP/dA = 0)$ $dC^2/dA^2 < 0$, but under fixed grips $(d\Delta/dA = d(PC)/dA = 0)$, only the less difficult condition, $dC^2/dA^2 < 2(dC/dA)^2/C$, has to be satisfied. In general tensile loads cause instability under both constant load and fixed grip conditions; bending loads (as in the double cantilever beam specimen) cause instability under constant load conditions, but the specimen may be stable under fixed grip conditions; specimens loaded in compression are usually stable under both fixed grips and constant load conditions.

Fixed grip conditions can never be achieved precisely, because there must be some stiffness associated with any loading arrangement. If C_m is the machine stiffness, then the deformation of the fracture specimen and machine can be schematically represented by a spring of compliance C_m and a specimen of compliance C stretched between two rigid abutments (see Figure 1.5). The indicated deflection, Δ_i, is given by

$$\Delta_i = \Delta + \Delta_m = P(C + C_m) \qquad (1.16)$$

where Δ_m is the machine displacement. The condition for fixed indicated deflection is $d[P(C + C_m)]/dA = 0$, and eqn 1.15 shows that stability is ensured if $dC^2/da^2 < 2(dC/da)^2/(C + C_m)$.

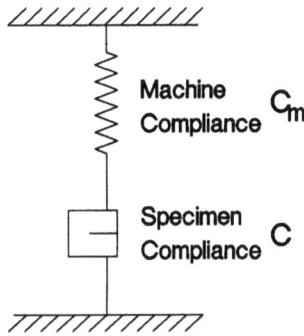

Figure 1.5 Equivalent load situation for a specimen loaded in a machine whose compliance is non-zero.

1.5   Linear elastic fracture mechanics (LEFM)

Although historically linear elastic fracture mechanics was first formulated in terms of energy, a stress approach has become more usual. Irwin (1957, 1958) and Williams (1957) realized that near a crack tip the stresses are inversely proportional to the square root of the distance from the crack tip. Under a pure opening mode, where near the crack tip the only displacements are normal to the crack surfaces, the only non-zero stresses near the crack tip (see Figure 1.6) can be written as

$$\left.\begin{array}{l} \sigma_x = \dfrac{K_{\mathrm{I}}}{\sqrt{2\pi r}}\cos(\theta/2)[1 - \sin(\theta/2)\sin(3\theta/2)] \\[2mm] \sigma_y = \dfrac{K_{\mathrm{I}}}{\sqrt{2\pi r}}\cos(\theta/2)[1 + \sin(\theta/2)\sin(3\theta/2)] \\[2mm] \tau_{xy} = \dfrac{K_{\mathrm{I}}}{\sqrt{2\pi r}}\sin(\theta/2)\cos(\theta/2)\cos(3\theta/2) \end{array}\right\} \qquad (1.17)$$

where K_{I}, the stress intensity factor, has the units of MPa$\sqrt{\mathrm{m}}$ and depends upon the geometry, applied stress, and crack length. For the classic Griffith

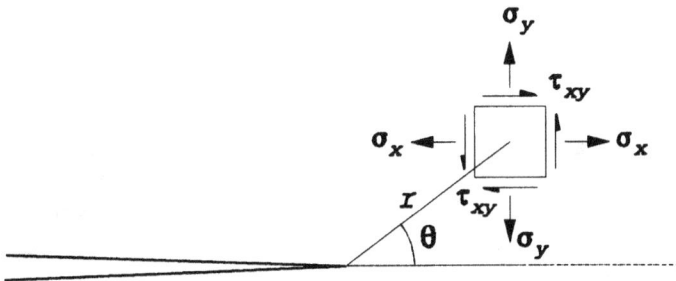

Figure 1.6 Stresses near a crack tip.

crack, the stress intensity factor is given by

$$K_I = \sigma\sqrt{\pi a} \tag{1.18}$$

For other geometries the stress intensity factor can be written as

$$K_I = \sigma_N \phi \sqrt{\pi a} \tag{1.19}$$

where σ_N is the nominal stress at the crack and ϕ is a geometric factor that depends upon the geometry of the specimen. The geometric factor, ϕ, is usually of the order of unity and can be found in stress intensity handbooks such as Tada *et al.* (1973) and Rooke and Cartwright (1976).

The pure crack opening mode (mode I), where the displacement is normal to the crack surfaces, is not the only possible opening mode. In general the displacement of the crack surface can be in any direction, and there are two other archetypal shearing modes: mode II where the only displacements near the crack tip are in the plane of the crack surface and normal to the crack front, and mode III where the only displacements are in the plane of the crack surface but parallel to the crack front. The three archetypal crack opening modes are schematically illustrated in Figure 1.7. The non-zero stresses near a crack tip under mode II and mode III opening are:

Mode II:

$$\left.\begin{aligned}
\sigma_x &= \frac{K_{II}}{\sqrt{2\pi r}}\sin(\theta/2)[2 + \cos(\theta/2)\cos(3\theta/2)] \\[2mm]
\sigma_y &= \frac{K_{II}}{\sqrt{2\pi r}}\sin(\theta/2)\cos(\theta/2)\cos(3\theta/2) \\[2mm]
\tau_{xy} &= \frac{K_{II}}{\sqrt{2\pi r}}\cos(\theta/2)[1 - \sin(\theta/2)\sin(3\theta/2)]
\end{aligned}\right\} \tag{1.20}$$

where K_{II} is the mode II stress intensity factor;

Mode III:

$$\left.\begin{aligned}
\tau_{xz} &= -\frac{K_{III}}{\sqrt{2\pi r}}\sin(\theta/2) \\[2mm]
\tau_{yz} &= \frac{K_{III}}{\sqrt{2\pi r}}\cos(\theta/2)
\end{aligned}\right\} \tag{1.21}$$

where K_{III} is the mode III stress intensity factor.

One of the attractions of the stress approach to LEFM is that it enables the principle of superposition to be used, provided the stress intensity factor is separated into its archetypal modes. An apparent problem with the stress approach is that the stresses right at the crack tip are infinite. Barenblatt (1959, 1962) discussed this problem and formulated the necessary hypotheses for the strict application of classic LEFM. Considering an ideal elastic-brittle

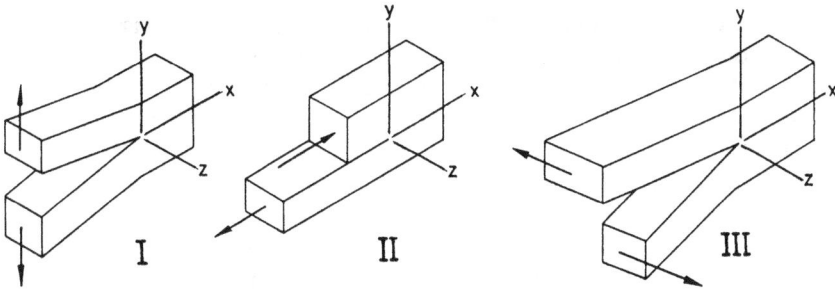

Figure 1.7 The three crack opening modes.

material, Barenblatt argued that very near the tip of a crack, where the separation of the crack faces is of the order of the equilibrium intermolecular distance, b, the forces of cohesion are important. These cohesive forces pull the two faces together in a similar fashion to the forces in the macroscopic FPZ shown in Figure 1.2. The cohesive forces cause a stress intensity factor K_r at the crack tip, which is negative, and adds to the stress intensity factor, K_a, due to the applied loads. Thus by the principle of superposition, the total stress intensity factor is given by

$$K_t = K_a + K_r \tag{1.22}$$

The infinite stress at the crack tip is removed if the total stress intensity factor is zero. In this case the stresses at the crack tip are continuous with the stresses in the FPZ and the faces meet in a cusp (see Figure 1.8). Barenblatt

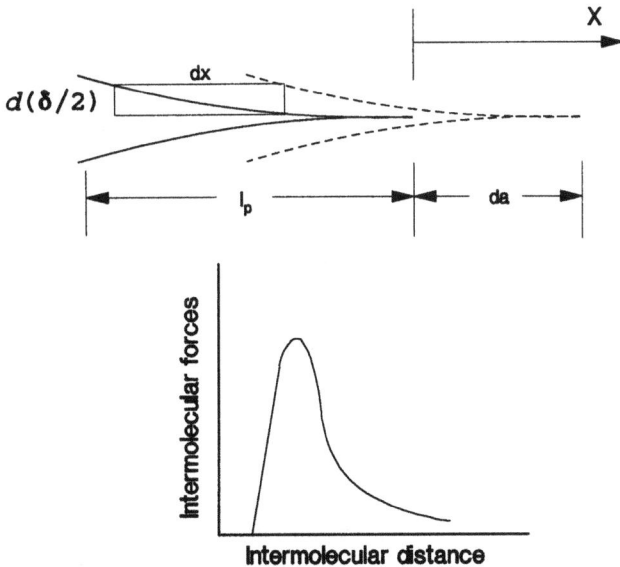

Figure 1.8 Crack in an ideal elastic-brittle material and the forces of cohesion.

(1959, 1962) showed that Griffith's theory of fracture and classic LEFM depends on two hypotheses: (i) the size of the FPZ is small compared with the size of the crack and other dimensions of the specimen; (ii) the shape of the crack surface in the FPZ (and consequently the cohesive forces) is the same for a given material during equilibrium crack growth.

The first of these hypotheses is well recognized but the second, which is very important for cementitious materials because they have large macroscopic FPZs, is often forgotten. During a crack extension of da, the work done against the forces of cohesion, dw_s, per unit width of the plate is given by

$$dw_s = \int_{-l_p}^{0} \sigma \frac{d\delta}{da} \, da \, dx \qquad (1.23)$$

where the integral is taken from a point where the forces of cohesion are negligible. From Barenblatt's second hypotheses

$$\frac{d\delta}{da} = -\frac{\partial \delta}{\partial x} \qquad (1.24)$$

and hence

$$\frac{dw_s}{da} = \int_{0}^{\infty} \sigma \, d\delta = 2\gamma_s = G_{If} \qquad (1.25)$$

Irwin (1957) showed that the mode I crack extension force, G_I, and stress intensity factor, K_I, are related by[4]

$$G_I = \frac{K_I^2}{E^*} \qquad (1.26)$$

Hence for equilibrium crack growth

$$G_{If} = \frac{K_a^2}{E^*} = \frac{K_r^2}{E^*} \qquad (1.27)$$

which formally shows that at equilibrium crack growth, the critical applied stress intensity, or plane strain fracture toughness, $K_{Ic} = K_a$, is a material constant and is related to the fracture energy, G_{If}, through eqn 1.26.

If Barenblatt's (1959, 1962) hypotheses are satisfied for a cementitious solid, then the fracture energy is given by eqn 1.25 as

$$G_{If} = \int_{0}^{\delta_f} \sigma \, d\delta \qquad (1.28)$$

---

[4] The mode I and mode II stress intensity factors are related to the respective crack extension forces by (Paris and Sih, 1965)

$$G_{II} = \frac{(1 - \nu^2)}{E} K_{II}^2$$

$$G_{III} = \frac{(1 + \nu)}{E} K_{III}^2$$

The crack extension forces for the three modes can be simply added to give the total energy release rate, because there is no interaction between the modes.

where δ_f is the crack tip opening displacement (CTOD) across the FPZ at the initiation of a continuous crack.

1.5.1 Weight functions of Bueckner and Rice

Very often in the analysis of the fracture behaviour of cementitious materials it is necessary to calculate the stress intensity factor due to stresses acting over the crack faces. In this case the concept of the weight function introduced by Bueckner (1970) and Rice (1972) can be used. If the stress on the crack face is $\sigma(x)$ then the stress intensity factor is given by

$$K_I = \frac{1}{\sqrt{W}} \int_0^a \sigma(x) m(a, x) \, dx \qquad (1.29)$$

where W is a characteristic dimension of the specimen and $m(a, x)$ is the weight function and

$$k_I = \frac{1}{\sqrt{W}} m(a, x) \qquad (1.30)$$

is the stress intensity factor for unit point forces, per unit thickness of the specimen, acting on both faces of the crack at position x. However, the weight function can be found by use of the reciprocity theorem from any known stress intensity factor for a particular geometry. The weight function method of calculating stress intensity factors is well described by Wu and Carlsson (1991) who show that eqn 1.29 is applicable to mixed boundary problems as well as static boundary problems. The general expression for the weight function is

$$m(a, x) = \frac{E^* \sqrt{W}}{2K_{ref}(a)} \frac{\partial \delta_{ref}(a, x)}{\partial a} \qquad (1.31)$$

where K_{ref} and δ_{ref} are the reference stress intensity factor and COD, respectively, whose geometry, including the boundary composition of the static and kinematic boundaries, is the same as the new load case.

1.5.2 Measurement of the fracture toughness

Provided the material is brittle enough the fracture toughness can be measured using any geometry for which the stress intensity factor is known. ASTM E399 (1990) describes the measurement of the plane strain fracture toughness, K_{Ic}, of metallic specimens. It will be seen that, for valid measurement of K_{If} in cementitious materials, specimens generally have to be very large and probably the most appropriate standard test geometry is the three-point-bend specimen (see Figure 1.9a). In large bend specimens tested with the notch downwards, either the bending moment at the notch due to the self-weight must be made zero by making the length of the

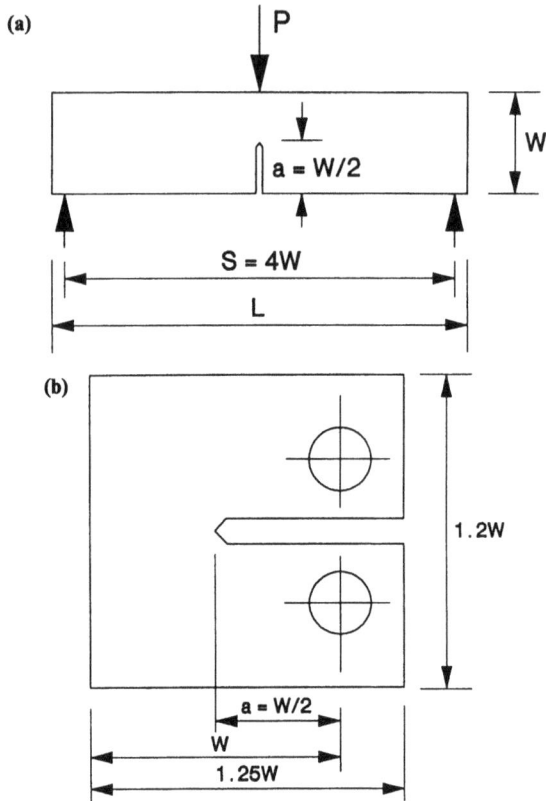

Figure 1.9 Fracture specimens: (a) three-point-bend, (b) compact tension specimen.

beam twice its span (or by providing counter-balancing weights) or allowance must be made for the stress intensity factor due to the self-weight. The other appropriate standard geometry is the compact tension specimen (see Figure 1.9b). Again large cementitious compact tension specimens must be counter-balanced to eliminate the effect of self-weight.

It is appropriate to record the load against the crack mouth opening displacement (CMOD) measured by a clip gauge, though in the case of the three-point-bend specimen an alternative would be to record the central displacement. The load-displacement curve can take on two forms (see Figure 1.10). It may be possible to calculate a valid K_{If} if the behaviour is brittle (Type I) using the procedure given in ASTM E399 (1990). For a cementious material it is appropriate to take the maximum load as an indication of crack initiation. However, it is impossible to calculate a valid K_{If} if the behaviour is ductile (Type II), because such a curve indicates that the FPZ is large.

For a valid measurement of fracture toughness the FPZ must be embedded in a K-stress field so that Barenblatt's (1959, 1962) first hypothesis applies.

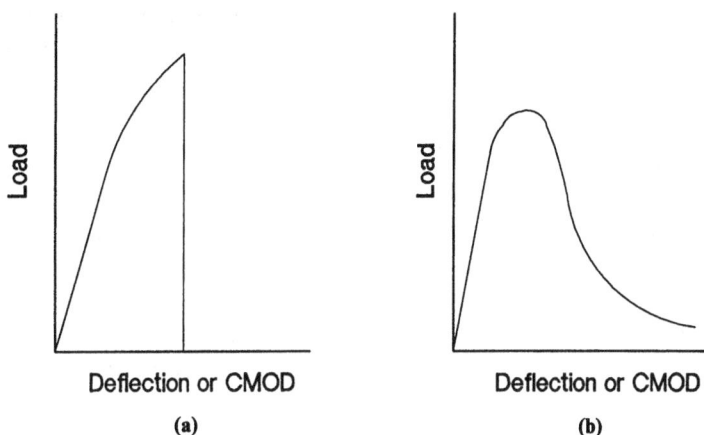

Figure 1.10 Load-deflection curves: (a) Type I—brittle behaviour, (b) Type II—ductile behaviour.

For common test geometries, the K-stress field is accurate to within 7% over a distance of 2% of the crack length ahead of the crack tip (Wilson, 1966). For metals the length, d_p, of the plastic zone or FPZ has been shown by Irwin (1960) to be approximately given by

$$d_p = \frac{1}{p\pi}\left(\frac{K_I}{\sigma_Y}\right)^2 = \frac{l_{ch}}{p\pi} \tag{1.32}$$

where $p = 1$ for plane stress and $p = 3$ for plane strain. The difference in the size of the plastic zone (FPZ) in metals, depending upon the stress state, is because plastic flow depends upon the degree of constraint which is low in thin metal sheets under essentially plane stress conditions and high in thick metal plates under essentially plane strain conditions. There are no constraint effects with cementitious materials and, if the FPZ is modelled as a region of constant stress (as Irwin (1960) did), the appropriate value in eqn 1.32 is $p = 1$ with the yield strength σ_Y replaced by the tensile strength of the cementitious material f_t. The requirement for a valid measurement of the plane strain fracture toughness, K_{Ic}, according to ASTM E399 (1990) is that the plastic zone is smaller than about 4% of the crack length which on substitution from eqn 1.32 gives

$$a > 2.5\left(\frac{K_{Ic}}{\sigma_Y}\right)^2 \tag{1.33}$$

Similar requirements necessary for the application of classic LEFM to cementitious materials are discussed in section 4.3.

In plane strain fracture toughness testing of metals, there is also the additional requirement that the thickness of the specimen shall also satisfy the inequality given in eqn 1.33. This requirement is to ensure that the fracture is essentially plane strain and the shear lips on the fracture edges are small. A limitation on thickness is not necessary for cementitious materials.

1.6 Crack opening displacement (COD)

Crack surfaces open under load and Wells (1961) suggested that in metals, fracture occurs when the crack tip opening displacement (CTOD) reaches a critical value. Such a criterion of fracture is attractive because it is independent of whether LEFM applies or not and has been tacitly assumed to apply to cementitious materials by many researchers. The concept is most readily applied in conjunction with the fictitious crack model of Hillerborg *et al.* (1976), which is discussed in detail in section 4.5. Although he did not call it a fictitious crack, the concept was first applied by Dugdale (1960) to the problem of a Griffith crack in an elasto-plastic solid. In thin plates the plastic zone is in the form of a thin flame-like zone that Dugdale assumed could be modelled as an infinitely narrow line zone extension, or fictitious crack extension, to the true crack across which a closing stress equal to the yield strength existed (see Figure 1.11). By these assumptions, Dugdale turned an elasto-plastic problem into a simpler elastic one. At the end of the plastic fictitious line crack, the elastic stresses must be continuous with the plastic ones. The problem is similar to Barenblatt's (1959, 1962) analysis of the

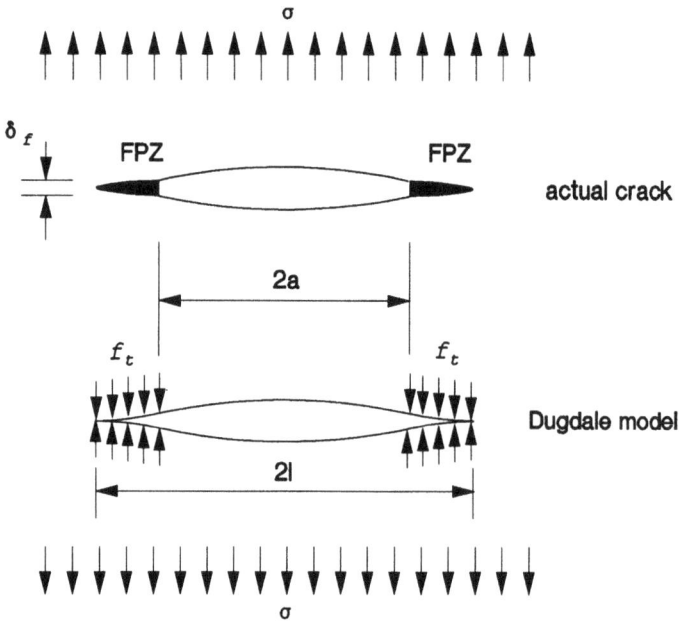

Figure 1.11 The Dugdale model.

cohesive forces at a crack tip. The stresses equal to the yield strength acting across the fictitious crack cause a stress intensity factor, K_r, that is negative. The applied stress, σ, causes a positive stress intensity factor, K_a, at the tip of the fictitious crack. For continuity in the stresses at the tip of the fictitious crack the total stress intensity factor $K_a + K_r$ must be zero as in eqn 1.22. The Dugdale model can be applied to cementitious materials if it is assumed that the stress within the FPZ is constant and equal to the tensile strength, f_t.

The stress intensity factor for a central crack loaded over its centre by a normal stress f_t is given by (Tada et al., 1973)

$$K_I = 2f_t \left(\frac{l}{\pi}\right)^{1/2} \sin^{-1}\left(\frac{a}{l}\right) \tag{1.34}$$

Combining this equation with that for a Griffith crack of length l closed by a stress $(f_t - \sigma)$ the stress distribution for the Dugdale model is obtained, and the total stress intensity factor is given by

$$K_I = -(f_t - \sigma)\sqrt{\pi l} + 2f_t \left(\frac{l}{\pi}\right)^{1/2} \sin^{-1}\left(\frac{a}{l}\right) = 0 \tag{1.35}$$

Thus

$$\frac{l}{a} = \sec \beta,$$

$$\tag{1.36}$$

where

$$\beta = \frac{\pi}{2}\left(\frac{\sigma}{f_t}\right)$$

Hence the extent of the FPZ, $d_p = l - a$, is given by

$$d_p = a[\sec \beta - 1] \tag{1.37}$$

If σ/f_t is small

$$\sec \beta \approx \frac{1}{(1 - \beta^2/2)} \tag{1.38}$$

and

$$\frac{d_p}{a} = \frac{\pi^2}{8}\left(\frac{\sigma}{f_t}\right)^2 \tag{1.39}$$

which is very similar to the estimate of Irwin (1960) for plane stress ($p = 1$) given in eqn 1.32.

The crack tip opening displacement, δ_t, can be calculated from the elastic field (Goodier and Field, 1963) and is

$$\delta_t = \left(\frac{8f_t a}{\pi E}\right) \ln(\sec \beta) \tag{1.40}$$

If δ_f is the critical crack opening displacement the specific fracture energy, G_{If}, is given by

$$G_{If} = \int_0^{\delta_f} \sigma \, d\delta = f_t \delta_f \qquad (1.41)$$

However, the specific fracture work is only equal to G_{If} if σ/f_t is very small because during crack growth, at constant applied stress, the size of the FPZ increases with crack length as given by eqn 1.37. In general the specific fracture work, w_f, is given by

$$w_f = -f_t \int_{x=a}^{x=l} \frac{\partial \delta(x,a)}{\partial a} \, dx \qquad (1.42)$$

Goodier and Field (1963) have evaluated eqn 1.42 and obtained

$$w_f = G_{If} \left[\frac{\beta \tan \beta}{\ln(\sec \beta)} - 1 \right] \qquad (1.43)$$

The normalized specific fracture work, w_f/G_{If}, is shown as a function of the normalized FPZ d_p/a, in Figure 1.12. Since the size of the FPZ increases with crack growth the specific fracture work, w_f, is always greater than the fracture energy G_{If} and the discrepancy increases with the size of the FPZ relative to the crack length.

This example of an approximate analysis of the fracture of cementitious materials using the Dugdale model demonstrates how the specific fracture work, w_f, is only identical to the fracture energy, G_{If}, if both of the

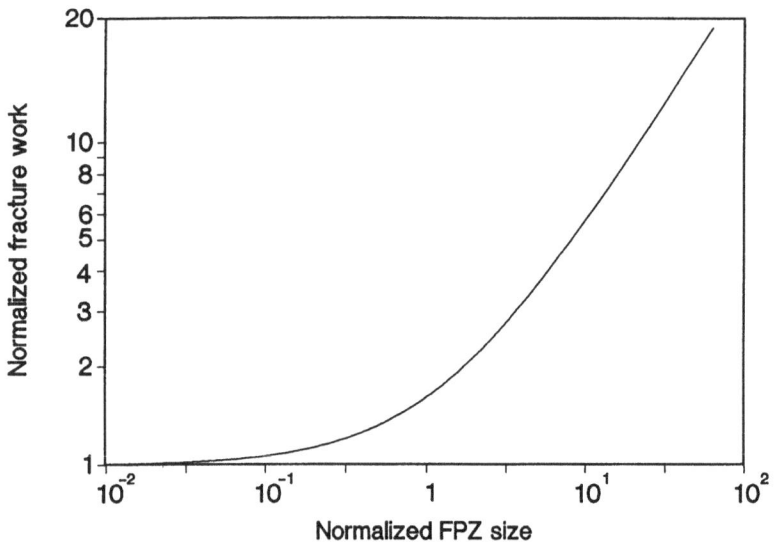

Figure 1.12 The specific fracture work as a function of the relative size of the FPZ.

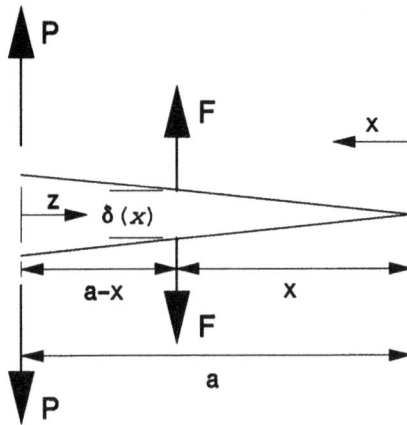

Figure 1.13 Calculation of the crack face displacements using Castigliano's theorem and a fictitious force.

hypotheses of Barenblatt (1959, 1962) are true. Thus, in general, G_{If} cannot be measured from the rate of fracture work.[5]

Fracture must occur when the CTOD reaches its critical value δ_f. In infinitely big specimens, such as presented in the Dugdale model, unstable fracture does initiate when the CTOD reaches its critical value, δ_f, but it will be seen in Chapter 4 that in finite specimens, if the FPZ is large, unstable fracture occurs before the critical value of the CTOD is reached.

1.6.1 The use of Castigliano's theorem to calculate the COD

If a more realistic stress-displacement relationship is used in the FPZ, instead of a constant stress, it is necessary to calculate the COD along the fictitious crack faces. The COD can be easily calculated from the stress intensity factor, which can be found in handbooks such as Tada *et al.* (1973) and Rooke and Cartwright (1976), by use of Castigliano's theorem (Tada *et al.*, 1973; Foote *et al.*, 1986a). Suppose the displacement $\delta(x)$ is required at a distance x from the crack tip due to the applied forces. Applying a fictitious force F at x (see Figure 1.13) the displacement is given by the application of Castigliano's theorem by

$$\delta(x) = \left[\frac{\partial \Lambda}{\partial F}\right]_{F=0} \tag{1.44}$$

---

[5] Although the rate of fracture work is not equal to the fracture energy, in a finite width specimen that fractures in a stable fashion the total work of complete fracture must equal the fracture energy multiplied by the fracture area if the stress-displacement relationship in the FPZ is unique for a particular material.

The strain energy stored, Λ, can be found from the crack extension force, G, and is given by

$$\Lambda = \int_0^A G\, dA + \Lambda_0 = \int_0^a \frac{K^2}{E^*} B\, da + \Lambda_0 \qquad (1.45)$$

where Λ_0 is the strain energy stored in the uncracked body and K is the sum of the stress intensity factor due to the applied forces, K_a, and the stress intensity factor, K_F, due to the fictitious force, F. Thus the displacement is given by

$$\delta(x) = \frac{2B}{E^*} \int_{(a-x)}^a K_a(z) \left[\frac{\partial K_F(z-a+x)}{\partial F} \right]_{F=0} dz \qquad (1.46)$$

A similar expression obtained if the applied force is a point force acting between x and the crack tip is

$$\delta = \frac{2B}{E^*} \int_{x-a}^x K_a(z-x+a) \left[\frac{\partial K_F(z)}{\partial F} \right]_{F=0} dz \qquad (1.47)$$

By integration the displacement due to a continuous distribution of stress $\sigma(x)$ along the crack faces can be found.

1.6.2 Measurement of the CTOD

For ductile metals the CTOD can be measured in deep notch bend specimens that yield completely before fracture from measurement of crack mouth opening displacements (CMOD), because the deformation is essentially the rotation of the halves of the specimen about a rigid hinge. Attempts have been made to measure the CTOD of fibre reinforced concrete by this method (Brandt, 1980). However, at best, this method can only give a rough approximation to the CTOD of cementitious materials, because the stress-strain relationship does not approximate to the rigid-plastic behaviour necessary for an accurate measurement of the CTOD.

1.7 The T-stress and higher order stress terms

The stress fields given by eqns 1.17, 1.20 and 1.21, give the singular part of the stress field at the tip of a crack. Williams (1957) showed that the stresses near a crack tip can be written in terms of the ascending power series

$$\sigma_{i,j} = \sum_{n=1}^{n=\infty} a_n f_{i,j}^n(\theta) r^{(n-2)/2} \qquad (1.48)$$

The first terms give the singular stress fields given by eqns 1.17, 1.20 and 1.21. These singular terms are the most important but the higher order terms, especially the second, also control some of the aspects of fracture (Cotterell,

1966). The most important of the higher order terms is the second, which under mode I opening gives what is called the T-stress. This stress field is simply $\sigma_x = T$, a constant, all the other in-plane stresses are zero. Under mode II crack opening, the second constant term is identically zero.

1.7.1 Mixed-mode fracture and crack paths

So far the discussion of fracture mechanics has been to mode I, though expressions for K and G for the other modes have been given. The direction for continuous crack growth in an ideal elastic-brittle material where one of the principal applied stresses is tensile, is such that the local stress field at the tip is mode I, which is often referred to as the criterion of local symmetry (Gol'dstein and Salganik, 1974). This result is consistent, in the sense that it follows by contradiction, with the various proposed mixed mode fracture criteria. However, the FPZ, if large, can have a significant effect on the crack and make it deviate from the ideal path. Rubinstein (1991) examined the experimental data on the propagation of a crack under tensile loading near to a hole in a polystyrene sheet (Chudnovsky *et al.*, 1987). Only near to the hole was the crack path affected (see Figure 1.14). The experimental

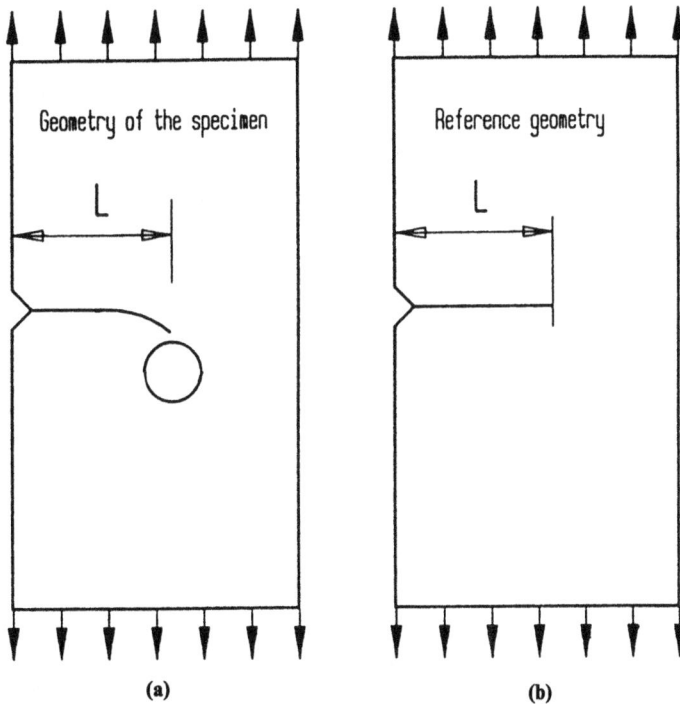

Geometry of the specimen

L

Reference geometry

L

(a)

(b)

Figure 1.14 Crack propagation near a hole. (a) Actual specimen. (b) Definition of the reference crack. (After Rubinstein, 1991.)

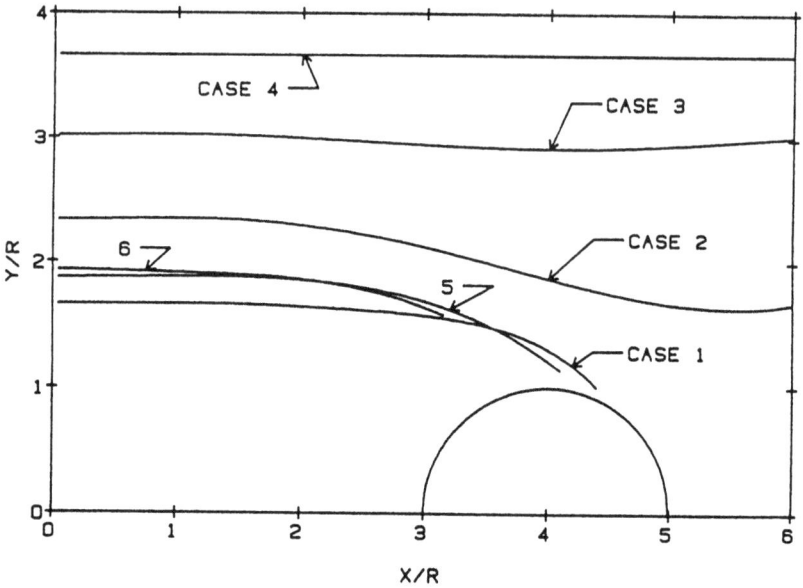

Figure 1.15 Experimental crack paths. (After Chudnovsky *et al.*, 1987.)

crack paths of cracks initiated near to a hole are shown in Figure 1.15. These crack paths are close to satisfying the criterion of local symmetry ($K_{II} = 0$) as Rubinstein (1991) showed by calculating the mode II stress intensity factor at the crack tip (see Figure 1.16). Although the maximum value of K_{II} is small,

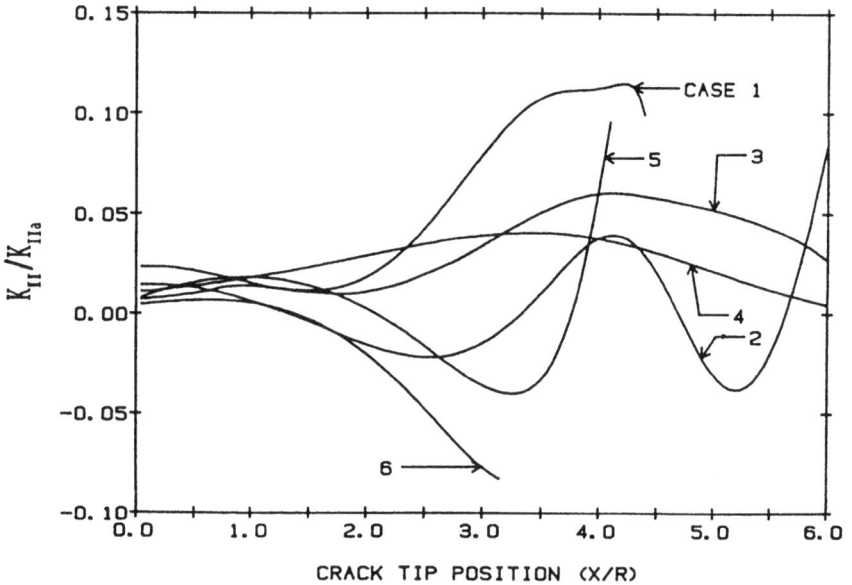

Figure 1.16 K_{II} variation along the crack paths near the hole. (From Rubinstein, 1991.)

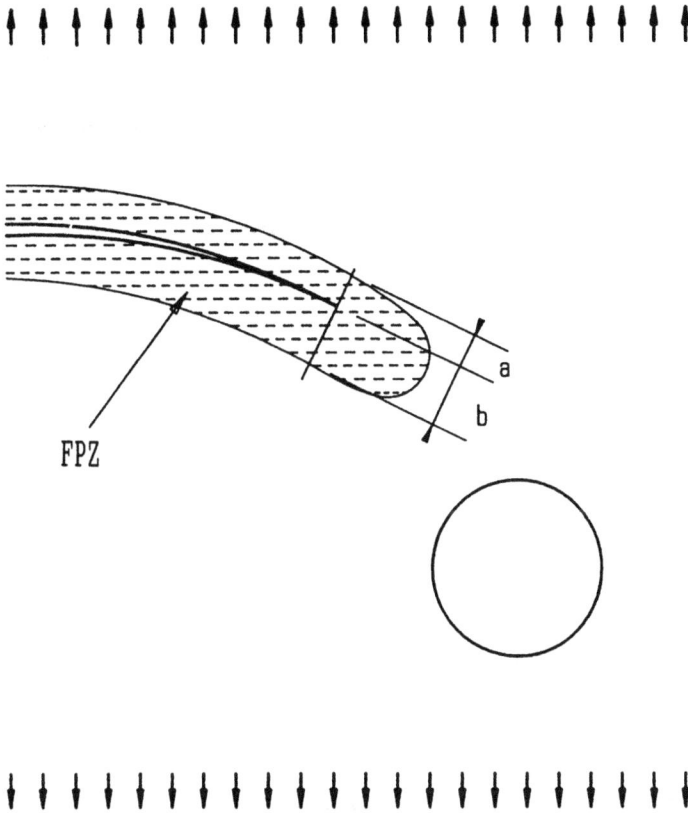

Figure 1.17 The FPZ surrounding the tip of a tightly turning crack. (After Rubinstein, 1991.)

only a little more than 10%, it is significant. The maximum deviation from the condition of local symmetry occurs in Case 1 which shows the greatest curvature. Rubinstein (1991) argues that the reason for the deviation in the crack path from the ideal is the FPZ surrounding the crack tip. While the crack path curvature is slight the FPZ is nearly symmetrical but, if the path turns sharply, the FPZ becomes asymmetrical and makes the path deviate from the ideal (see Figure 1.17). In polystyrene, the FPZ size is only of the order of 0.5 mm (Berry, 1964). In cementitious materials the FPZ is very much larger and even more significant deviations from ideal paths can be expected. One example of strong deviation from the ideal crack path is provided by the 'shear specimen' of Bažant and Pfeiffer (1986) which will be discussed further in Chapter 4. Melin (1989) has compared the crack path in a PMMA specimen, whose FPZ size is only of the order of 25 μm, with that obtained by Bažant and Pfeiffer in a concrete specimen (see Figure 1.18). Real crack paths can be expected to deviate from the ideal crack path if the radius of curvature of the ideal crack path is comparable with the size of the FPZ.

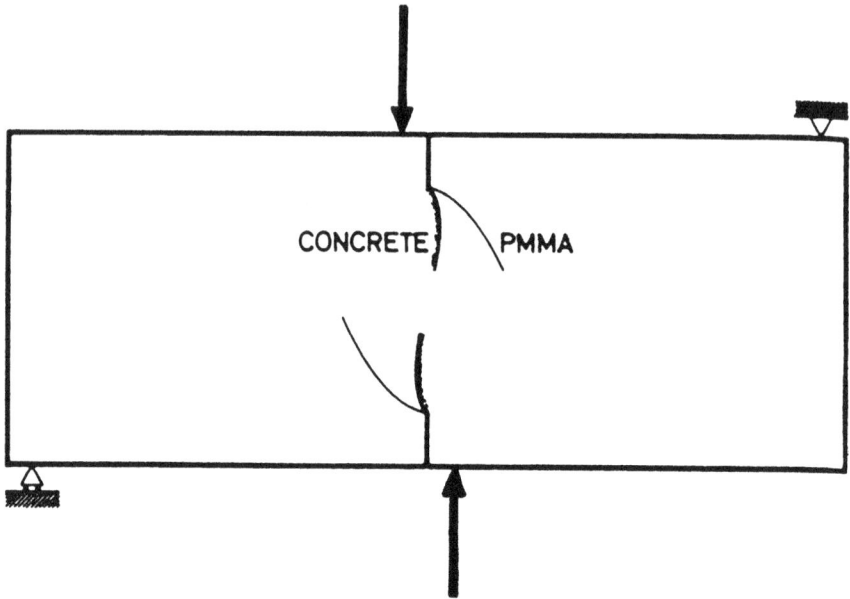

Figure 1.18 Comparison between the crack paths in PMMA and concrete specimens. (After Melin, 1989.)

While the criterion of local symmetry is accepted as the criterion for the crack path in an ideal elastic-brittle material, there is still controversy over the initial direction of crack growth from a crack tip under mixed mode loading conditions. This controversy almost certainly is due to the very small, but finite, FPZ of the brittle polymers that are used to model an ideal elastic-brittle material. The original criterion for crack direction is the maximum tangential stress (MTS) criterion proposed by Erdogan and Sih (1963). In this criterion crack growth is assumed to occur in a radial direction perpendicular to the direction of maximum tension. The stress intensity factor at the tip of a kink formed according to the MTS criterion has, to a first order, a pure mode I stress intensity factor (Cotterell and Rice, 1980). It was quickly realized that the FPZ could affect the crack direction and a better fit to the experimental data was obtained by taking, not just the singular stress field, but also the *T*-stress. In this modification of the MTS criterion the direction of crack propagation is taken as the maximum tangential stress at a critical distance, which can be identified with the tip of the FPZ (Williams and Ewing, 1972; Finnie and Saith, 1973; Ewing and Williams, 1974; Ewing *et al.*, 1976; Streit and Finnie, 1979). Wu (1978) assumed that the maximum tangential strain controlled the crack path. Palaniswamy and Knauss (1978) postulated that a crack would grow in the direction of maximum energy release. To obtain the direction of maximum energy release rate, Palaniswamy and Knauss (1978) calculated the energy

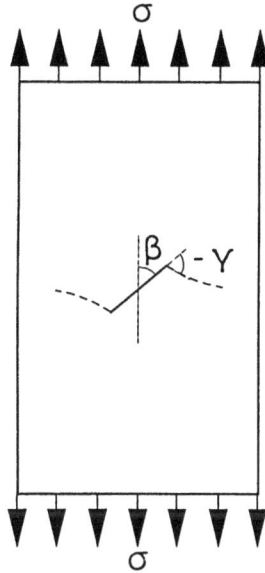

Figure 1.19 The CNT specimen used for crack path studies.

release rate for a kinked crack in terms of the length of the kink, and then let the length of the kink tend to zero. Sih (1973a) proposed that a crack develops in the radial direction of the minimum strain energy density. However, Sih (1973a) then goes on to state that propagation occurs when the strain energy density reaches a critical value which is self-contradicting. With the strain energy density criterion the crack extension direction is a function of Poisson's ratio. Finnie and Weiss (1974) conducted experiments on cross-rolled beryllium sheets that effectively had a Poisson's ratio of zero and found that the MTS criterion gave the better prediction than the strain energy density criterion.

The most common geometry used to examine the criteria of crack direction is the centre notched tension geometry (CNT) with the crack at an angle, β, to the applied load (see Figure 1.19). Using the mode mixity parameter, $\lambda = K_{II}/K_{I}$, the MTS criterion gives the crack direction as

$$\sin\gamma + (3\cos\gamma - 1)\lambda = 0 \tag{1.49}$$

where for the geometry shown in Figure 1.19, $\lambda = \cot\beta$. Mahajan and Ravi-Chandar (1989), in their excellent review of mixed-mode fracture, point out that there is considerable similarity between the various criteria if $\lambda < 0.37$ or $\beta > 15°$. The CNT specimen is not really suitable for studying larger mode mixities, because as $\beta \rightarrow 0°$, $\lambda \rightarrow \infty$, but both K_I and K_{II} tend to zero and hence the general stress level is high for $\beta < 15°$. A better geometry for studying crack directions at high mixities is the compact shear specimen

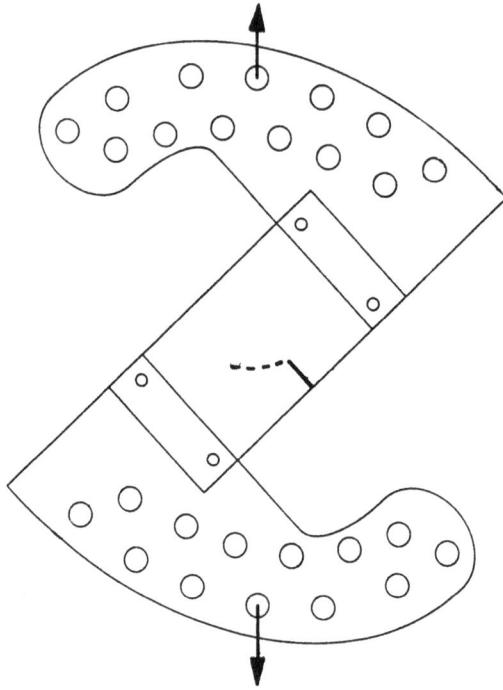

Figure 1.20 Schematic illustration of the CTS specimen. (After Mahajan and Ravi-Chandar, 1989.)

(CTS) shown in Figure 1.20. With this specimen the mode mixity needs to be measured directly at the crack tip by photoelasticity or the method of caustics. Experiments by Mahajan and Ravi-Chandar (1989) on PMMA and a birefringent brittle polymer, Hormalite-100, using the CTS geometry clearly showed that the crack direction was accurately predicted by the MTS criterion. However, none of the criteria gave a good prediction of the critical load because of strain rate effects and changes in fracture mode.

Mode II crack propagation in a brittle material is possible if both principal stresses are compressive (Melin, 1986; Broberg, 1987). However, even if both principal stresses are compressive a kink can occur with a tensile K_I providing the ratio between the principal stresses is large enough. Under a combined hydrostatic pressure, p, and a shear stress, τ, the direction of maximum shear stress is given by (see Figure 1.21)

$$\theta = 0 \qquad \text{if } p/\tau \leq 0$$

$$\theta = \tfrac{1}{2}\sin^{-1}(p/\tau) \quad \text{if } 0 < p/\tau \leq \mu/(1+\mu^2)^{1/2} \qquad (1.50)$$

$$\theta = \tfrac{1}{2}\tan^{-1}\mu \qquad \text{if } \mu/(1+\mu^2)^{1/2} < p/\tau < (1+\mu^2)^{1/2}/\mu$$

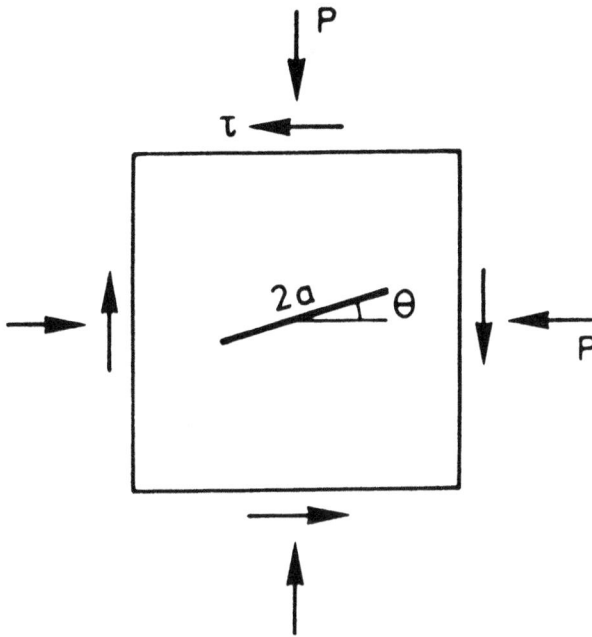

Figure 1.21 A crack under compressive principal stresses. (After Melin, 1986.)

where μ is the coefficient of friction between the closed crack surfaces (Melin, 1987). There are two options for crack propagation: the crack can propagate under mode II in the direction of the crack, or a kink can form with a tensile mode I stress intensity factor. The kink angle for maximum K_I occurs at about 70° but the maximum value is only weakly dependent on the angle. Melin (1986) has calculated the maximum mode II stress intensity factor at the tip of a favourably aligned crack, K_{IImax}, and the maximum mode I stress intensity factor at a kink making an angle $\alpha = 70°$. The ratio $\kappa = K_{IImax}/K_{Imax}$ is plotted in Figure 1.22 as a function of p/τ. If $p/\tau < 2.4$, K_I is tensile, which is quite close to the limit for all crack growth at $p/\tau = 2.69$ when friction prevents all motion between the crack faces. However, the ratio, κ, increases rapidly if $p/\tau > 2$. Broberg (1987), experimenting with cracks in PMMA, found that the crack could be made to propagate in mode II without kinking if $p/\tau > 2$, the experimentally determined coefficient of friction being 0.46. Thus pure mode II fracture is possible if the ratio, $\kappa = K_{IImax}/K_{Imax}$, is large enough. It must be noted that in these compression experiments, PMMA was quite ductile and that substantial plastic flow occurred. In concrete and rock, mode II type fractures are possible under compressive load even without any confining pressure because of the development of a large FPZ.

If a crack is under mixed mode I and III, the crack front cannot advance smoothly because it wants to rotate about an axis perpendicular to the crack front to achieve a mode I fracture. The crack front as a whole cannot

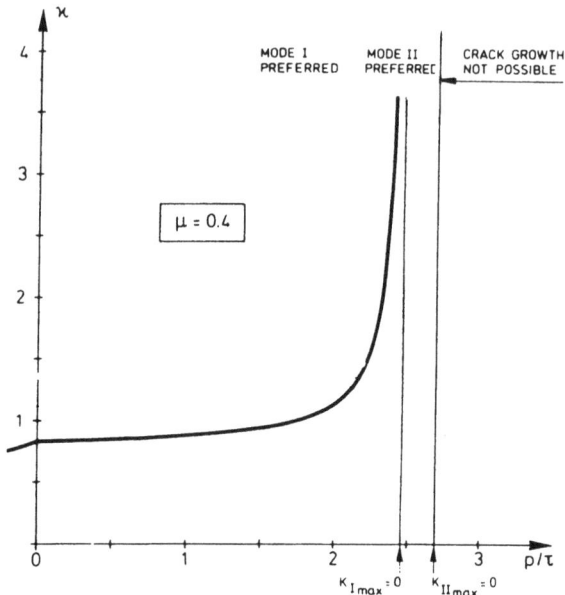

Figure 1.22 The ratio $\kappa = K_{IImax}/K_{Imax}$ as a function of p/τ for $\mu = 0.4$. (After Melin, 1986.)

accomplish this rotation and breaks up into a series of sub-fronts. Broberg (1987) shows that pure mode III fracture is possible, in PMMA, if there is sufficient compressive stress perpendicular to the crack surface. However, in these experiments he did not use closed cracks, but open notches and, therefore the compressive stress could generate a negative K_I, which would effectively eliminate the possibility of mode I fracture.

The ideal mode I crack path is not always stable. The crack trajectories from fracture of a biaxially loaded PMMA containing a central Griffith crack are shown in Figure 1.23 (Radon et al., 1977). When the transverse stress parallel to the crack is less than the stress normal to the crack, the trajectory is almost identical to the ideal path normal to the maximum stress. However, if the transverse stress is equal to or greater than the normal stress, then the crack deviates from the line of symmetry. Cotterell and Rice (1980) showed that the stability of a mode I crack path in an ideal elastic-brittle material is determined by the T-stress. If the T-stress is negative the crack path is stable, that is, after any small deviation from the mode I direction, the crack tends to return to its original path. For a classic Griffith crack under normal tension, the T-stress is negative and equal to the applied stress, hence the fracture path is very close to the ideal path.[6] The transverse stress simply adds to the T-stress, so that if the transverse stress

---

[6] This statement is true for slowly propagating cracks. If the crack velocity approaches about one-third of the speed of shear wave propagation, the fracture can branch. The dynamic effects on fracture will not be discussed.

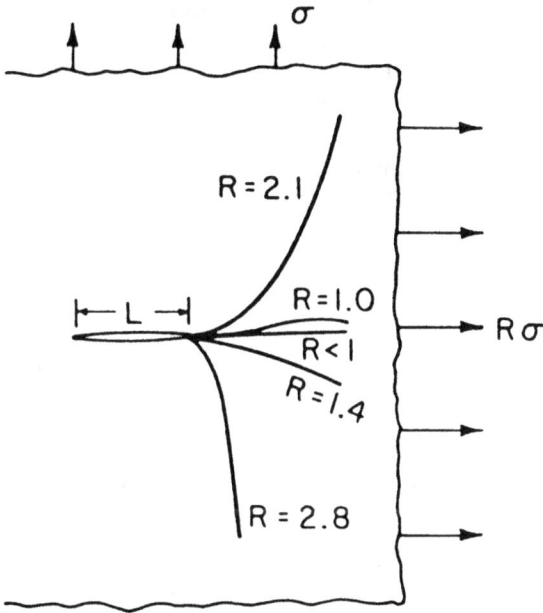

Figure 1.23 Crack trajectories from a biaxially loaded Griffith crack. (After Radon *et al.*, 1977.)

is greater than the normal stress, the T-stress is positive and the crack path is unstable. Naturally in real materials the presence of a FPZ can modify the criterion stated above. In Cotterell and Rice's (1980) analysis, an infinitesimal deviation was considered. Sumi *et al.* (1985) have considered terms higher than the T-stress and finite deviations. The net result is that there is a grey area around a zero T-stress where, in a real material, the crack path stability cannot be determined from the T-stress alone.

The nominal stresses near a crack tip under mode I loading can be used as a practical guide to crack path stability. For example, in the compact tension specimen simple engineer's theory of bending can be used to calculate the nominal stresses parallel and normal to the crack, if the nominal stress normal to the crack is greater than the stress parallel to the crack, then the crack path is likely to be stable (Cotterell, 1970). The double cantilever bend specimen has the disadvantage that its crack path is inherently unstable because of the high bending stress parallel to the crack and symmetrical crack propagation can only be achieved by providing guiding grooves or super-imposing a longitudinal compressive stress.

1.8 Crack growth resistance

So far it has been assumed that the fracture toughness of a material is a constant independent of crack growth, but that is the exception. In general, resistance to fracture and the fracture toughness increase with crack growth. Kraft *et al.* (1961) originated the concept of crack growth resistance to explain the fracture behaviour of thin ductile metal sheets. In this case the crack growth resistance is due to the development of shear lips. However, there are many mechanisms whereby crack growth resistance develops (Mai, 1988). With cementitious materials, the crack growth resistance is due to the development of the FPZ and the bridging of the crack by aggregate. Fibre reinforced cementitious materials have a very large-scale crack growth resistance due to the pull-out of the fibres. In this section only the general mechanical consequences of crack growth resistance will be discussed; the details of the fracture resistance mechanisms will be left until the next chapter.

The fracture toughness of a material exhibiting crack growth resistance increases with crack growth. If the FPZ, and any trailing bridging zones caused by fibres not pulled out completely, is small compared with the crack length and the remaining ligament, then the fracture toughness is only a function of the crack growth, Δa, and can be considered a material property. Under these circumstances FPZ is in a K-stress field, and the crack growth resistance, $K_R(\Delta a)$, is defined as the value of the applied stress intensity factor, K_a, calculated at the tip of the continuous crack,[7] that causes a stable crack extension of Δa. The crack growth resistance, K_R, increases from its initiation value, K_i, with crack growth, to a plateau value, K_{If}. A schematic crack growth resistance curve is shown in Figure 1.24. In smaller sized specimens, where the FPZ is not in a K-stress field, the crack growth resistance is both size and geometry dependent and the concept of a crack growth resistance curve is not particularly useful.

Crack growth resistance curves can also be based on the energetic concept of fracture. In this case the fracture resistance, $G_R(\Delta a)$, is defined as the value of the applied crack extension force, G, that causes a stable crack extension of Δa. The fracture resistance rises from an initiation value, G_i, to a plateau value, G_{If}, if the FPZ is small compared with the specimen dimensions and Barenblatt's (1959, 1962) hypotheses hold. The plateau value in this case is the true fracture energy and K_{If} is given by

$$G_{If} = \frac{K_{If}^2}{E^*} \tag{1.51}$$

However, since Barenblatt's second hypothesis cannot hold while the crack growth resistance is developing, because the crack shape must change with crack growth, in general the fracture resistance, G_R, is not exactly the same as the energy required to produce a unit area of fracture surface. However, from a practical viewpoint the distinction between G_R and the true fracture energy is unimportant.

---

[7] The definition of the crack tip varies according to the material. In cementitious materials the crack tip is usually defined as the tip of the visible continuous matrix crack. If there is fibre reinforcing, fibres will bridge the crack behind the crack tip.

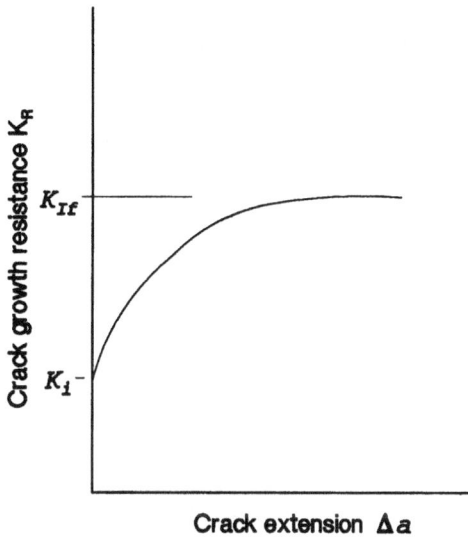

Figure 1.24 Crack growth resistance curve.

Crack growth resistance confers extra stability to crack growth. However, the maximum plateau value can only be utilized in very large specimens. Consider the fracture of a classic Griffith specimen with crack growth resistance (see Figure 1.25). The crack extension force is given by eqn 1.1 and increases linearly with the crack length, a. Hence, in Figure 1.25, G_I is

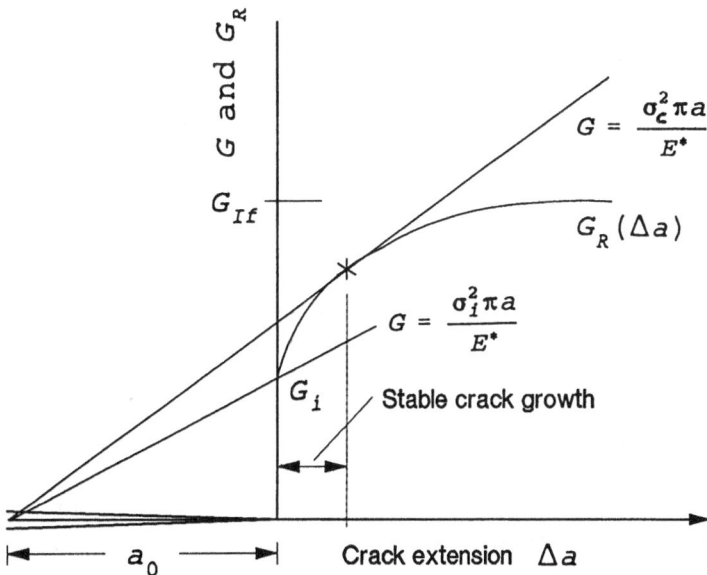

Figure 1.25 Fracture of a Griffith specimen that has crack growth resistance.

represented by a straight line whose slope is proportional to σ^2. When the stress increases to a value, σ_i, a crack can start to grow, but if the material has a crack growth resistance, the crack growth is stable because the work requirement for fracture increases with crack growth. Therefore, unstable fracture does not occur immediately. As the stress is increased the crack can propagate stably so that

$$G(a) = G_R(\Delta a) \tag{1.52}$$

However, when the crack extension force curve makes a tangent to the crack growth resistance curve, the stress reaches a critical value, σ_c, and the potential energy released is greater than that required for further growth. Unstable fracture occurs at this critical stress. The critical condition is given by

$$\frac{\partial G(\sigma, a)}{\partial a} = \frac{dG_R(\Delta a)}{d\Delta a} \tag{1.53}$$

The critical crack growth, Δa_c, increases with the size of the initial crack and the critical fracture resistance increases correspondingly. The energy approach was chosen to illustrate the concept of stable crack growth because, for a Griffith crack, G_I is a linear function of a, but the argument is similar for the stress approach using K and K_R.

A crack growth resistance confers stability to all geometries and its modifications to the stability discussed in section 1.4 are dealt with by Atkins and Mai (1985).

1.9   Non-linear fracture mechanics (NLFM)

In theory LEFM can be applied to any material provided the specimen is large enough, but for many materials the size of the specimen is larger than can be tested in a laboratory, or the component, whose strength prediction is required, is too small. NLFM was originally developed to cope with medium and low strength metals where the plastic zone is too large to be dominated by the K-stress field. In this case it is argued that there is an inner FPZ within the plastic zone where the essential fracture work is performed. Hence, NLFM is really about separating the essential work of fracture, which is a material property, from the larger plastic work which depends upon the size and geometry of the specimen. In cementitious materials, though the FPZ is large, there is no inner kernel that can be separated because by definition the material is linearly elastic outside of the FPZ. Hence, the basic rationale for NLFM does not really apply to cementitious materials, but the computational and experimental procedures of NLFM can be useful. The fundamentals of NLFM have been well reviewed by Hutchinson (1983).

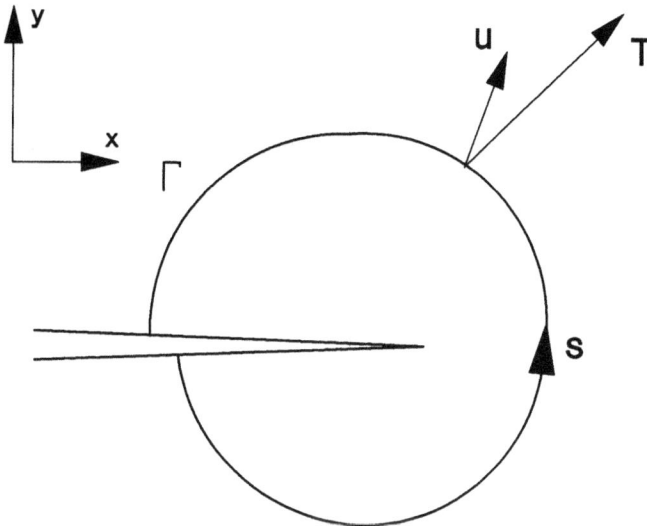

Figure 1.26 The *J*-integral.

1.9.1 The J-integral

For a non-linear elastic solid the line integral

$$J = \int_\Gamma W\,\mathrm{d}y - T_i\frac{\partial u_i}{\partial x}\,\mathrm{d}s \qquad (1.54)$$

taken anticlockwise around a crack tip (see Figure 1.26) along a path Γ that starts on one crack surface and ends on the other, where W is the strain energy density function, T_i are the components of the traction vector, u_i are the components of the displacement vector, and $\mathrm{d}s$ is an element along the path of the integral, is a constant independent of the path taken provided it is outside of the FPZ (Rice, 1968). The *J*-integral can be applied to metals deforming plastically if the loading is near proportional so that a deformational theory of plasticity which is indistinguishable from a non-linear elastic deformation, provided there is no unloading, approximates to the exact incremental theory. The *J*-integral applies exactly to cementitious materials that are essentially elastic outside of the FPZ.

If the FPZ is narrow and the *J*-integral is taken around its edge of the FPZ at the tip of a crack under mode I loading, then the first term in eqn 1.54 is small and can be neglected. Along the bottom of the FPZ, $\mathrm{d}s = \mathrm{d}x$ and the displacement vector $v = -\delta/2$, and along its top, $\mathrm{d}s = -\mathrm{d}x$ and $v = \delta/2$. The stress vector, $T_i = \sigma$, is the stress at the edge of the FPZ. Thus the *J*-integral at fracture initiation reduces to

$$J = J_{\mathrm{Ic}} = \int_0^{\delta_\mathrm{f}} \sigma\,\mathrm{d}\delta = G_{\mathrm{If}}, \qquad (1.55)$$

where δ_f is the crack opening displacement across the FPZ at crack initiation and the critical value of the J-integral, J_{Ic}, is identical to the fracture energy, G_{If}. The expression given in eqn 1.55, unlike that of eqn 1.28, apparently does not have the limitation placed upon it that the length of the FPZ must be small compared with the crack and remaining ligament lengths. However, if the FPZ is restricted in its length because the specimen is small, then the FPZ is short and fat and the first term in the J-integral, eqn 1.54, will not be negligible. Hence, the J-integral is also subject to the limitations imposed by Barenblatt's (1959, 1962) hypotheses.

An alternative interpretation of the J-integral is that it is the rate of change in potential energy of the system with crack growth (Begley and Landes, 1972). Consider a crack extension da (see Figure 1.27). The change in potential energy within the contour is

$$\frac{d\Pi}{dA} = \iint_s \frac{\partial W}{\partial a}\, dS - \oint_c T_i \frac{\partial u_i}{\partial a}\, ds \qquad (1.56)$$

where the double integral is taken over the area S enclosed by the contour Γ. Providing the crack propagation is self-similar

$$\frac{\partial}{\partial a} = -\frac{\partial}{\partial x} \qquad (1.57)$$

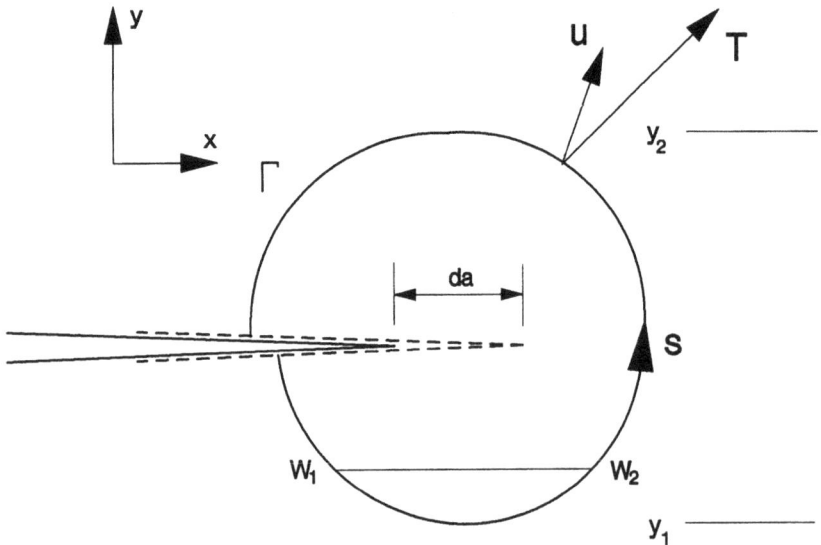

Figure 1.27 The energetic interpretation of the J-integral.

Hence

$$-\frac{d\Pi}{dA} = \iint_s \frac{\partial W}{\partial x} \, dx \, dy - \oint_\Gamma T_i \frac{\partial u_i}{\partial x} \, ds$$

$$= \int_{y_1}^{y_2} (W_2 - W_1) \, dy - \oint_\Gamma T_i \frac{\partial u_i}{\partial x} \, ds$$

$$= \oint_\Gamma W \, dy - T_i \frac{\partial u_i}{\partial x} \, ds \tag{1.58}$$

and

$$J = -\frac{d\Pi}{dA} \tag{1.59}$$

For a linear elastic material $G = J$. The derivation of eqn 1.59 clearly shows that Barenblatt's (1959, 1962) hypotheses are applicable to NLFM. Thus there is a minimum specimen size for determination of a valid fracture toughness J_{Ic}. For metallic specimens of high constraint (such as deep notch bend specimens) the ligament, b, at the notched section should satisfy the inequality

$$b > 25 \frac{J_{Ic}}{\sigma_Y} \tag{1.60}$$

but for low constraint (such as centre notch tension specimens) the condition is more onerous (Hutchinson, 1983)

$$b > 175 \frac{J_{Ic}}{\sigma_Y} \tag{1.61}$$

There is no constraint effect with cementitious materials and discussion of the appropriate limitation for these materials is deferred until Chapter 4.

The J-integral can also be expressed in terms of the complementary strain energy, Ω, under constant load conditions or the strain energy, Λ, under fixed grips by

$$J = -\frac{d\Pi}{da} = \left(\frac{\partial \Omega}{\partial A}\right)_P = -\left(\frac{\partial \Lambda}{\partial A}\right)_\Lambda \tag{1.62}$$

1.9.2   Measurement of J_{Ic}

There are three methods that are commonly used to evaluate J_{Ic}, all of which are based on the energetic interpretation of J:

(i) A direct application of eqn 1.59 can be made if the fracture is stable. Two specimens with slightly different crack lengths are loaded up to crack initiation and record the load and load-line deflection (see Figure 1.28). J_{Ic} is then given directly by $-\Delta\Pi/\Delta A$, where $-\Delta\Pi$ is the area OAB in Figure 1.28.

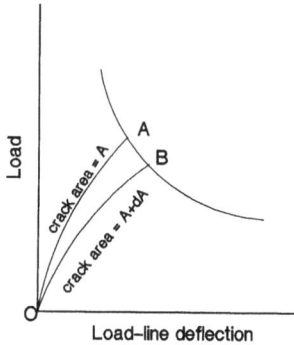

Figure 1.28 Direct calculation of J from stable load-deflection curve.

(ii) A number of specimens of different crack lengths are loaded to various displacements and the strain energy, Λ, plotted against the crack length for constant values of the displacement, u (see Figure 1.29). J is then given by the slope of the lines of constant displacement (see Figure 1.29a) through eqn 1.62 and J can be plotted against the deflection Δ (see Figure 1.29b). J_{Ic} is the value of J at the displacement at which a crack initiates.

(iii) The third method enables J to be determined from a single $P - \Delta$ plot and is the basis of the ASTM standard methods for determination of J_{Ic} for metals (ASTM, 1989). In this method, J is related to the work done on the specimen through the equation (Turner, 1979)

$$J = \frac{\eta \Lambda}{Bb} \tag{1.63}$$

where Λ is the strain energy stored (which is the same as the work done), b is the remaining ligament and η is a function of the geometry. The values of η

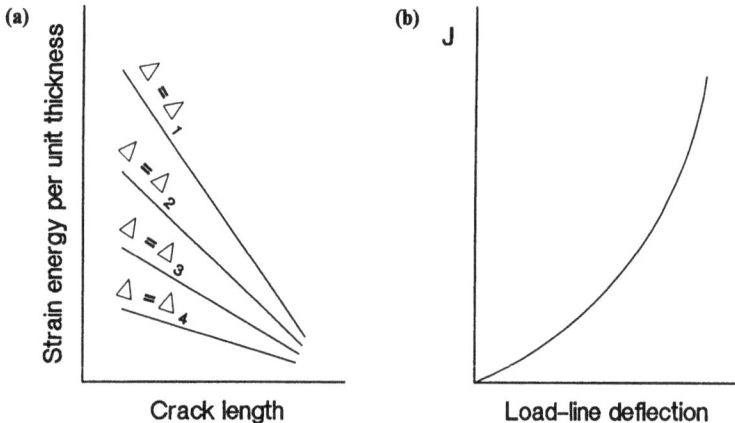

Figure 1.29 Calculation of J from multiple specimens. (a) Strain energy as a function of crack length for constant deflection. (b) J-integral as a function of deflection.

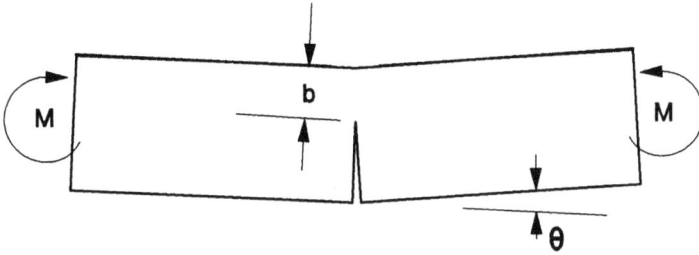

Figure 1.30 The deformation of a deeply notched bend specimen.

for deep notched specimens where the deformation is dominated by the deformation at the notch are particularly easily obtained. For a deep notched bend specimen (see Figure 1.30) $\eta = 2$, this result was first obtained by Rice et al. (1973) from dimensional arguments. If the deformation is confined to the ligament then the bending moment, M, at the notched section is a function of the rotation, θ, of the halves of the beam given by

$$M = Bb^2 f(\theta) \tag{1.64}$$

where B is the width of the beam and b is the ligament length. Thus,

$$\Lambda = \int_0^\theta M \, d\theta = \int_0^\theta Bb^2 f(\theta) \, d\theta$$

and

$$J = -\left(\frac{\partial \Lambda}{\partial A}\right)_\theta = 2 \int_0^\theta bf(\theta) \, d\theta \tag{1.65}$$

Substituting for $f(\theta)$

$$J = \frac{2}{Bb} \int_0^\theta M \, d\theta = \frac{2\Lambda}{Bb}, \tag{1.66}$$

since up to initiation b is constant. For general geometries η is given reasonably accurately by the linear elastic value (Turner, 1973)

$$\eta = \frac{b\phi^2 a}{\int_0^a \phi^2 a \, da} \tag{1.67}$$

1.9.3 Crack growth resistance J_R

The crack growth resistance of a material can be measured in terms of $J_R(\Delta a)$ the value of the applied J that produces a crack extension of Δa. Provided the FPZ is small compared with the remaining ligament, the crack growth resistance $J_R = G_R$ may be obtained from eqn 1.63. However, if there is appreciable crack growth, then eqn 1.63 is inaccurate because b

is not constant and should be replaced by (Hutchinson and Paris, 1977)

$$J = \frac{\eta}{Bb} \int_0^\Delta P \, d\Delta - \int_0^a \frac{\gamma J}{b} \, da$$

where

$$\gamma = \eta - 1 - \frac{b}{\eta W} \frac{d\eta}{d(a/W)} \tag{1.68}$$

There is a problem in applying the J-integral to the crack growth of metallic specimens because the material behind the crack tip must unload during crack extension and this unloading is completely non-proportional. Therefore, the J-integral can only be applied where the loading due to the increase in J_R dominates the unloading due to crack extension (Hutchinson and Paris, 1977; Hutchinson, 1983). However, there is no such problem with cementitious materials and the restrictions on the measurement of a valid J_R are the same as those for a valid J_{Ic} discussed in section 1.9.2.

1.10 Summary

The fracture of cementitious materials can only be characterized by a single parameter if the FPZ is small and embedded in a stress field dominated by the singularity at the crack tip. In this case LEFM is sufficient because outside of the FPZ a cementitious material is essentially elastic. However, since the FPZ of a cementitious material is usually large, laboratory sized specimens are in general too small for the valid application of classic LEFM. Small cementitious specimens fracture before the FPZ is fully developed and the fracture stress is not a direct function of the fracture energy, G_{If}.

Outside the FPZ, a cementitious material is essentially an elastic material. Thus it is possible to use LEFM if the FPZ is adequately modelled. In this case the simplest criterion for crack initiation is that the CTOD reaches a critical value, δ_f. This application of LEFM to cementitious materials is discussed in Chapter 4.

It is not strictly necessary to use the J-integral concept for cementitious materials, because they are essentially linear elastic and J is hence identical with G. However, there are situations where the J-integral is useful.

2 Fracture mechanisms in cementitious materials

2.1 Introduction

In the microscopic sense most engineering materials are not homogeneous but have domains with different microstructures. The properties, whether physical, chemical or mechanical, measured on these materials are no more than the average of these domains. Thus, cementitious materials, such as pastes, mortars and concretes, are heterogeneous materials with very complex microstructures that increase in complexity from cement pastes to concretes. In cement pastes, hydrated cements form from the reaction of cement particles with water, and the microstructure consists of both unhydrated cement and hydrated products (see Figure 2.1). The hydrated cement consists of a hierarchy of needle-like colloidal particles of calcium silicate hydrate (C–S–H) surrounding and intergrowing with calcium hydroxide (CH) crystals (see Figure 2.2). In mortars and concretes there is an additional phase of aggregates of various sizes and shapes which becomes bounded by the cement paste matrix as it hardens. Many of these aggregates are chemically stable rock materials such as river gravels or crushed particles. Whilst the shapes and sizes of the aggregates do affect the mechanical properties they are not as important as the interfacial bond strength between the paste and the aggregates.

For material modelling, Zaitsev and Wittmann (1981) consider cementitious materials as a four-level system: the nano, micro, meso and macro levels. At the nano-level, typical of the hardened cement paste with capillary pores, the characteristic size is about 500 nm. The micro-level refers to cement paste with large pores as the main inhomogeneity and the characteristic size is about 0.5 mm. The size of the meso-level, characterized by the structure of mortar, is taken as typically 10 mm. Finally, at the macro-level of concrete, the coarse aggregates are the main inhomogeneities and the characteristic size is at least four times the aggregate size which for a typical concrete is about 100 mm. Wittmann (1983) has argued that the appropriate structural, materials engineering and science models are different for each microstructural level. However, this rather complicated characterization may not be strictly necessary. Modeer (1979) suggested that it is simpler to consider all cementitious materials as a two-phase composite with a homogeneous phase and a particle phase. Thus, in cement paste the matrix is the hydrated cement gels and the unhydrated cement particles the reinforcement; in mortar the two phases are cement paste and fine aggregates, while in concrete the matrix is mortar and the reinforcement is

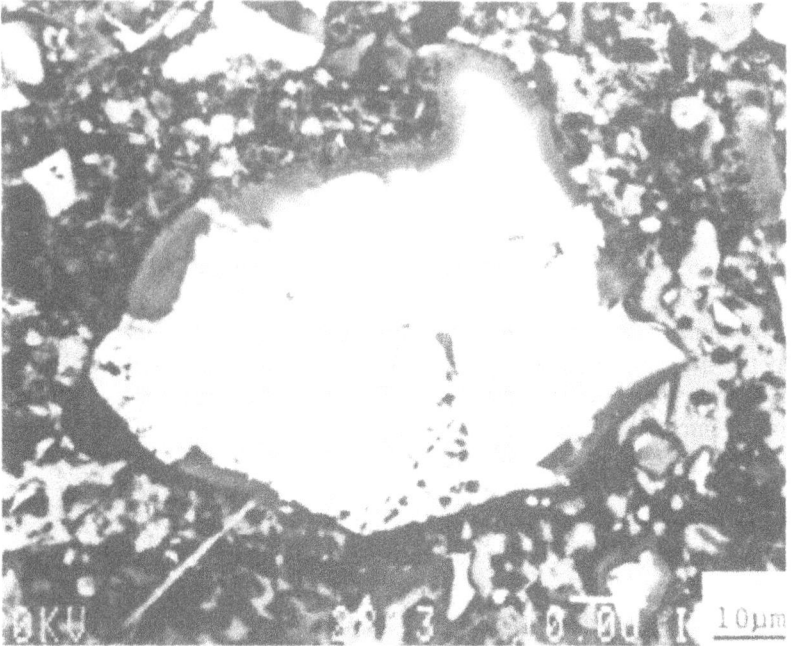

Figure 2.1 Large anhydrous core. BEI polished surface. (After Baldie, 1985.)

the coarse aggregates. In this way the properties of concrete, mortar and paste can be considered as the averages of the individual properties of the phases and the interfacial bonds between the phases.

When fibres are added to reinforce a cementitious matrix the interphases formed between the fibre and matrix are significant in determining the effective stress transfer between them. Various types of fibres have been used and these include both high and low Young's modulus fibres such as steel, asbestos, glass, synthetic (e.g. aramid, polyethylene, polypropylene, acrylic, nylon, polyvinyl alcohol, carbon) and natural fibres (e.g. cellulose, sisal and coconut). Strengthening of the matrices occurs not only because the fibres are able to take up the applied load transferred from the matrix material, but more importantly because the fibres can bridge the pre-existing matrix pores or microcracks thus delaying critical failure. This latter aspect means that even low modulus fibres (and not necessarily high modulus fibres as required by conventional composite mechanics) can strengthen brittle cementitious materials.

2.2 Cementitious materials

When a plain cementitious specimen is loaded in tension a FPZ forms right across the section at maximum load (see section 1.2). A similar FPZ also

Figure 2.2 Small hexagonal crystals of calcium hydroxide and outer product of CSH needles. SEI fracture surface. (After Baldie, 1985.)

develops at the tip of a notch or a crack. The critical size of the FPZ is dependent upon both the material properties and the geometry and size of the specimen. Only in extremely large specimens can the FPZ develop to its full extent. The most important material property that determines the size of the FPZ is the stress-displacement relationship for the FPZ. The width of the FPZ is reasonably constant except in very small specimens, and the stress-displacement relationship of the strain-softened FPZ is very nearly a material property independent of geometry and size. The magnitude and shape of the stress-displacement curve depends upon: the size and shape of the aggregates, the matrix properties, the interfacial bond between the matrix and reinforcing phase, as well as on the failure mechanism. Because it is difficult to identify precisely where the FPZ ends, various methods have been proposed in the past decade to measure the extent of the FPZ. These experimental methods are discussed in section 3.2.

2.2.1 Fracture process zone (FPZ) and strain-softening characteristics

The development of the FPZ in concrete during a direct tensile test and the associated stress-displacement relationship has been obtained by Petersson (1985) using a very stiff testing machine and a short-gauged tensile specimen. These requirements must be satisfied, otherwise the fracture process is unstable. The development of the FPZ in such a direct tension test is shown in Figure 2.3 and the corresponding stress-displacement curve is shown in Figure 2.4. Widening of the FPZ begins from photograph 3 where the width is about 15 μm and increases to about 170 μm in

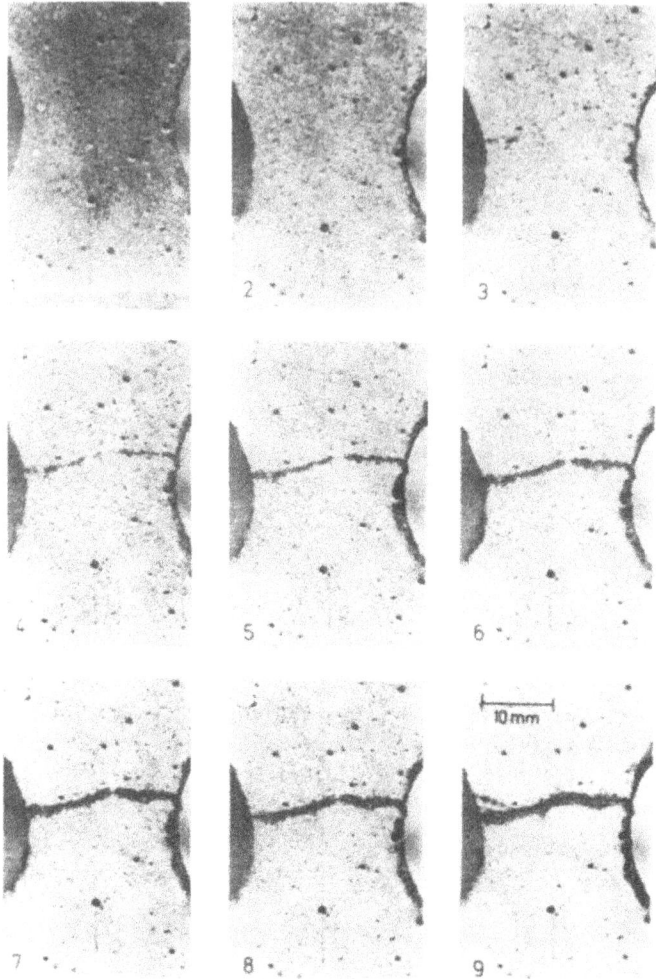

Figure 2.3 The development of a FPZ during a tensile test. The surface has been cast against a steel mould. (After Petersson, 1985.)

photograph 9. Up to the maximum stress f_t there are no signs of a FPZ; in fact the first visible signs of the FPZ appear in photograph 3 where the stress carried is already very much less than f_t. The reason for the delayed appearance of a FPZ is because in a plain tension specimen there is no focus for the formation of a FPZ and at first microcracking occurs over a diffuse area. In the presence of a notch or crack, this initial diffuse development does not occur and a concentrated FPZ forms directly. Although there is some curvature in the stress-displacement curve up to the maximum stress, it is usually assumed that the displacement up to this point is elastic and uniform over the entire gauge length. Since the FPZ is narrow compared with

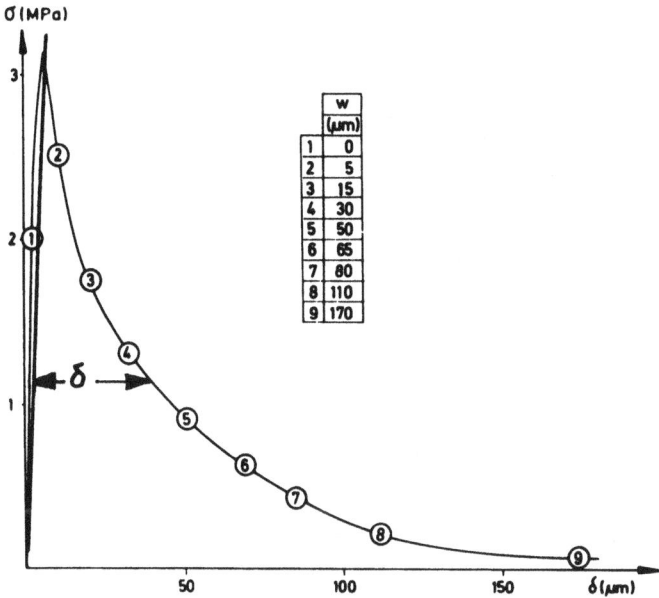

	w (μm)
1	0
2	5
3	15
4	30
5	50
6	65
7	80
8	110
9	170

Figure 2.4 The stress-displacement curve. The numbers on the graph correspond to the photographs in Figure 2.3. w is widening of the FPZ. (After Petersson, 1985.)

the gauge length, the elastic contribution from the pupative FPZ can be neglected and the opening of the FPZ, δ, measured from the elastic loading line as shown in Figure 2.4. The stress-displacement curve is shown schematically in Figure 2.5. There are three regions in this schematic stress-displacement curve. Up to about 60% of the ultimate strength, f_t, the specimen is elastic (Region A). In Region B the deformation is non-linear because of dispersed cracking. At a critical stress the damage in a cementitious material becomes localized and a FPZ starts to form. It is only from this point that the displacement becomes virtually independent of the gauge length. Hence, in Figure 2.5 the displacement is measured from the elastic loading line passing through the point of maximum load. There is elastic recovery in the body of the specimen as the stress decreases in Region C where the FPZ strain softens as the bridges pull out or fracture. A visible continuous crack forms when the local displacement across the FPZ reaches a critical value, δ_f. The stress-displacement relation for the FPZ is obtained by subtracting the elastic displacement of Region A from the total displacement as shown in Figure 2.4. Of note is the long tail to the stress-displacement curve where locked bridging aggregates are pulled out against frictional forces. The long tail makes accurate measurement of the fracture energy, G_{If}, difficult (see section 3.5).

It may be noted that in cementitious materials the FPZ, albeit analogous to the plastic yielding zone in ductile materials, is not affected by the state of

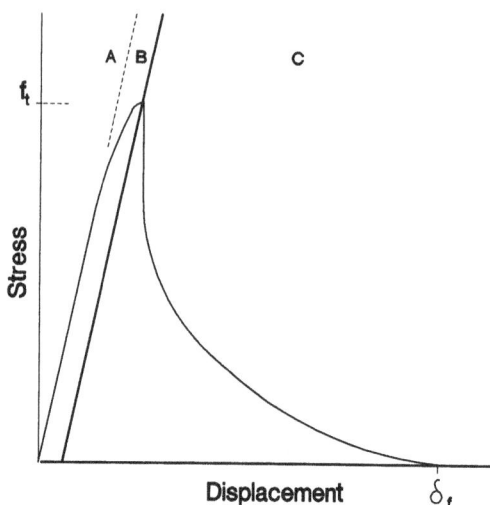

Figure 2.5 Schematic stress-displacement curve for a cementitious material.

stress and is therefore independent of the thickness of the specimen provided the specimen is more than three or four times the thickness of the aggregate. The reason is that the loss of strength is caused by microcracking and the pull-out of interlocking aggregates and there is little shear deformation. Hence, there is very little difference in the FPZ between plane stress and plane strain.

2.3 Cementitious fibre reinforced composites

When fibres, whether continuous, or discontinuous, are added to the brittle cement matrix materials, the fracture toughness and strength of the resultant composites are significantly improved. The fibres alone are capable of sustaining a stress greater than that which causes the matrix to crack if the volume fraction of the reinforcing fibres is greater than a critical value, v_c (Aveston *et al.*, 1971). In this case the matrix breaks up into a number of parallel cracks, the spacing being determined by the distance over which the stress transferred from the fibre to the matrix builds up to the critical value (Aveston *et al.*, 1971). The critical volume fractions are in the range of 0.3–0.8% for aligned continuous steel, glass and polypropylene fibres (Hannant, 1978). The stress-strain behaviour for this kind of cementitious composite we term Type I and Figure 2.6 shows the curves for carbon fibre and steel wire reinforced composites of this type. A schematic stress-displacement curve for Type I composites is shown in Figure 2.7; note that since the deformation is not confined to a single FPZ or FBZ, the displacement is not separated into deformation across the FPZ/FBZ and the bulk

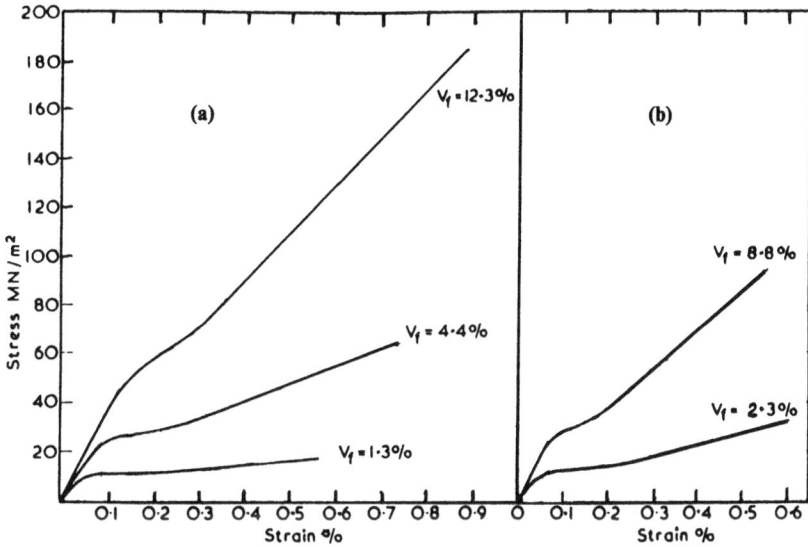

Figure 2.6 Stress–strain curves for (a) carbon fibre reinforced cement, (b) steel wires. (After Aveston *et al.*, 1974.)

elastic displacement. In Region A the behaviour is essentially elastic though there is some diffuse microcracking. A defined FPZ forms in Region B, causing non-linearity. If the fibres have a higher elastic modulus compared with the matrix, Region B may not be detectable. At a critical composite stress, σ_{cmc}, the matrix cracks and a continuous matrix crack forms in the FPZ. Since the volume fraction is greater than the critical value, the stress carried by the bridging fibres increases with further straining and leads to the development of multiple cracks (Region C) until the spacing reaches a minimum distance depending upon the bond strength between the fibres and the matrix (Aveston and Kelly, 1973; Hannant *et al.*, 1983). The stress in Region C may remain constant or increase (as discussed below). In Region D further straining causes the stress to increase almost linearly with a less steep slope than in Region A, but greater slope than in Region C, as the fibres stretch. Region D ends in either the fibres breaking,[1] or if the fibres are shorter than the critical length at which the fibres can just develop their fracture strength, the fibres gradually pull out (Region E). The ultimate strength of cementitious matrices reinforced with continuous wires or fibres of steel carbon or polypropylene can be very much greater than the stress, σ_{cmc}, at which the matrix first cracks (Aveston *et al.*, 1974; Keer, 1984; Majaumdar and Walton, 1984; Swamy and Hussin, 1989; Li *et al.*, 1993).

---

[1] If the Weibull distribution for the strength of the fibres has a high modulus there can be no significant fibre pull-out. However, if the Weibull modulus is low, fibres can break away from the matrix crack causing some fibre pull-out.

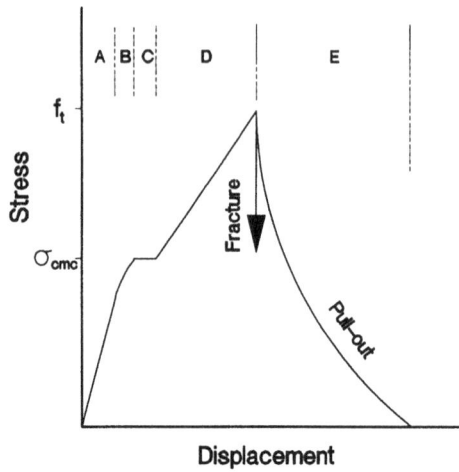

Figure 2.7 Schematic stress-displacement curve for a Type I composite with multiple matrix cracking.

There is some controversy as to whether the stress in Region C is constant or increases. The stress during multiple cracking of the matrix can rise for two reasons:

(i) There is a significant variation in the strength of the matrix. Since the Weibull modulus for cementitious materials is of the order of 10 (see Chapter 6), comparatively large matrix strength variations are possible.
(ii) The fibres debond easily so that there is no longer strain compatibility between the fibres and the matrix and the fibres take more than their fair share of the load. Hence, the load has to rise before more matrix cracks can be generated. Few fibres bond well to cementitious matrices.

Examination of the stress-strain curves of a variety of Type I composites suggests that the stress in Region C is more nearly constant if the volume fraction of the fibres is close to the critical value for multiple cracking and that Region C becomes less distinguishable from Region D as the volume fraction increases.

The stress-strain curve for a cementitious composite is also difficult to measure and some of the experimental techniques used may introduce errors. Li *et al.* (1993) report a stress-strain curve for steel fibre reinforced cementitious composite ($v = 1.53\%$, $d_f = 0.4\,\text{mm}$) where the change in slope from Region C to Region D is very small (see Figure 2.8) and the test method may have affected the curve. They aligned 30 steel wires and cast epoxy resin at the ends to form the anchorages and then cast the

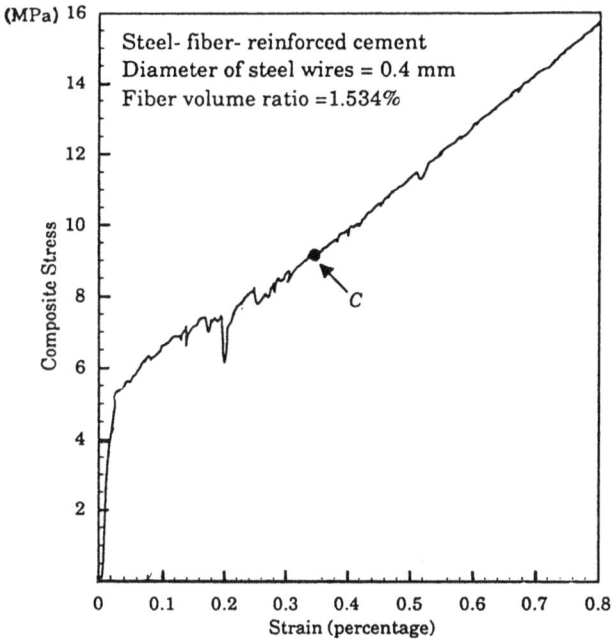

Figure 2.8 Stress–strain curve for steel wire reinforced cement, $v = 1.53\%$, $d_f = 0.4$ mm, C marks the end of multiple cracking. (After Li *et al.*, 1993.)

Portland cement in the test section. By this method the wires were very securely anchored in the epoxy resin which bonds well to steel and poorly bonded in the test section. It is suggested that under these circumstances the wires might have been more highly stressed than would be expected from the rule of mixtures even before the matrix cracked so that the deformation of the specimen may have been very much controlled by the steel wires throughout the experiment with some slippage taking place between the wires and the cement matrix even before the matrix cracked. Composites that are Type I behave in many ways like elastic-plastic materials and fracture mechanics as such is not needed to describe the behaviour of structures made from these materials. Type I fracture behaviour is most common for cementitious materials reinforced with continuous fibres or networks. If the volume fraction of the fibres is less than v_c then the stress sustainable by the composite drops either when the matrix cracks or, if there is some stable debonding, just after the matrix cracks. This kind of composite we term Type II. Most Type II composites are reinforced by discontinuous fibres which usually pull out rather than fracture. Multiple cracking does not occur in Type II composites and fracture in a tension specimen occurs essentially on a single narrow FPZ. The behaviour of Type II composites can be best described in terms of a stress-displacement curve where the bulk elastic displacement is separated from that acting

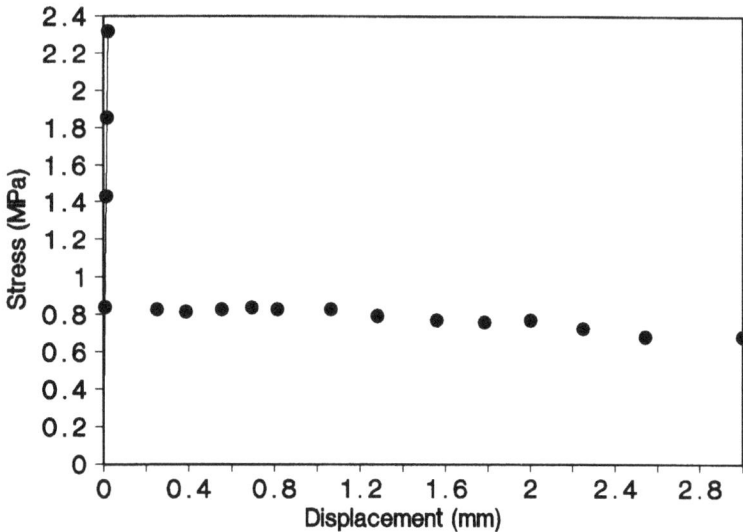

Figure 2.9 Stress-displacement curve for a steel–fibre–concrete reinforced with straight wires 0.565 mm in diameter and 30 mm long; fibre volume fraction 1.5%. (After Lim *et al.*, 1987b.)

across the FPZ/FBZ. There are two sub-types. In Type IIA composites the fibres are long and the fibre pull-out stress is virtually constant for large CODs. The stress-displacement curve for a steel–fibre–concrete reinforced by smooth steel wires with a volume fraction of 1.5%, shown in Figure 2.9 is typical of a Type IIA composite. A Type IIB composite has short fibres and the stress drops more sharply with fibre pull-out. Typical Type IIB composites are shown in Figure 2.10.

A schematic stress-displacement curve for Type IIA composites is shown in Figure 2.11. The initial behaviour is elastic (Region A). At higher stresses the load-deflection curve becomes non-linear (Region B) primarily because of microcracking in the FPZ. The microcracks are partially stabilized by the fibres and the maximum strain achieved before the FPZ becomes localized is larger than the corresponding strain in the unreinforced matrix. This increase in matrix strain can occur even if the modulus of the reinforcing fibres is less than that of the matrix. The fibre debonding can be either completely unstable or the fibres may debond stably at first in which case the partial debonding can contribute to the non-linearity of Region B and a continuous matrix crack forms before the tensile strength, f_t, is reached. On reaching the fracture strength, f_t, of the composite the stress drops as friction takes the load. There can be a slight recovery in load (Region C) as fibres that are debonded in the vicinity of the matrix crack stretch elastically before pulling out against friction. If the fibres are long, the stress remains almost constant over COD of the order of 5 mm or more.

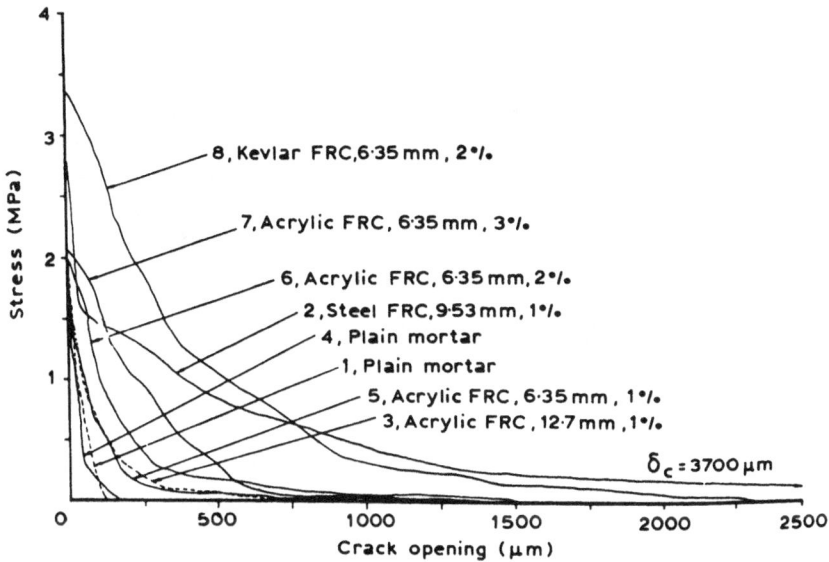

Figure 2.10 Stress-displacement curves for various fibre reinforced cementitious materials. (After Li and Ward, 1989.)

With short fibres (Type IIB) the schematic stress-displacement curve superficially looks the same as that for an unreinforced cementitious material (Figure 2.5), but the COD, δ_f, which is of the order of 0.1 mm for unreinforced material is equal to half the fibre length for fibre reinforced composites. After the essentially elastic Region A, diffuse microcracking

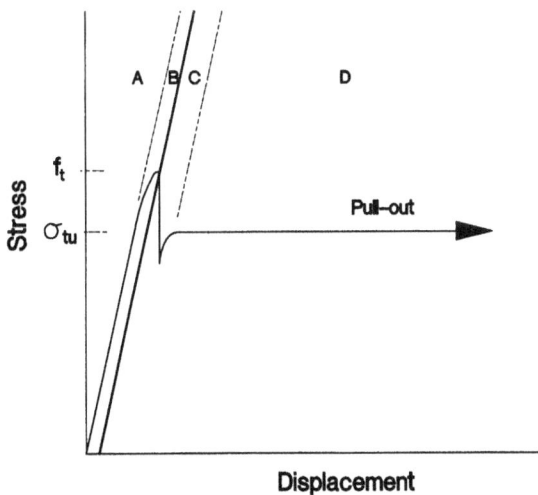

Figure 2.11 Schematic stress-displacement curve for a Type IIA composite with long fibres.

and possibly stable partial debonding of fibres causes the non-linearity in Region B. If there is some stable debonding, the matrix cracks completely at a stress, σ_{cmc}, which is less than the strength of the composite. Once debonding becomes unstable the stress drops. If the fibres are poorly bonded so that friction provides the main shear transfer from the fibres to the matrix, the drop in stress, though steep has a finite slope, whereas for well bonded fibres there is a sudden drop in stress as friction takes over. In Region C the fibres pull out against friction and the stress decreases.

2.3.1   The crack tip fibre bridging zone (FBZ)

When a Type II cementitious structure containing a notch is loaded, a FPZ initially forms at its tip. With increase in load the CTOD reaches the critical value, δ_m, of the matrix and a fracture bridging zone (FBZ) starts to grow. When the FBZ is fully developed the bridging fibres either pull out or fracture (see Figure 2.12). The division between the FPZ and the FBZ is the tip of the continuous matrix crack which is often difficult to locate exactly and the measurements of the sizes of the two zones can be inaccurate. Generally, the FBZ in cement composites depends on the aspect ratio of the fibres and

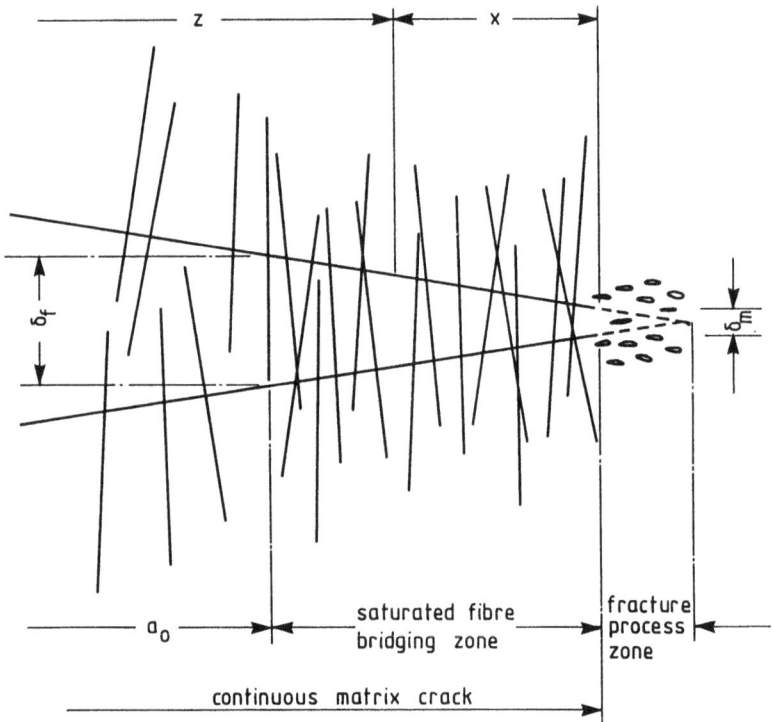

Figure 2.12 The FPZ and FBZ in a notched Type II fibre reinforced composite.

on the geometry and size of the specimen and the loading configuration. The magnitudes of the FBZ in common cement composites can vary considerably: in asbestos cement with a fibre length of about a couple of millimetres the FBZ is of the order of 30 mm in a double-cantilever-beam (DCB) specimen (Lenain and Bunsell, 1979); in a steel fibre reinforced concrete with a fibre length of 6.35 mm the FBZ, in a DCB specimen, is more than 600 mm long (Visalvanich and Naaman, 1983); in a glass fibre cement with chopped strand glass mat the FBZ is about 15 mm (Patterson and Chan, 1975).

2.4 Fracture mechanisms in cementitious materials and fibre composites

In this section the micro-failure mechanisms of cementitious materials and their fibre composites are discussed. Emphasis is placed on the role of interfaces on strength and fracture energy absorption. The development of macro-defect-free (MDF) cement pastes in the last 10 years by ICI in the UK, which give both high strength and high toughness, and the mechanisms involved in this new class of material, are described.

2.4.1 The micromechanisms of failure in concretes and the new cement pastes

The micro-failure mechanisms in cement paste, mortar and concrete have been studied with great detail by Diamond and Bentur (1985) using a scanning electron microscope equipped with a chamber that permits saturated specimens to be observed in a wet environment by detecting back-scattered electrons. Similar studies have also been conducted by Tait and his co-workers (1986, 1990) using the same technique but with a double-torsion specimen which allows stable crack growth to occur. In mortar and concrete, failure is characterized by crack deviation, branching and multiple cracking. Crack deviation, where the crack meanders in the paste matrix between aggregates or propagates around the aggregates and along debonded paste/aggregate interfaces, is shown in Figure 2.13. Branches are more predominant in concrete than in mortar and multiple sub-division into three or more branches is not uncommon as is shown in Figures 2.14–2.16. In the case of mortar, multiple cracking occurs where the crack leaves a debonded aggregate or around an air void (see Figure 2.17). In concrete, multiple cracking often occurs between adjacent sand grains rather than around isolated ones as in mortar. The interaction of the propagating crack with the inhomogeneities is of particular interest. In dried mortar specimens the crack runs along the interface between the sand grain and the paste matrix resulting in debonding. However, in undried specimens the crack always avoids the interface even though it may run along it for some distance. In concrete the interaction of the crack tip with sand

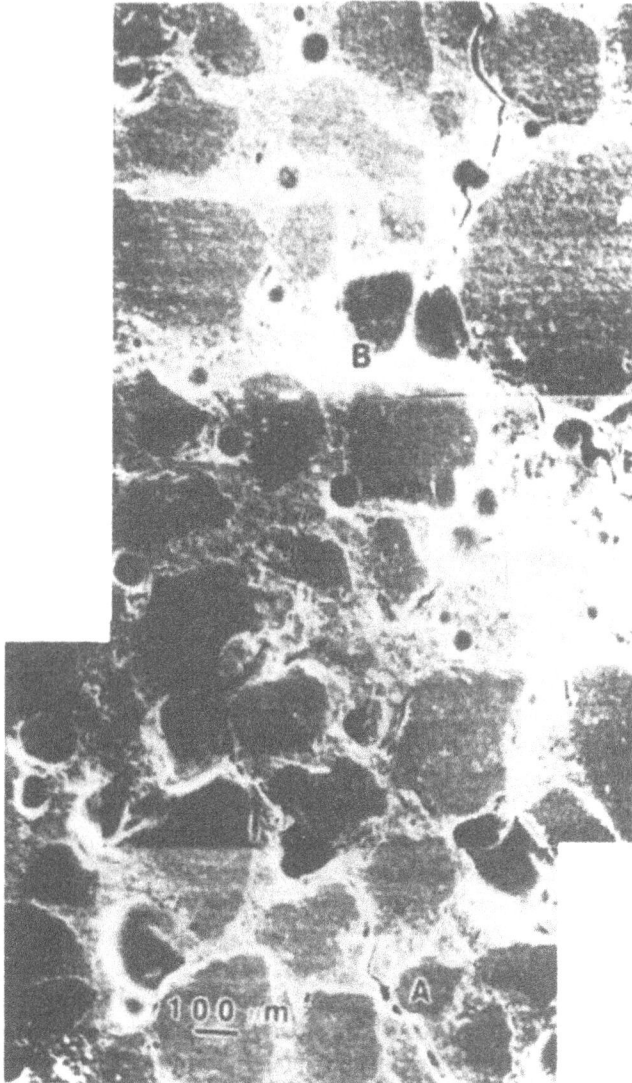

Figure 2.13 Typical portion of a crack in air dry mortar. (After Diamond and Bentur, 1985.)

grains is much the same as in mortar but, in addition, multiple radial cracks are often found between adjacent grains caused by shrinkage misfit (see Figure 2.18). However, the presence of large coarse aggregates produces some interesting features. Occasionally, the cracks run into the large aggregates but more often they are arrested in front of these aggregates thus activating another main crack in a new path. In more recent work where the crack was made to propagate stably in a double-torsion specimen,

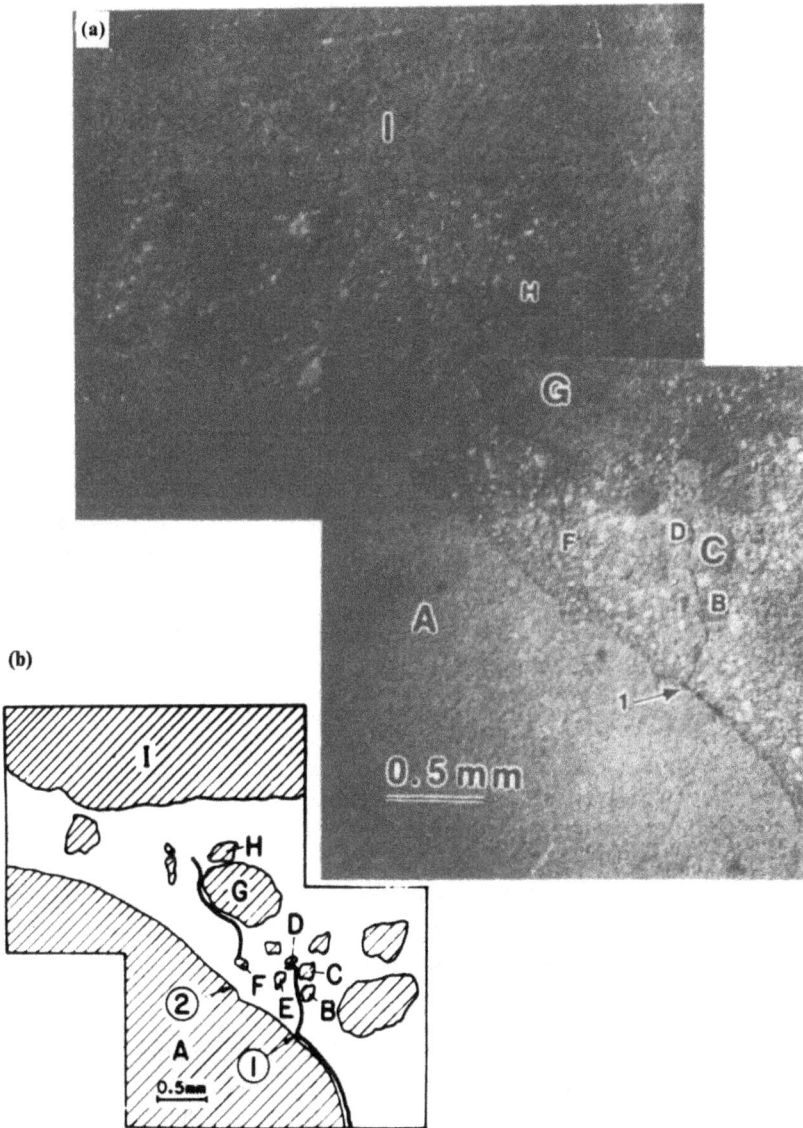

Figure 2.14 The crack tip zone in concrete during the first stage of loading. (After Diamond and Bentur, 1985.)

Tait *et al.* (1990) found that for mortar there was a substantial microcrack zone at the main crack tip much like 'tree branches'. Eventually one of these branches becomes the main crack with its own system of microcracks. In so doing the previously formed microcracks are unloaded and a good illustration of this feature is given in Figure 2.19.

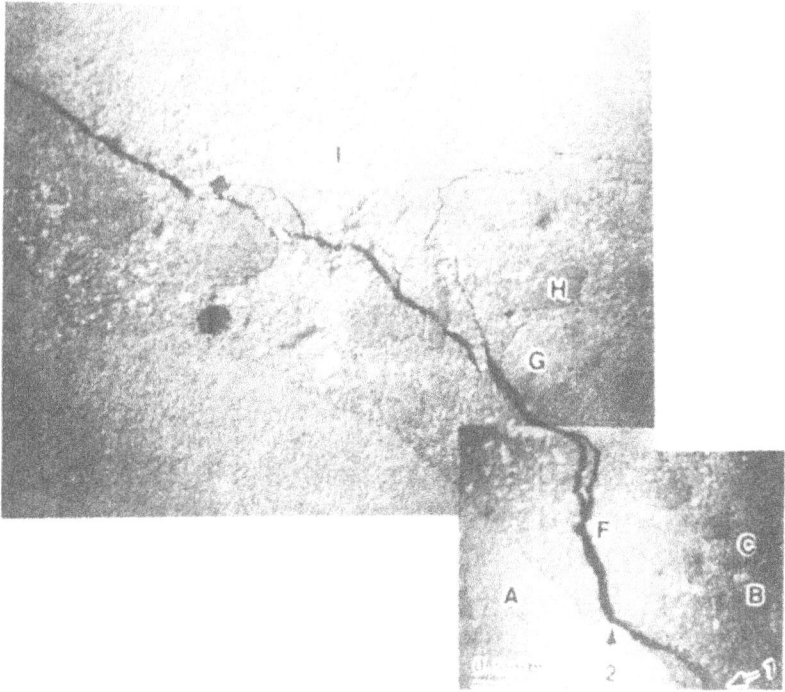

Figure 2.15 The same zone as in Figure 2.14 after the second loading stage. (After Diamond and Bentur, 1985.)

The failure processes are complicated. The high fracture energy absorptions, which are associated with crack deviations, crack branching, and multiple microcracking, make concrete a much tougher material than paste. Whilst the microcracking zone can be reasonably treated as a FPZ it is very difficult to distinguish a traction-free main crack behind this region. Furthermore, the main crack is never straight and the FPZ is somewhat diffused. All these features seem to cast some doubts on the fracture mechanics modelling of concrete and mortar using a single equivalent traction-free crack (Tait *et al.*, 1990). Hillerborg's fictitious crack model (Hillerborg *et al.*, 1976; Petersson, 1985) and Bažant's crack band model (Bažant and Cedolin, 1979; Bažant and Oh, 1983; Bažant and Lin, 1988; Bažant *et al.*, 1988; Bažant and Ozbolt, 1990), which is discussed in Chapter 4, are better equipped to deal with these materials.

In ordinary Portland cement paste, Bailey and his co-workers (Higgins and Bailey, 1976; Bailey *et al.*, 1986) have used a 'diffuse illumination' optical microscopy technique to reveal a small FPZ at the notch tip. They show that the FPZ is synonymous with a discontinuous or tied-crack, within which the silicate hydrate fibrils are pulled apart. Unfortunately, this model is inconsistent with recent microstructural observations on polished

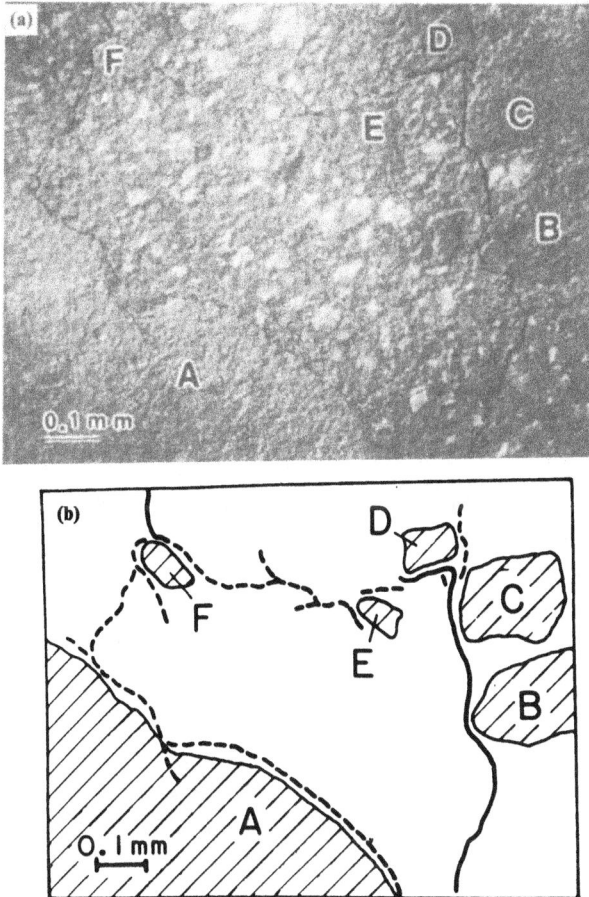

Figure 2.16 Higher magnification (75×) of the field between aggregates A and F in Figure 2.14 during the first stage of loading, showing a network of cracks of width 1 μm or less between the aggregates and along the aggregate surfaces. (After Diamond and Bentur, 1985.)

epoxy-impregnated cement paste carried out by Struble *et al.* (1989) using SEM with back-scattered electron imaging. They show quite conclusively that the crack-resistance curve behaviour in cement paste is primarily due to the discontinuous crack path in which the gaps between offset segments form crack bridges behind the advancing crack front. These unbroken gaps across the offsets behave similarly to the localized grain bridging in coarse-grained alumina (Swanson *et al.*, 1987). It is these bridging unfractured segments rather than the interlocking silicate hydrate fibrils that provide the microstructural basis for the so-called 'tied-crack' model. *In-situ* studies in the SEM show that the main crack is generally straight but higher magnification reveals that it is composed of many linked short

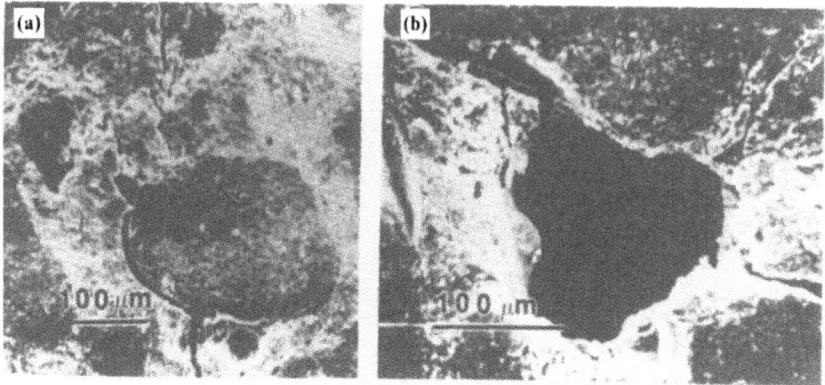

Figure 2.17 Debonding and multiple cracking in air-dry mortar, (a) around a sand grain, (b) around a void. (After Diamond and Bentur, 1985.)

segments about 60 μm in length. Restricted crack branching is visible and more importantly the crack is discontinuous in several places near its tip as shown in Figure 2.20. This observation confirms the comment of Struble *et al.* (1989) that unbroken segments exist in the wake of the propagating crack tip. Higgins and Bailey (1976) found that the fracture toughness, K_c,

Figure 2.18 Multiple radial cracks between adjacent sand grains in concrete. (After Diamond and Bentur, 1985.)

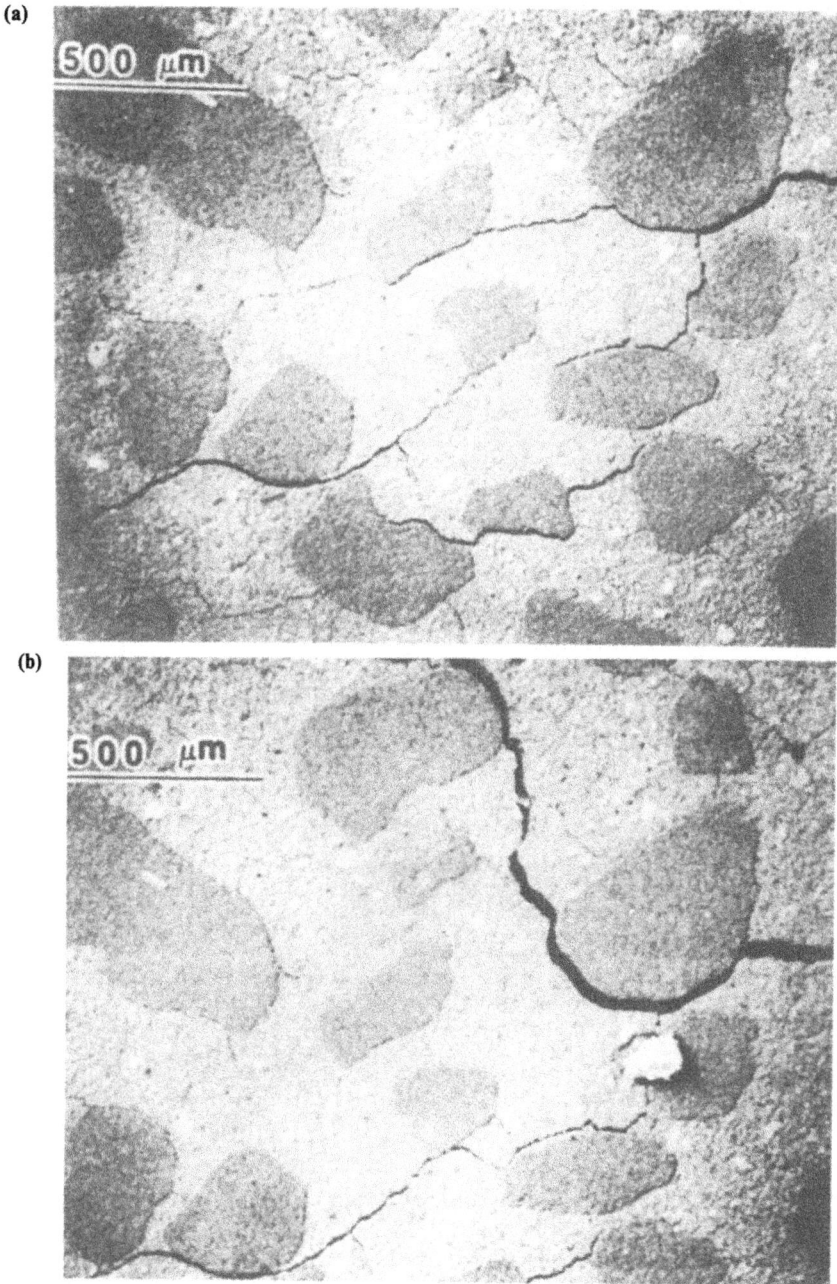

Figure 2.19 A crack system in mortar, the crack has propagated from top right to bottom left. (a) A crack bridge has developed. (b) On increasing the load, the main crack has propagated upwards allowing the initial cracks to close. (After Tait *et al.*, 1990.)

(a)

(b)

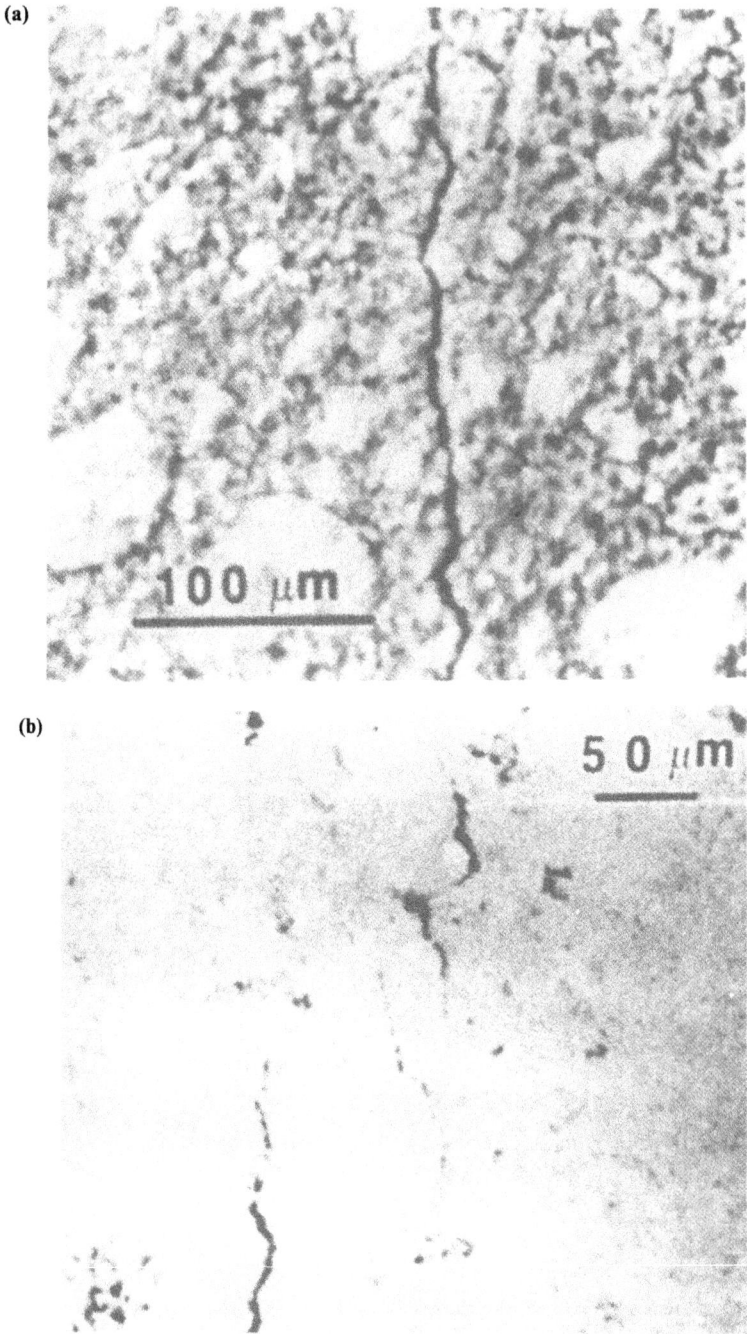

Figure 2.20 Typical cracks in cement paste. (a) Continuous crack. (b) Discontinuous cracks. (After Struble *et al.*, 1989.)

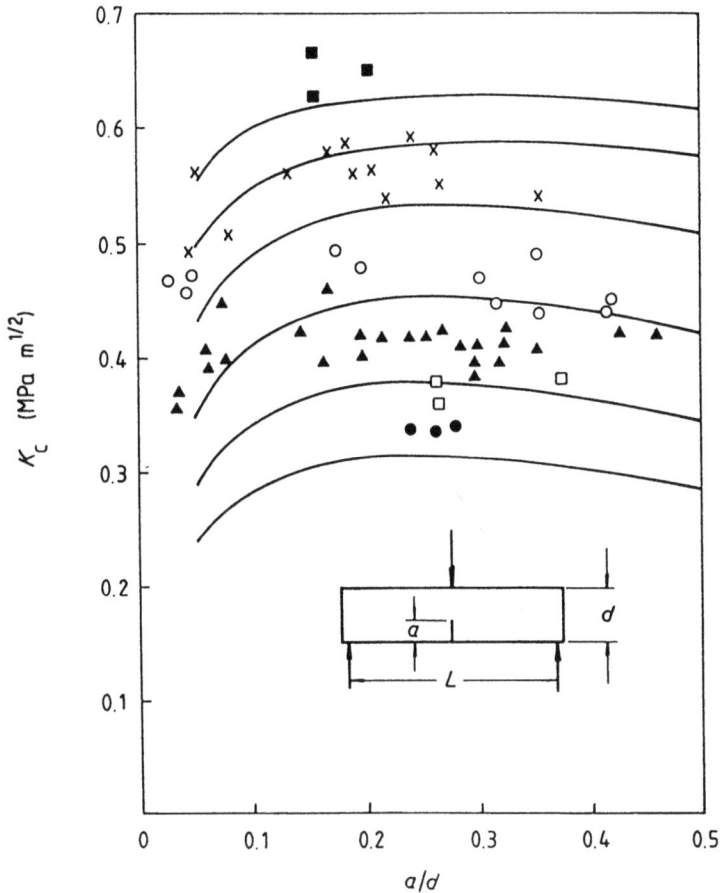

Figure 2.21 The critical stress intensity factor, K_c, as a function of the initial notch length for various cement paste specimen sizes with $L = 5d$ (experimental data from Higgins and Bailey, 1976; d (mm) = 5 ●, 8 □, 14 △, 28 ○, 56 ×, 110 ■: theoretical curves from Cotterell and Mai, 1987).

for their Portland cement pastes was not a constant but increased with the size of the notched beam specimens. They concluded that the Griffith fracture criterion is invalid without realizing that the 'tied-crack' dimension is much larger than the very small zone of silicate hydrate fibrils and is of the order of the unbroken ligaments. Cotterell and Mai (1987) have actually shown that these K_c results are in good agreement with the predictions of a crack-resistance curve model (see Figure 2.21). It was further shown that the FPZ is dependent on the size of the beam. Thus the crack-resistance is size dependent as is shown in Figure 2.22, and there is no unique crack-resistance curve. The R-curve is only unique if the size of the beam is much larger than the FPZ.

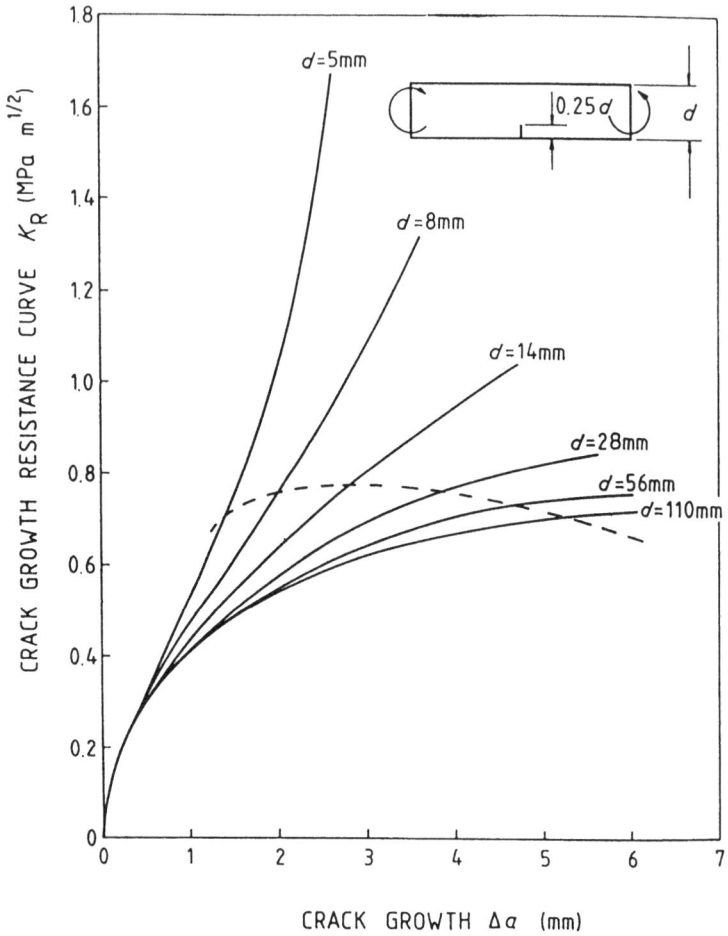

Figure 2.22 Crack growth resistance curves for different size cement paste specimens with $a/d = 0.25$, instability locus - - - -.

The source of weakness in cement pastes is associated not with the total porosity but with the size and shape of the largest pore. Thus, by reducing the size and amount of these large pores the strength of cement pastes can be improved. Birchall and co-workers (Birchall *et al.*, 1981, 1982; Kendall *et al.*, 1983, 1985) have made use of this concept to develop what is commonly called macro-defect-free (MDF) cement paste. The MDF paste was prepared using calcium aluminate cement premixed with a polyvinyl alcohol–acetate copolymer which are passed through a two-roll mill with a high shearing rate to reduce the size and level of porosity in the pastes. Typical properties of these MDF cement pastes in 2–3 mm thicknesses are: bending strength (three-point bend and span 100 mm) of 145 ± 5 MPa with a Weibull modulus of between 20 and 30, Young modulus, E, of 52 ± 1 GPa, fracture toughness, K_{Ic}, of 3.3 ± 0.1 MPa$\sqrt{m}$ and fracture energy, G_{If}, of 400 ± 60 J/m^2. According to Birchall and his co-workers, the improved tensile strength of the new pastes is entirely caused by the reduction in pore size ($2c$) according to the equation

$$\sigma_c = \left[\frac{E_0 G_{If0}(1 - v_p)^3 \exp(-\alpha v_p)}{\pi c} \right]^{1/2} \qquad (2.1)$$

where E_0 and G_{If0} are values for the Young's modulus and the fracture energy when the volume fraction, v_p, of the porosity is zero; α is a constant. In this view the polymer largely acts as a rheological aid to reduce the pore size, though it is understood that there is a chemical interaction between polymer and cement (Sinclair and Groves, 1985; Roger et al., 1985) which alters both E_0 and G_{If0}. Poon and Groves (1987) confirm that E_0 and G_{If0} are strength limiting factors. A totally different argument for the strength improvement in polymer-modified cement pastes was presented by Eden and Bailey (1984a,b, 1985a,b). They did not agree with the pore size reduction theory and suggested that the polymer must have played a significant role in the strength development in these polymer-modified pastes. For the Portland (calcium silicate) MDF cements they proposed a 'tied-crack' model in which the tensile strength is controlled by the pulling apart of the hydrate fibrils with the polymer acting as an adhesive interface (Eden and Bailey, 1984b, 1988). It is the variation of the interfacial shear strength with polymer content and water absorption that determines the magnitude of the fracture stress according to the Griffith equation. This is a physically attractive model which relates the macroscopic tensile strength to the microscopic interfacial shear strength properties. However, Bailey and co-workers realized that the 'tied-crack' model could not be applied to the calcium aluminate MDF cements because they could not detect the development of an FPZ (akin to the tied-crack) ahead of the machined notch. Consequently, they believed that the high strength and high fracture toughness obtained in these MDF cement pastes must have come from the large deformation of the polymer. But again there is no experimental evidence to support this proposed mechanism for calcium aluminate MDF cements.

Recently, Mai et al. (1990) have shown that in the ICI calcium aluminate MDF cements a crack-resistance curve behaviour does exist similar to that observed for unmodified paste. The saturated FPZ in these MDF cements is about 3 mm long in a double-cantilever-beam specimen. Toughening mechanisms giving rise to the R-curve have been identified, using both optical and scanning electron microscopy, in the crack wake. The predominant mechanism is due to unbroken cement ligaments bridging the crack faces. The evolution and break-up of these bridges are shown in Figure 2.23.

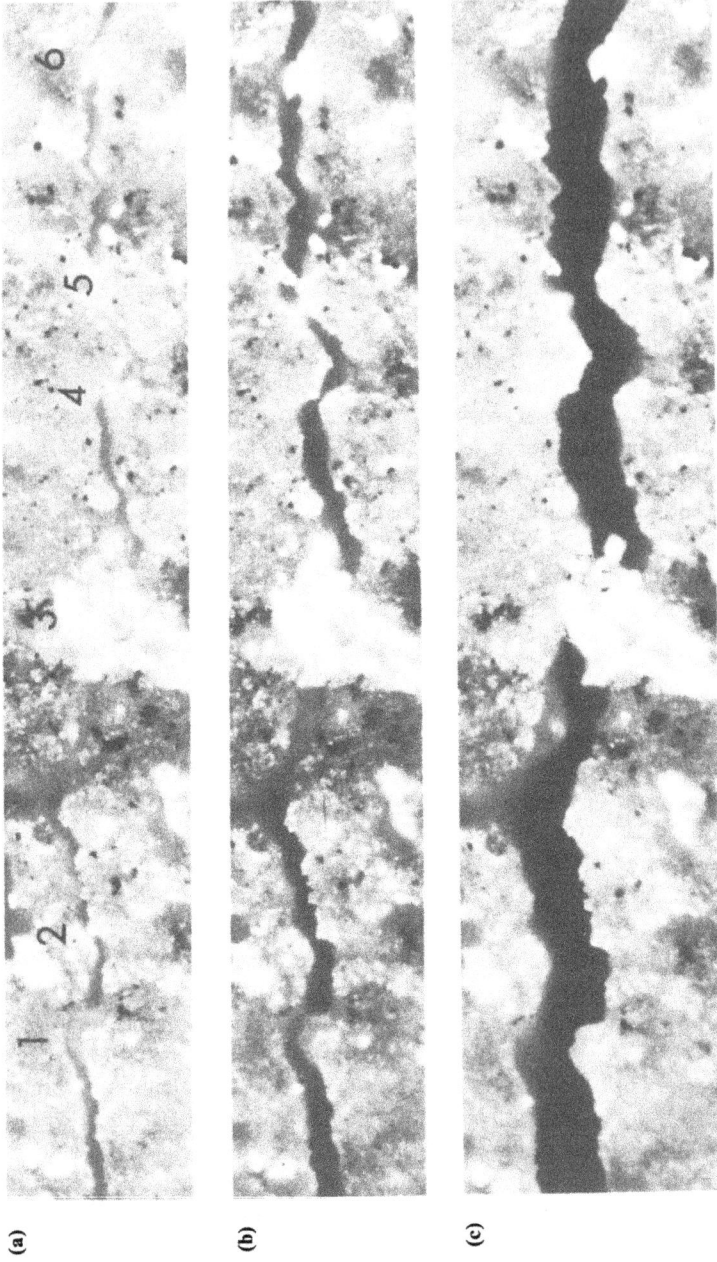

Figure 2.23 Evolution and break-up of crack bridges in MDF paste, ×500.

Similar bridges by untorn ligaments are shown in Figure 2.24e, and by isolated grains in Figure 2.24f. Other less significant contributions are due to frictional interlocking of adjacent grains on the fracture plane (see Figures 2.24a, 2.24b) and tearing of polymer fibrils (see Figure 2.24g). It is concluded from these results that the much enhanced total fracture energy of the polymer modified cement paste comes equally from the improved intrinsic fracture toughness G_{Ifo} and from the fracture work necessary to pull the crack bridges apart. The role of the polymer is, therefore, more than just a rheological processing aid.

2.4.2 The micro-fracture mechanisms in fibre reinforced cements

The total fracture energy of a Type II fibre reinforced cement composite comes from many different energy absorbing mechanisms. It is the understanding of these micromechanisms and the material parameters that control them that will enable the design of stronger and tougher fibre cements. There have been many studies on the fracture micromechanisms in resin-based fibre composites (e.g. Harris, 1980; Kim and Mai, 1991). Such mechanisms include fibre–matrix interface debonding, stress relaxation caused by fibre failure, fibre pull-out, and fractures of fibre and matrix. Because the failure strain of the reinforcing fibres is usually larger than that of the cementitious matrix material, the matrix cracks before the fibres fracture and the major contribution to the total fracture energy comes from work involved in pulling out the fibres from the matrix. An example of fibre reinforced cementitious composites is the high strength steel fibre reinforced concretes. The total fracture energy in this case is given by the sum of the fibre pull-out work and the matrix fracture work including the fibre–matrix debonding work. The work of fibre pull-out gives by far the largest contribution to the fracture energy. For many glass and polymeric fibre reinforced cementitious materials the fibre length is often longer than the critical value l_c so that those fibres which are embedded to a depth of more than $l_c/2$ will break first before they are pulled out; those fibres embedded to a depth of less than $l_c/2$ are simply pulled out. The total fracture energy is now the sum of the work of pulling out of fibres over a length of l_c, the energy absorbed by stress redistribution, the work of fracture of the matrix and fibres, and the work of fibre debonding. Generally, the stress redistribution term is small and can be ignored. Also the fracture work for high strength brittle fibres such as glass, asbestos, carbon, etc. is negligible. However, for ductile fibres like Kevlar and polypropylene, both of which are used for reinforcing cementitious matrices, the work of fracture can be quite substantial.

Fibre reinforcement in cementitious matrices is usually randomly orientated and this affects the fibre pull-out contribution to the fracture work. The pull-out term will have to be multiplied by an orientation efficiency

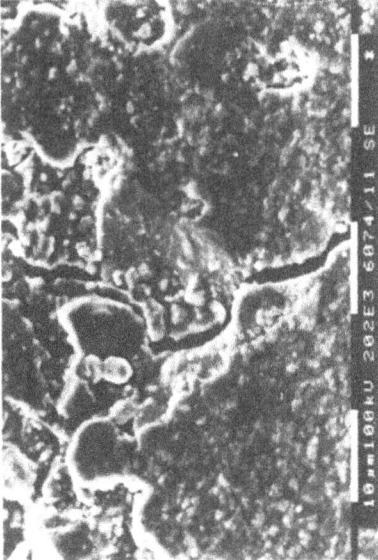

(a)

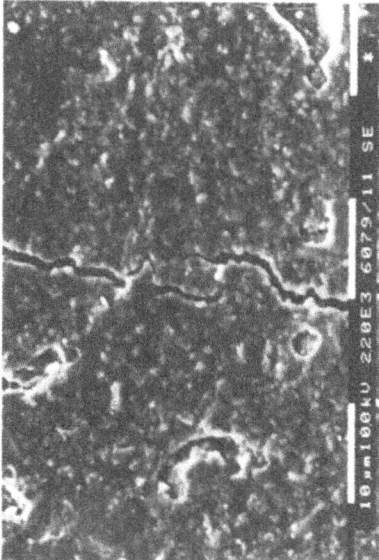

(b)

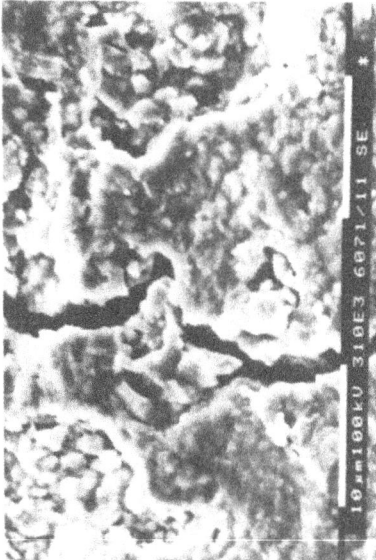

(c)

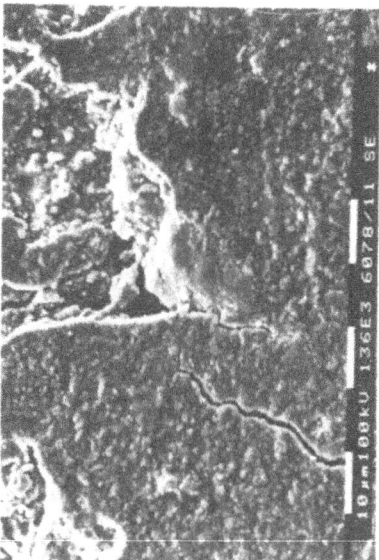

(d)

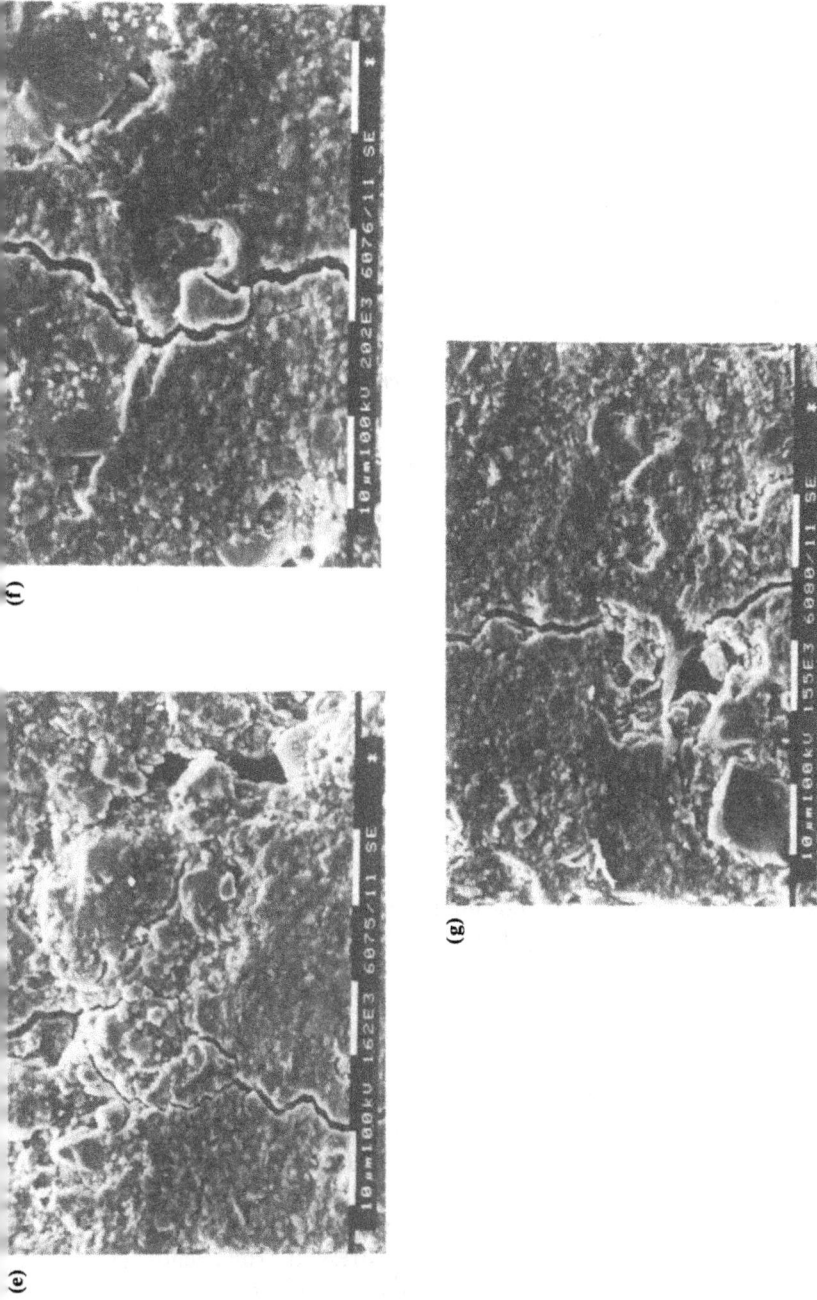

Figure 2.24 Crack bridging mechanisms in MDF paste showing (a) and (b) frictional interlocking; (c) and (d) offset cracks bridged by untorn ligaments; (e) and (f) grain-localized bridging; and (g) polymer fibril bridging.

factor depending on the degree of randomness. While this has an adverse effect of reducing the pull-out work the fact that the fibres are randomly orientated across a fracture plane can introduce some toughness enhancement in certain types of fibre cements such as steel fibre concretes where the fibre length is less than l_c. The toughness enhancement occurs because the steel fibres lying at an angle to the fracture plane are plastically sheared and extra work is dissipated (Harris et al., 1972). A number of investigators have commented on the beneficial toughening effect due to fibre orientation in steel fibre concretes (Helfet and Harris, 1972; Naaman and Shah, 1975; Morton, 1979). Unfortunately, there is no such toughness enhancement in more flexible fibres such as glass, asbestos and cellulose. However, there can be a snubbing effect (Li, 1990) with these fibres.

2.4.3 The role of interfaces in controlling strength and toughness

In all types of composites, the strength and toughness are largely controlled by the interfacial properties. Generally, a strong bond is required for high composite strength and a weak bond is needed to achieve large composite toughness. The nature of the interface is such that there is a physical transition zone or what is commonly called an interphase layer between the cement paste and the reinforcement. The microstructural features of this zone are different from those of the bulk matrix material away from it. For mortars and concretes, this interface layer is less compact with a higher amount of calcium hydroxide crystals oriented with their c-axis normal to the aggregate surface (Larbi, 1993). The interphase layer is the weakest link in those cementitious composites that have a high density of fractures and microcracks. This interphase is also affected by the water–cement ratio. The heterogeneity and microstructural defects increase with the amount of water leading to reduced compressive strength. Larbi (1993) has shown that by adding silica fume, fly ash and metakaolinite to cement paste produces a denser and thinner interphase layer with a smaller amount of calcium hydroxide crystals. This in turn enables a better stress transfer between paste and aggregate particles so that the strength is improved.

In fibre cements there is also an interphase region which is sufficiently distinct from the bulk matrix material. The interphase region for a steel fibre reinforced cement consists of a thin duplex film 1–2 μm thick in direct contact with the steel fibre and a 10–30 μm thick layer of calcium hydroxide crystals surrounded by a highly CSH porous layer (Diamond and Bentur, 1985) as shown in Figure 2.25. Cracks propagate in this porous layer parallel to the fibre. During the pull-out of the fibres this porous layer can densify causing the frictional stress to drop (Bentur et al., 1985). The weak porous layer also acts as the source for the well known Cook–Gordon (1964) debonding mechanism in deflecting a propagating

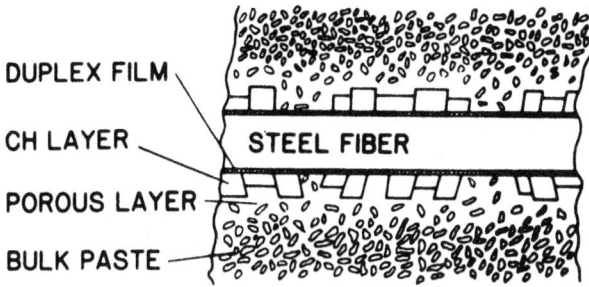

DUPLEX FILM

CH LAYER

POROUS LAYER

BULK PASTE

STEEL FIBER

Figure 2.25 Schematic description of the microstructure of a steel fibre-cement paste interface. (After Diamond and Bentur, 1985.)

crack along the porous layer (see Figure 2.26). The crack will eventually fracture the fibre before being arrested again in another porous layer. In principle this mechanism should increase the fracture toughness of the steel fibre cement composite.

The role of the interface on the mechanical properties of glass cements is more striking than that in steel fibre cements. At an early stage under normal cure conditions the bond at the interface is predominantly mechanical (Vekey and Majumdar, 1970). Pores gather at the interface and the bond strength is very low. Kim *et al.* (1993) have inferred from pull-out

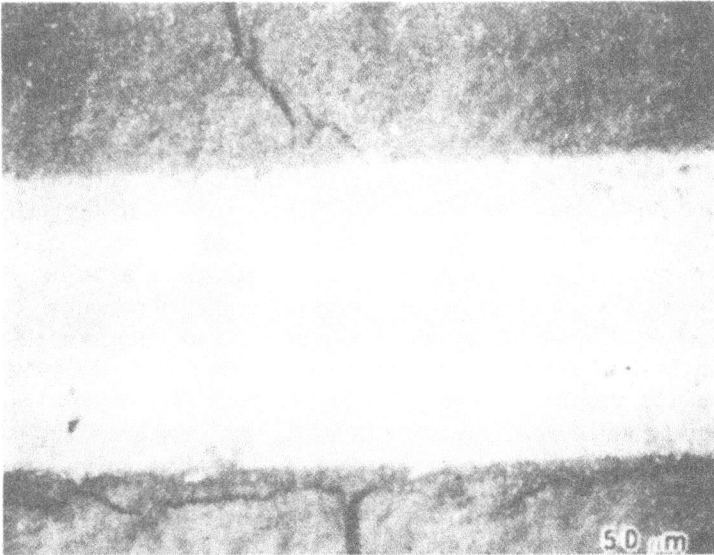

Figure 2.26 Arrest of a crack near the cement paste/steel fibre interface by the Cook–Gordon mechanism. (After Diamond and Bentur, 1985.)

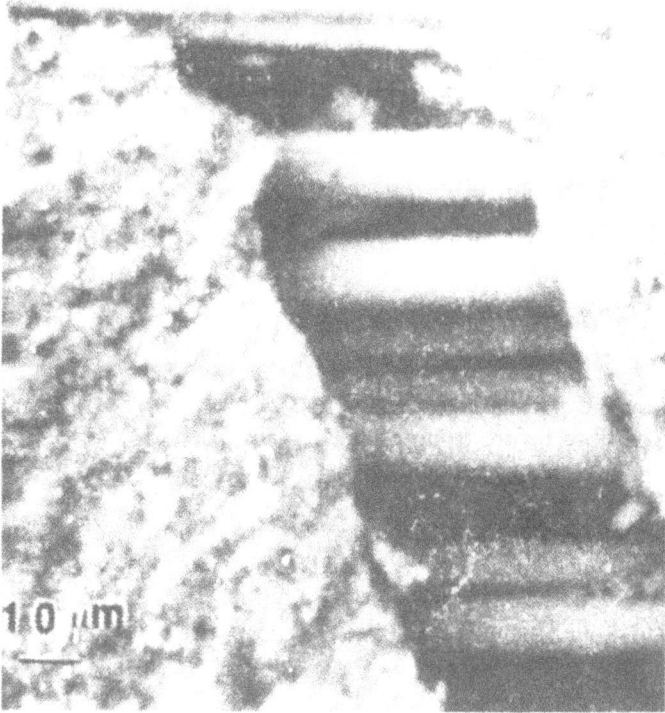

Figure 2.27 Pulled out glass filaments bridging a crack in a glass fibre reinforced cement prepared with Cem FIL-2 glass fibre strand, cured for 14 days in lime water. (After Majaumdar and Walton, 1984.)

tests on single fibres that the interfacial fracture energy, G_{IIb}, after a 3 day normal cure is only $0.03 \, J/m^2$. Even after 14 days in lime water, there is true debonding along the interface exposing the clean surfaces of the fibre (see Figure 2.27). With age the CH layer forming at the interface, when the cement is hydrated, becomes progressively more crystalline and less porous. The interfacial bond strengthens as a result of the chemical reaction and also because of an increase in fibre/matrix contact area. Accelerated ageing in lime water at elevated temperatures gives the same result.[2] When the interface is completely mature the interfacial bond energy is about $20 \, J/m^2$ (Kim *et al.*, 1993). The fracture surface of a glass fibre reinforced cement paste that has been subject to accelerated ageing is shown in Figure 2.28. The fibres are broken and there are cement hydration products between the empty spaces of adjacent fibres.

---

[2] One day of accelerated curing at 50°C in lime water is equivalent to 100 days in water at 10°C.

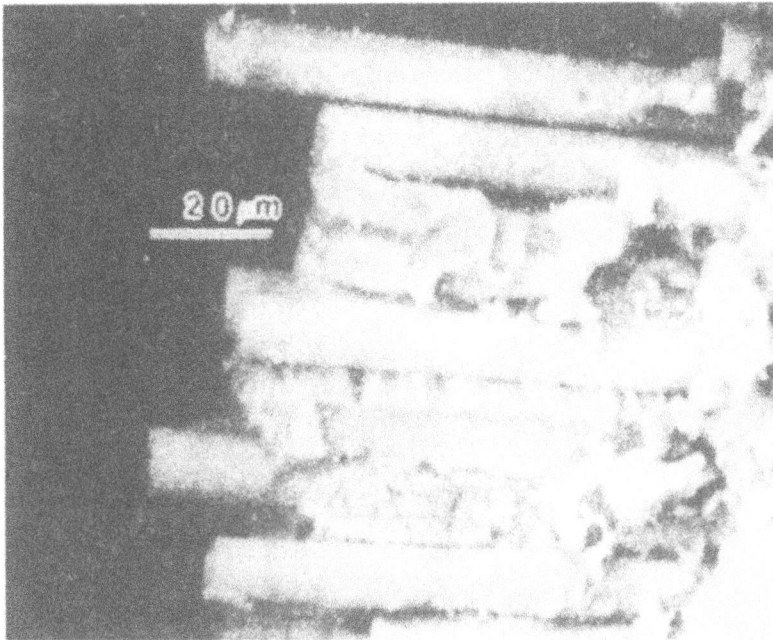

Figure 2.28 Broken glass filaments observed at a crack in a glass fibre reinforced cement prepared with Cem FIL-1 glass fibre strand, after accelerated ageing. The spaces between the filaments are filled with hydration products. (After Majaumdar and Walton, 1984.)

2.5 Summary

The fracture of cementitious materials is typified by the presence of a fracture process zone (FPZ) in which all kinds of micro-failure mechanisms take place. Such processes include microcracking, crack deviation, crack branching and cement-aggregate interface debonding which all contribute to the fracture energy.

The realization that the sizes and shapes of the pores in cement pastes are strength-limiting factors has led to the development of polymer modified cement pastes using the polymer both as a processing aid to reduce the sizes of the pores and a toughening agent to enhance its specific work of fracture. These new MDF cement pastes have rising R-curve characteristics. An effective way to make high strength cements is to engineer the interfacial properties between the paste and the aggregates. The addition of silica fume and fly ash, etc., is useful in changing the structure and properties of the interphase layer to increase the strength of concrete.

The fracture of cementitious fibre composites is characterized by the development of a fibre bridging zone (FBZ) in the wake of the continuous tip of the matrix crack and matrix FPZ. The precise position of the crack tip that divides these two zones is hard to identify. There are many micro-failure mechanisms that take place in the FBZ including fibre–matrix interface debonding, fibre fracture, fibre pull-out, and shear yielding of the fibre. Many of these processes depend on the nature and properties of the fibre–matrix interfaces that can vary between different composites. Design of new tough and strong cement composites therefore depends on optimal control of the interfacial properties.

3 Fracture parameters for cementitious materials

3.1 Introduction

In the previous chapter it is shown that the failure of unreinforced cementitious materials is characterized by the development of a fracture process zone (FPZ) at the notch tip. For fibre reinforced composites, there is an additional fibre bridging zone (FBZ) formed immediately behind the FPZ. The dimensions of the fracture process zone and the fibre bridging zone are difficult to determine precisely and it is also difficult to distinguish between the two zones. Consequently, many experimental methods have been developed to enable the sizes of the FPZ and the FBZ to be measured in cementitious materials and fibre reinforced composites.

To predict the crack-resistance behaviour and the fracture strength of cementitious structures it is necessary to know the constitutive relationships of material in the FPZ and the FBZ. For the unreinforced cementitious materials this constitutive relationship is best described by a strain-softening characteristic in terms of a closure stress and a crack opening displacement (or strain) in the FPZ. Experimental techniques to determine these relationships for mode I and mixed mode fractures and models for these relationships are discussed.

The most important parameter to define the mechanical behaviour of cementitious materials is the fracture energy. This parameter is usually assumed to be a material constant, but does vary slightly with size. The standard RILEM (1985) method of measuring the fracture energy is presented. The stress-displacement relationship of fibre reinforced cement composites, both Type I composites, whose fibre volume fraction is greater than the critical value and whose fibres alone can bear a higher stress than that necessary to crack, and Type II where the stress sustained at matrix cracking is the maximum value, are discussed.

3.2 Experimental techniques for measurement of the FPZ and FBZ

3.2.1 Measurement of fracture process zone in cementitious materials

A comprehensive review of the experimental techniques and methodologies to detect the shape and size of the fracture process zone at the notch tip in unreinforced cementitious materials has been given by Mindess (1991a,b). For convenience, these may be classified into direct and indirect methods. The direct methods involve either surface measurements such as (a) optical microscopy, (b) scanning electron microscopy, (c) resistance strain gauges, and (d) interferometry techniques, or measurements through the specimen interior which include (a) X-ray techniques, (b) mercury penetration measurements, (c) dye penetrants, (d) ultrasonic pulse velocity, (e) infrared vibrothermography, and (f) acoustic emission. The indirect methods are those of (a) compliance measurements and (b) multi-cutting techniques.

Numerical methods can also be used to estimate the FPZ as a function of specimen size and geometry and crack growth, if the closure stress–crack face separation relationship (or strain-softening) is known, based on fracture mechanics analysis (for example, see Cotterell and Mai, 1987). Similar numerical techniques can be applied to fibre reinforced cements. For both analyses it is necessary to know the stress–crack opening relationships for the FBZ and the FPZ. The fracture mechanics procedure for the prediction of the FPZ and FBZ as a function of specimen geometry and size as well as crack length is also given by Cotterell and Mai (1988b).

Not all of the experimental techniques mentioned above are useful to detect the size of the FPZ in cementitious materials. For example, X-ray (Slate and Hover, 1984), mercury penetration measurements (Schneider and Diederichs, 1983), dye penetrants (Swartz and Go, 1984) and resistance strain gauges (John and Shah, 1986) are not always sensitive enough to define the process zone. Mixed results have also been reported with a variety of techniques based on both optical and scanning electron microscopy. Some investigations confirm the existence of the FPZ but others do not, even though the same technique is used. In optical microscopy these methods include diffuse illumination (Eden and Bailey, 1986), thin sections with impregnated epoxy containing a fluorescent dye (Knab et al., 1984, 1986; van Mier, 1989). In scanning electron microscopy, in-situ observations of specimens in the chamber of the SEM (Mindess and Diamond, 1982a,b; Diamond and Bentur, 1985; Tait and Garrett, 1986), back-scattered electron imaging (Baldie and Pratt, 1986; Knab et al., 1986), and replica technique (Ringot et al., 1987; Bascoul et al., 1989a,b) have also been used. Similarly, averaging techniques based on total damage such as the infrared vibro-thermography (Luong, 1986) and ultrasonic pulse velocity (Alexander and Blight, 1986; Chhuy et al., 1986; Alexander, 1988; Reinhardt and Hordijk, 1988; Berthaud, 1989; Alexander et al., 1989) have also yielded mixed results in measuring the size of the FPZ.

Mindess (1991a,b) has also suggested that the compliance measurement technique is not a good method to determine either the 'total' (Kobayashi et al., 1985) or the 'effective' crack length (Karihaloo and Nallathambi, 1989). Although the laser holographic and speckle interferometry techniques are the most sensitive, the definition of the FPZ is often in terms of some limiting strain and it is not always possible to specify the FPZ

(Ferrara and Morabito, 1989). However, Moiré interferometry has been used successfully to measure the FPZ of concrete by Du *et al.* (1987, 1989) and Raiss *et al.* (1989) though the values seem rather large. Acoustic emission measurements can detect the growth of the FPZ and the evolution of damage in a specimen during loading. Berthelot and Robert (1987) found a microcracked zone (i.e. FPZ) ahead of the continuous crack and that this zone grew with crack extension. Maji and co-workers (Maji and Shah, 1988; Maji *et al.*, 1990) also observed acoustic emission events, after the maximum load, to occur both ahead of (which indicates a frontal process zone) and behind (which confirms the existence of ligament bridging) the visible crack tip.

3.2.2 The multi-cutting technique to measure the FPZ in cementitious materials

A simplified view of a critical or saturated FPZ in an unreinforced cementitious material is shown in Figure 3.1. The continuous crack is traction-free. Within the FPZ the effective Young's modulus is reduced from that of the undamaged material E to E' due to the presence of the microcracks. The closure stress associated with the bridging grains and the localized damage is a maximum f_t at the tip of the FPZ and decreases to zero at the continuous crack tip where the crack opening displacement is δ_f. For the same continuous crack length, a, the compliance of a saw cut specimen $C(a)$ is smaller than that

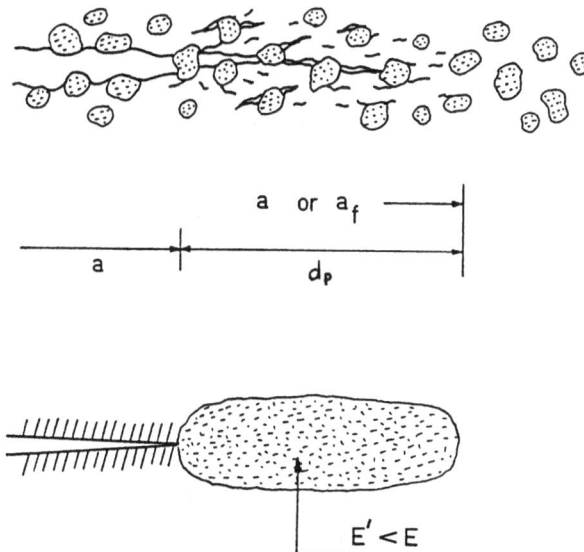

Figure 3.1 Schematic representation of a fracture process zone (FPZ) in a cementitious material. (After Wittmann and Hu, 1991.)

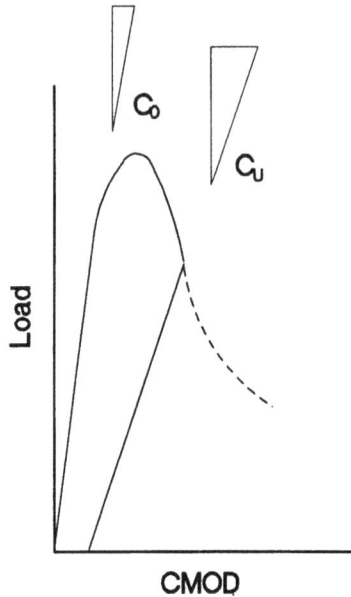

Figure 3.2 Increase in compliance from C_0 for the initial crack length to C_u with the growth of the crack and FPZ. (After Hu and Wittmann, 1989.)

of a natural crack containing an FPZ, $C_u(a)$ (see Figure 3.2). The compliance for a saw cut notch can be obtained experimentally using small loads or obtained from elastic solutions. In the latter case the compliance is readily obtained from the stress intensity factor by the method of Castigliano as described in section 1.6.1.

The multi-cutting technique was first used by Knehans and Steinbrech (1982) to determine the crack wake bridging effect in a coarse-grained alumina. Similar methods have been employed to estimate the FPZ size of mortars and cement pastes (Hu and Wittmann, 1989, 1990, 1991, 1992a; Wittmann and Hu, 1991). Essentially, the bridging zone is consecutively removed by a small amount by saw cutting and the compliances measured after each removal step. Before any bridges are removed there is no change from the original compliance, C_u. However, the compliance, C_p, will increase if any bridging ligaments are removed by cutting. When the whole FPZ has been cut through the compliance, C_p, is equal to the elastic compliance. A schematic illustration is shown in Figure 3.3, where d_p is the size of the saturated bridging zone or critical FPZ.

Compliance measurements made on wedge-opening-loaded (WOL) mortar specimens after saw cutting along the FPZ by Hu and Wittmann (1992a) are shown in Figure 3.4. The WOL specimen dimensions were $200 \times 197 \times 15\,\text{mm}$ with $a/W = 0.4$. Grooves $2\,\text{mm}$ deep were machined on both sides to guide the fracture. The water/cement and sand/cement

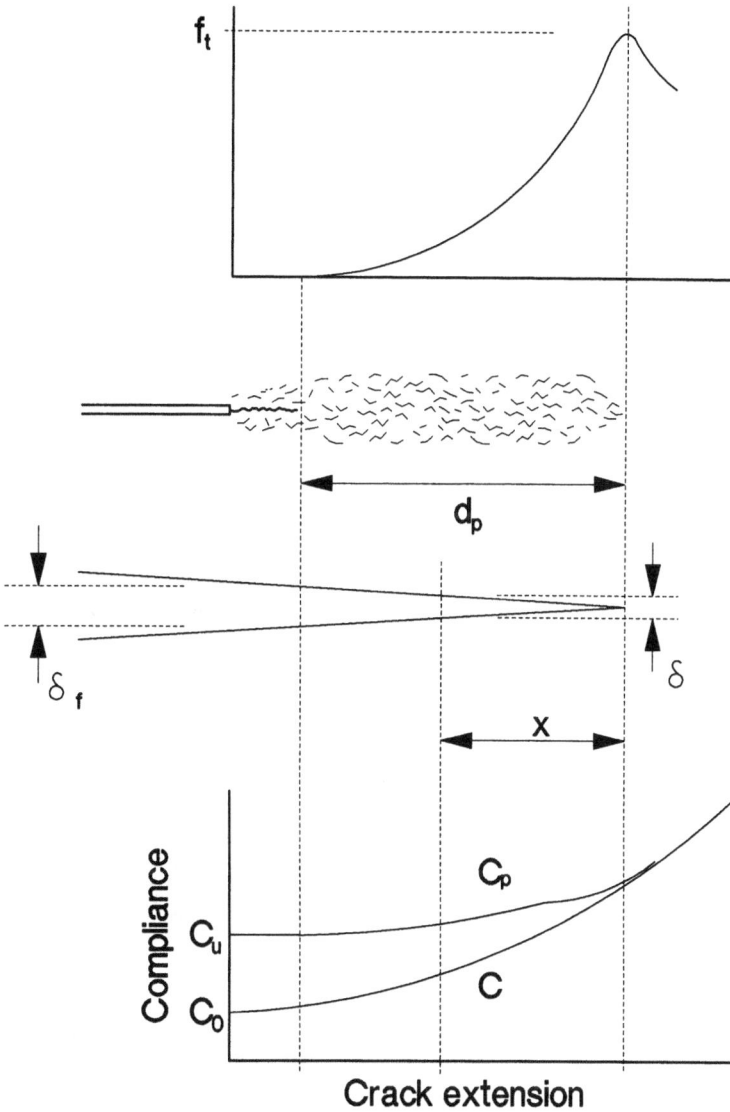

Figure 3.3 Schematic distribution of stress in the FPZ and the change in compliance due to saw cutting into the FPZ. (After Hu and Wittmann, 1989.)

ratios for the specimens were 0.4 and 1.5, respectively. The specimen was loaded to increase the initial compliance from 0.15 to 0.9 m/MN. There was no change in the saw cut compliance until the saw cut was greater than about 32 mm. Hence the continuous crack was about 32 mm long. With longer saw cuts there was an increase in compliance. A curve fitted to the saw cut compliances for $\Delta a > 32$ mm intersects with the

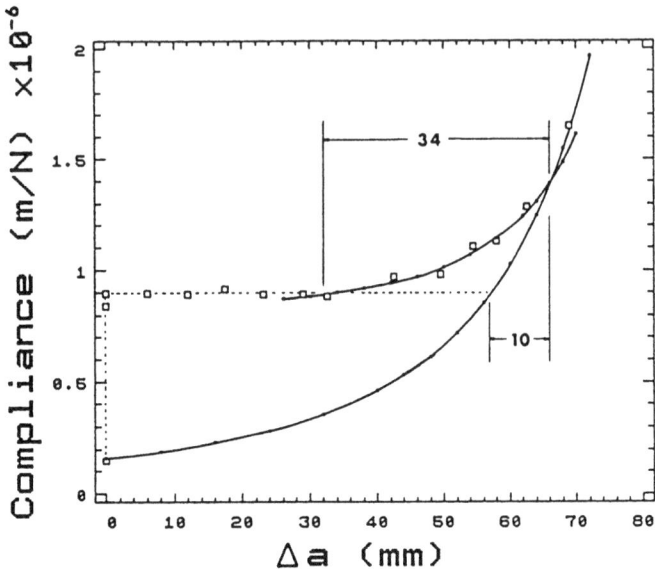

Figure 3.4 Use of saw cutting technique to determine FPZ in a WOL mortar specimen. (After Hu and Wittmann, 1989.)

theoretical compliance at a crack length of 66 mm giving the FPZ length as 34 mm.

The size of the FPZ depends upon the specimen size and geometry as well as on the material properties of the cementitious material as can be seen from the multi-cutting experiment shown in Figure 3.5 for the same mortar specimen used in the experiment shown in Figure 3.4. The initial a/W was again 0.4. In the first experiment the length of the FPZ was found to be 43 mm. After the first saw cutting experiment the specimen was reloaded to extend the FPZ. The compliance curve for the second saw cutting experiment intersects the elastic compliance with the theoretical compliance curve to give a much smaller FPZ of 12 mm. The shortening of the FPZ as the continuous crack grows can be predicted theoretically (see, for example, Cotterell and Mai, 1987, 1988b).

In using the multi-cutting technique, some precautions are required to avoid any additional damage or crack growth due to saw cutting and subsequent reloading. For concretes and mortars with relative large FPZs, a fast cutting speed and a slow feeding rate plus reloading to only half the critical level required for further crack extension are sufficient. However, for cement pastes with much less pronounced FPZs, although the saw-cutting technique may, in principle, be applied, in practice it is not easy because the accuracy of removal of the bridging zone has to be within a millimetre. The same difficulty is experienced in the case of FPZs in quasi-brittle ceramics.

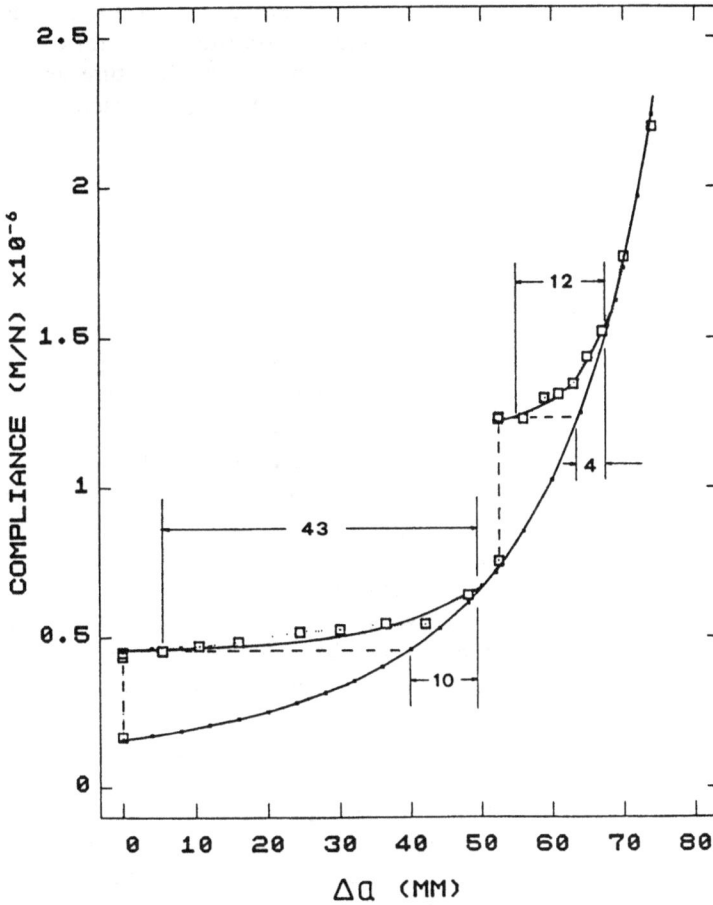

Figure 3.5 Repetition of saw cutting on one specimen after the previous FPZ has been removed. (After Hu and Wittmann, 1992.)

3.3 Measurement of the mode I strain-softening relationship for cementitious materials

The strain-softening relationship for cementitious materials can be expressed in terms of either a stress-strain relationship or a stress-displacement relationship measured at the edge of the FPZ. For mode I fracture these two approaches are very similar because the width, w, of the FPZ is a quasi-material constant. Since the FPZ is narrow it can usually be assumed that the elastic contribution to the displacement is negligible.

The stress-displacement relationship for the FPZ of a cementitious material can in theory be obtained directly from a tension test because, though the initial microcracking activity is dispersed, the deformation becomes localized in a FPZ after the maximum load has been reached. The most important parameter that can be obtained from the mode I stress-displacement curve is the mode I fracture energy, G_{If}, which is the specific work of fracture necessary to develop and completely fracture the FPZ. Thus the mode I fracture energy is given by

$$G_{\text{If}} = \int_0^{\delta_f} \sigma \, d\delta \tag{3.1}$$

3.3.1 The direct tension method

In the direct tension method the specimens used to obtain the stress-displacement curves can either be plain (Petersson, 1985; Guo and Zhang, 1987) or have shallow notches on both edges to locate the FPZ (Reinhardt, 1984; Gopalaratnam and Shah, 1985; Reinhardt et al., 1986; Hordijk et al., 1987; Rots, 1988). The tensile area of the specimens must be large enough to contain a reasonable number of aggregates. Testing can be performed in a servo-hydraulic machine driven by LVDTs mounted on the specimens (Reinhardt, 1984; Gopalaratnam and Shah, 1985; Reinhardt et al., 1986; Hordijk et al., 1987) or in a universal testing machine with heavy springs in parallel with the specimen to stabilize the fracture (Petersson, 1985; Guo and Zhang, 1987). Apart from the problem of stabilizing the fracture, the main problem in making direct tensile tests is that the FPZ does not establish itself instantaneously right across the specimen. Hordijk et al. (1987) have made a detailed study of the mode of failure in the direct tension test and found that the deformation was often asymmetrical especially if the specimen was long. If the deformation was asymmetrical a 'bump' was introduced in the stress-displacement curve which was partly associated with local instability referred to as 'snap-back'. This behaviour has been modelled using finite elements (Rots, 1988).

Because the direct tension test is not as simple as first appears, indirect methods of obtaining the stress-displacement curve have been proposed.

3.3.2 The J-integral method

The J-integral can be used to obtain the stress-displacement curve from a compact tension test (Li, 1984; Li et al., 1987). If in the J-integral the integration is performed around the edge of the FPZ, which has a thickness t, the integral reduces to

$$J \approx \int_0^a \sigma \frac{\partial \delta}{\partial a} \, dx + \tfrac{1}{2} f_t \epsilon_e t = - \int_0^a \sigma \frac{\partial \delta}{\partial x} \, dx + \tfrac{1}{2} f_t \delta_e \tag{3.2}$$

where ϵ_e and δ_e are the elastic strain and displacement in the FPZ at the ultimate strength of the material which yields

$$J(\delta) = \int_0^\delta \sigma \, d\delta \qquad (3.3)$$

When a continuous crack forms, $\delta = \delta_f$ and $J(\delta_f) = J_{Ic}$ can be identified with G_{If}. J can be calculated experimentally using the energy interpretation given by eqn 1.59. If the load/load-line deflection $(P - \Delta)$ for two compact tension specimens of slightly different crack lengths a and $a + da$ are obtained (see Figure 1.28) the J-integral for a given deflection, Δ, is given by

$$J(\Delta) = \frac{\text{Area} \, (OAB)}{B \, da} \qquad (3.4)$$

The stress-displacement relationship is given by

$$\sigma(\delta) = \frac{\partial J(\delta)}{\partial \delta} \qquad (3.5)$$

and hence combining eqns 3.4 and 3.5

$$\sigma(\delta) = \frac{1}{B \, da} \frac{\partial \, \text{Area} \, (OAB)}{\partial \delta} \qquad (3.6)$$

Since in the experimental method only finite differences are measured, the CTOD, δ, measured simultaneously with the load-point deflection, is taken as the average of the CTODs obtained from the two specimens. The CTOD has to be measured from two points on the specimen a finite distance apart and thus an elastic displacement not associated with the FPZ is included in the measurement. This elastic component can be subtracted from the measured COD to give the true CTOD across the FPZ.

The main problem with Li's (1984, 1987) method is that the stress-deflection curve has to be obtained from two specimens. Since cementitious materials are very inhomogeneous and no two supposedly identical specimens behave identically the same, there can be considerable scatter in the stress-displacement curves. Rokugo et al. (1989) have proposed using a single specimen for the determination of the J-integral. However, the details of their proposition are not very clear. The similarity arguments used to obtain J for a deep notched bend specimen (see section 1.9.2) rely on the FPZ being small compared with the ligament. If the FPZ is not small compared with the ligament length, as is usually the case with cementitious materials, the similarity argument fails. The η factor used by Rokugo et al. (1989) is not stated explicitly.

3.3.3 Indirect method using a notch bend specimen

An alternative indirect method that uses a load-deflection curve obtained from a notched bend test has been suggested by Chuang and Mai (1989). In this method the strain distribution across the notched section is assumed to be linear and global equilibrium of forces and moments are satisfied. A disadvantage of this method, which is discussed in more detail in section 4.2, is that the width of the FPZ as well as the stress-strain relationship affects the load-deflection curve. However, the width of the FPZ can be inferred from the fracture energy G_{If}.

3.3.4 Compliance methods

Hu and Wittmann (1989, 1990, 1991, 1992a) have developed a compliance method, combined with cutting through the FPZ, to determine the stress-displacement relationship. The method is explained using the compact tension specimen, which is a suitable geometry, but other geometries may be used. An elastic compliance calibration curve, C, is obtained as a function of the crack growth from some datum crack length, either experimentally from a lightly loaded specimen with saw-cut cracks, or from theoretical compliance expressions and an effective Young's modulus that normalizes the compliance for the datum crack length. The specimen is then loaded to produce a FPZ. In this illustration of the method, it is assumed that the specimen is loaded until the FPZ is fully developed. The crack line is now sawn through to increase the datum crack length in stages, measuring the new compliance C_p at every stage, as described in section 3.2.2. If the first cutting stages do not cut into the FPZ, the compliance remains constant at its initial uncut value, C_u. On further cutting the compliance increases as the FPZ is penetrated. It is assumed that the crack faces remain straight.[1] Thus, at a distance x from the tip of the FPZ (see Figure 3.3), the COD, δ, is given by

$$\frac{x}{d_p} = \frac{\delta}{\delta_f} \tag{3.7}$$

and the fracture energy, G_{If}, is given by

$$G_{If} = \int_0^{\delta_f} \sigma \, d\delta = \frac{\delta_f}{d_p} \int_0^{d_p} \sigma \, dx \tag{3.8}$$

If the FPZ translates with no change in shape the second of Barenblatt's (1959, 1962) hypotheses is valid. This restriction, which is not mentioned by Hu and Wittmann (1991), means that strictly the method is only accurate for large specimens. The problem would not be nearly as accurate with the double-cantilever-beam specimen (see section 4.5.2). Consider the load-crack mouth opening (P-CMOD) relationship schematically illustrated in Figure 3.6, where P_u is the load without cutting into the FPZ and P_x is the load after part of the FPZ has been cut away keeping the CMOD constant.

---

[1] This assumption has been shown to be quite accurate (Foote *et al.*, 1986; Cotterell *et al.*, 1988, 1992; Cotterell and Mai, 1988a) and is discussed in section 4.5.2.

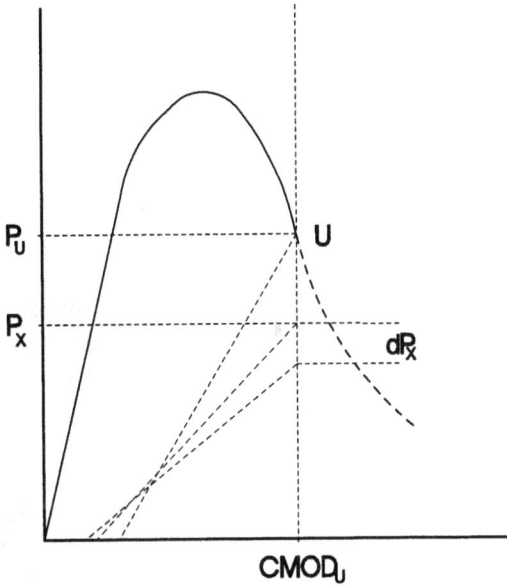

Figure 3.6 Load-CMOD during partial removal of the FPZ. (After Hu and Wittmann, 1991.)

If the residual deflection is neglected the load, P_x, after cutting into the FPZ leaving a length x intact, is given by

$$P_x = \frac{\text{CMOD}_u}{C_p} \tag{3.9}$$

The change in the load due to cutting, dP_x, depends upon the bridging stress, σ, removed. Hu and Wittmann (1991) introduce a coefficient $k(a, x)$, dependent on the length, a, of the combined true crack and FPZ and x, the remaining uncut FPZ, defined by

$$k(a, x) = \frac{\sigma(x)}{\left(\dfrac{1}{B}\dfrac{dP_x}{dx}\right)} \tag{3.10}$$

Using eqns 3.8 and 3.9, Hu and Wittmann (1991) obtained the expression

$$G_{\text{If}} = k\frac{\delta_f \text{CMOD}_u[C_p(0) - C_p(d_p)]}{Bd_p C_p(0)C_p(d_p)} \tag{3.11}$$

for the fracture energy, G_{If}, by assuming that dk/dx is small and can be neglected. Solving for k and substituting in eqn 3.10 yields the expression for the bridging stress

$$\sigma(x) = G_{\text{If}}\left(\frac{d_p}{\delta_f}\right)\frac{C_p(0)C_p(d_p)}{[C_p(0) - C_p(d_p)]}\frac{d}{dx}\left(\frac{1}{C_p}\right) \tag{3.12}$$

the tensile strength, f_t, is given by

$$f_t = G_{If} \left(\frac{d_p}{\delta_f}\right) \frac{C_p(d_p)}{C_p(0)} \frac{1}{[C_p(0) - C_p(d_p)]} \frac{dC}{da} \qquad (3.13)$$

because

$$\left[\frac{dC(a)}{da}\right]_a = -\left[\frac{dC_p(x)}{dx}\right]_0 \qquad (3.14)$$

The difficulty with eqn 3.14 is that the absolute magnitude of the bridging stresses depends upon the critical CTOD, δ_f, which is difficult to measure (Hu and Wittmann, 1992a). However, the ratio of the bridging stress to the tensile stress does not depend upon δ_f and is given by

$$\frac{\sigma}{f_t} = -\frac{d}{dx}\left(\frac{1}{C_p(x)}\right) \bigg/ \frac{d}{da}\left(\frac{1}{C(a)}\right) \qquad (3.15)$$

If the tensile strength, f_t, is obtained from an independent tensile test then eqn 3.15 can be used to give the absolute bridging stresses. The closure stress in the FPZ of a WOL specimen obtained by their multi-cutting technique and eqns 3.13 and 3.15 are shown in Figure 3.7. The average values of the important parameters are: G_{If} = 21.8 N/m, δ_f = 0.02 mm, and f_t = 3.71 MPa. Even though the FBZ varies from 12 to 43 mm, Figure 3.7 shows that there is a reasonable small scatter in the curve for the bridging stress. The accuracy of the method has been demonstrated by Alvaredo et al. (1989) who used this stress-displacement relationship to calculate the theoretical

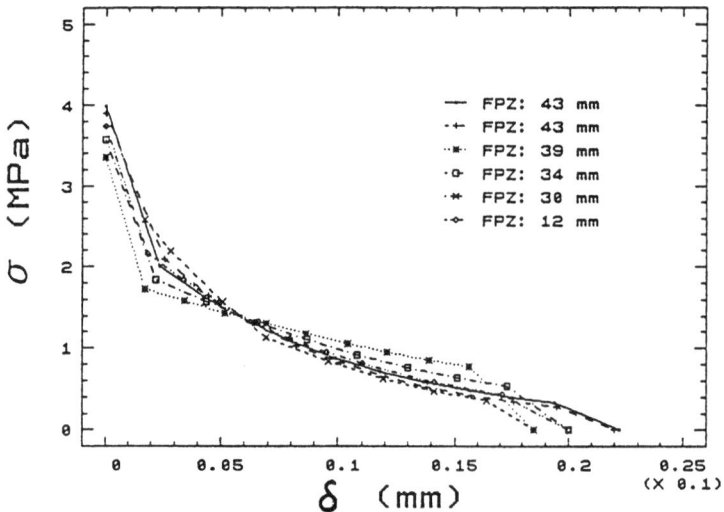

Figure 3.7 Strain-softening curves derived from six mortar specimens by the multi-cutting technique. (After Hu and Wittmann, 1992.)

load-deflection curve which was in good agreement with the experimental results.

To avoid the cumbersome procedure of multi-cutting, Hu and Mai (1992a,b) have extended the method described above so that saw cutting is not required. The method hinges on the realization that the unloading compliance, C^*, of a partially grown FPZ is equivalent to a saturated FBZ that has been partly cut through. The argument leads to the same expression as eqn 3.15. Hu and Mai (1992a) derived a function, ϕ, given by

$$\phi = \frac{C(a)}{C^*(a)} \frac{[C(a) - C^*(a)]}{dC/da} \qquad (3.16)$$

where Δa is the size of the partial FPZ. Assuming that the stress-displacement relationship can be modelled by the power law

$$\frac{\sigma(x)}{f_t} = \left[1 - \frac{\delta(x)}{\delta_f}\right]^n \qquad (3.17)$$

where $n > 0$, Hu and Mai (1992a) showed that the function, ϕ, can be expressed in terms of the size of the fully developed FPZ, d_p, the size of the partially developed FPZ, Δa, and the exponent, n, by

$$\phi = \frac{d_p}{(n+1)} \left[1 - \left(1 - \frac{\Delta a}{d_p}\right)^{n+1}\right] \quad \text{for } \Delta a < d_p$$

$$= \frac{d_p}{(n+1)} \qquad\qquad\qquad \text{for } \Delta a > d_p \qquad (3.18)$$

A plot of ϕ against Δa can be constructed using eqn 3.16, because the compliance is easily obtained from experiment. A schematic relationship of ϕ against Δa is shown in Figure 3.8. The plateau value of the curve in

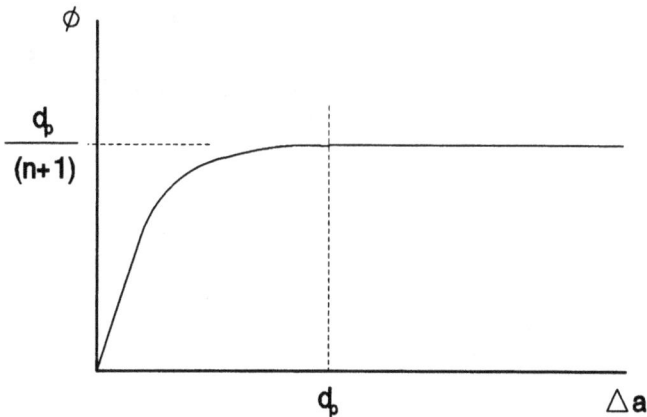

Figure 3.8 Schematic variation of ϕ with Δa.

Figure 3.8 gives the ratio $d_p/(n+1)$, and the size of the fully developed FPZ, d_p, is the crack extension necessary to reach the plateau value. Hence, the exponent n can be determined and, if f_t is determined separately, the bridging stress is known. An application of this method to a fibre cement composite is given in section 3.8.

3.3.5 Comment on the methods of determining the stress-displacement relationship

All of the methods of determining the stress-displacement relationship for the FPZ described above have their limitations and the best method is to assume a suitable form for the relationship as described in section 3.4 and then to model the load-deflection curve and find the stress-displacement parameters that give the best fit to the experimental data. This method is described in section 4.5.

3.4 Modelling the mode I strain-softening relationship for cementitious materials

Under symmetrical loading the principal stresses in the FPZ are essentially aligned to the line of symmetry. It is often useful to consider the strain as composed of two parts: an elastic strain, ϵ^{co}, derived from the uncracked material and a cracked component, ϵ^{cr}, due to the micro-cracking activity within the FPZ. It is assumed that strain-softening does not produce a strain parallel to the FPZ. In modelling the behaviour of the FPZ local cracked-strains cannot be used or localization will occur when the model is used to predict load-deflection curves for cementitious specimens. For this reason the cracked strain is usually taken as the average through the thickness of the FPZ though other non-local definitions of the cracked strain are discussed in section 4.4. Provided the width of the FPZ is constant, the stress-displacement relationship measured at the edge of the FPZ is equivalent to the stress-strain relationship.

The simplest approximation to the stress-displacement curve is the linear approximation

$$\frac{\sigma}{f_t} = 1 - \frac{\delta}{\delta_f} \tag{3.19}$$

This approximation is sufficiently accurate for many purposes, but cannot accurately predict the post-ultimate load behaviour of laboratory specimens, and a variety of more complicated expressions have been suggested.

The power law expression given by eqn 3.17, which gives zero slope at final separation or alternatively the power law

$$\frac{\sigma}{f_t} = 1 - \left(\frac{\delta}{\delta_f}\right)^k \tag{3.20}$$

with the index k in the range $0.2 < k < 0.4$, has been suggested by Reinhardt (1984) for concrete. However, a problem with the power law of eqn 3.20 is that it has a finite slope at final separation which makes its use to numerically simulate the load-deflection curve for notched concrete beams little better than the simpler linear law (Rots, 1986). Exponential relationships (Reinhardt, 1984; Gopalaratnam and Shah, 1985; Reinhardt et al., 1986) are better at modelling the load-deflection curves of specimens. Since the exponential expressions are asymptotic to zero at infinity, they are either truncated or have an additional linear term to bring them to zero at a finite critical crack opening displacement. A typical expression (Reinhardt et al., 1986) is

$$\frac{\sigma}{f_t} = \left[1 + \left(\frac{C_1\delta}{\delta_f}\right)\right]^3 \exp\left(-\frac{C_2\delta}{\delta_f}\right) - (1 + C_1^3)\left(\frac{\delta}{\delta_f}\right)\exp(-C_2) \qquad (3.21)$$

Ingraffea and his coworkers (Ingraffea and Gerstle, 1984; Ingraffea and Saouma, 1984; Ingraffea et al., 1984; Ingraffea and Panthaki, 1985) have used a semi-reciprocal expression with a linear term to make the stress zero at a finite displacement. However, the bilinear stress-displacement relationship, shown in Figure 3.9, has the advantage of simplicity and has enough parameters to enable the load-deflection curve to be accurately predicted by numerical simulation. The bilinear curve is, in the authors' opinion, the best expression if high accuracy is required. Petersson (1985) found that the parameters that gave a good fit to the stress-displacement curve for a range of concretes were $s = 0.33$ and $v = 0.22$. Wittmann and his co-workers (Roelfstra and Wittmann, 1986; Wittmann et al., 1987) found that with the parameters $s = 0.12-0.19$ and $v = 0.16-0.22$ they obtained the best fit to

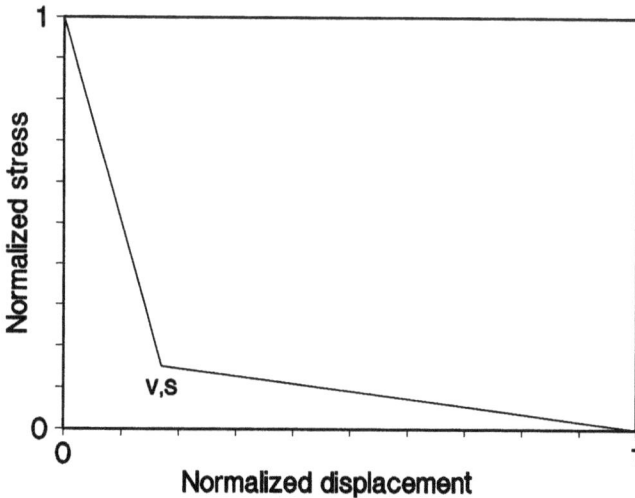

Figure 3.9 Bilinear stress-displacement relationship.

the load-deflection curves for notched concrete beams under three-point loading. The new draft CEB-FIP Model code (1990) recommends $s = 0.15$ (see section 8.1). For mortar, v is in the same range as it is for concrete, but s appears to be much smaller and in the range 0.02–0.11. With a bilinear stress-displacement curve the mode I fracture energy is given by

$$G_{If} = \tfrac{1}{2} f_t \delta_f (s + v) \tag{3.22}$$

Liaw *et al.* (1990) have proposed a trilinear stress-displacement curve where for very small crack opening displacements the stress in the FPZ remains constant at the ultimate strength f_t. However, with regard to the inherent scatter in results obtained from fracture tests on cementitious materials this refinement, which makes the stress-displacement curve more complicated, is not justified.

Smith (1994) has suggested that the stress-displacement relationship can be approximated by two regions of constant stress (see Figure 3.10). This simple stress-displacement relationship captures the essential features of strain-softening and only the crack displacement at two points is needed to determine the stress in the FPZ. The stress-displacement of Smith (1994) suggests an even simpler relationship. It is the long tail to the stress-displacement curve that causes the large FPZ. Thus there may be situations where the high stresses that can be sustained at small CTOD are modelled by a critical stress intensity factor, K_{Ic}, at the tip of the FPZ and the long tail in the stress-displacement relationship is modelled by a constant stress region. Then in the terms of Smith's stress-displacement relationship

$$K_{Ic} = \sqrt{E f_t v} \tag{3.23}$$

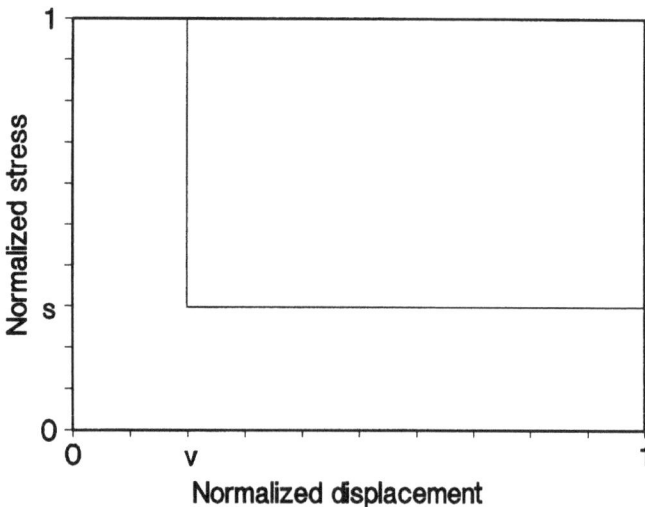

Figure 3.10 Piece-wise softening law.

A very similar technique is suggested for fibre reinforced cementitious materials in section 3.9.5 where the FPZ is modelled by a fracture toughness of the reinforced matrix and only the bridging fibre stress is modelled explicitly.

3.5   Measurement of mode I fracture energy

Although the most direct way of determining the mode I fracture energy is by means of a uniaxial tension test, the test is not easy because of stability problems. It is much easier to perform bending tests on notched specimens because these are stable in reasonably stiff testing machines. If the mode I fracture energy is a material property then it is simply given by the total work required to fracture a notch bend specimen divided by the ligament area. For these reasons a three-point notch bend test has been chosen by RILEM (1985) as a draft recommendation for measuring fracture energy.

The RILEM standard specimen is a rectangular bar notched at its centre to half the depth of the beam. The depth of the beam is at least six times the aggregate size. The beam is tested under three-point bending either under closed-loop servo control or in a stiff testing machine to ensure that the fracture is stable with the notched surface downwards. For the 100 mm deep beam, which is the smallest recommended, the required machine stiffness to ensure stability is about 10 kN/mm. To obtain the work done the load is plotted against the load-point deflection. The work done by the beam's own weight may not be negligible and the work measured from the load-deflection curve has to be corrected before the fracture energy can be calculated. In theory a correction to the work done by the applied load could be avoided if the weight of the beam is compensated by using a beam whose length is twice the span or by using weights applied to the ends of the beam. However, as Petersson (1985) has stated, there will then be a long tail to the load deflection curve and a small discrepancy in the balancing system can give rise to a significant error in the work done. Hence, a beam only slightly longer than the span is used and a correction is applied. The load-deflection curve for an uncompensated beam can be constructed from that for a compensated beam (see Figure 3.11), by shifting the zero load of the uncompensated beam to a load P_0 on the compensated beam's curve where the load P_0 gives rise to the same bending moment at the notched section as does the self-weight of the beam. Hence, if Mg is the weight of the beam between the supports then

$$P_0 = \frac{Mg}{2} \tag{3.24}$$

If Δ_0 is the deflection at fracture in the uncompensated beam then the area A_2 of Figure 3.11 is given by

$$A_2 = \frac{Mg\Delta_0}{2} \tag{3.25}$$

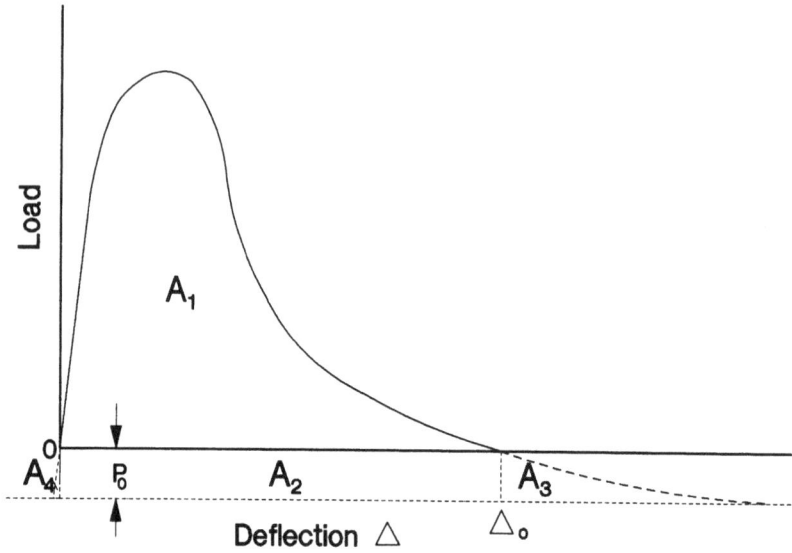

Figure 3.11 Compensated load-deflection diagram for a three-point bend test to measure the fracture energy, G_{If}.

A continuous crack will have propagated and the FPZ will have nearly reached the back surface of the beam by the time that specimen is finally broken by its own self-weight. Petersson (1985) assumed that the compression zone had degenerated almost to a point on the back surface of the specimen during the final stages of fracture, so that the halves of the beam effectively rotate about that point as rigid bodies (see Figure 3.12). During propagation the CTOD at the tip of the continuous crack is δ_f, and the angle of rotation of the beam, θ, is approximately δ_f/b, where b is the

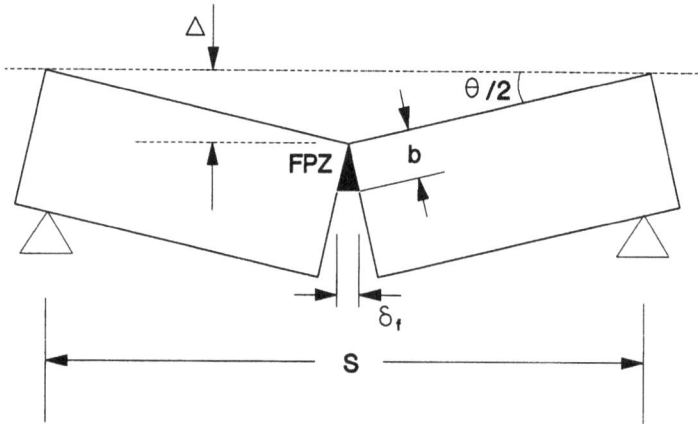

Figure 3.12 Beam during the last stages of fracture.

remaining ligament. Consequently the remaining ligament in the beam, b, is inversely proportional to the deflection, Δ, of the beam. Both the closing force in the FPZ and its moment arm about the point of rotation are proportional to the ligament length, b. Therefore, in the tail of the load-deflection curve, the load is given by

$$P = P_0 \left(\frac{\Delta_0}{\Delta} \right)^2 \qquad (3.26)$$

The work done in the tail of the compensated specimen (area A_3 in Figure 3.11) is given by

$$A_3 = \int_{\Delta_0}^{\infty} P \, \mathrm{d}\Delta \qquad (3.27)$$

which on substituting from eqn 3.26 gives

$$A_3 = \frac{Mg\Delta_0}{2} \qquad (3.28)$$

The small elastic work performed by the beam's self-weight, A_4, is negligible and hence the corrected expression for the mode I fracture energy is

$$G_{\mathrm{If}} = (A_1 + Mg\Delta_0)/Bb \qquad (3.29)$$

Hilsdorf and Brameshuber (1984) examined the effect of size on the fracture energy testing concrete and mortar beams 100, 400 and 800 mm deep. There was very little difference in the mean value of G_{If} between the two smaller concrete beams, but the fracture energy for the largest beam was 20% larger (see Table 3.1). There is much less difference in the fracture energy for the mortar beams and in fact the largest beam has the smallest value, but this value is doubtful because there may have been some instability in the fracture. For the 32 mm aggregate of the concrete used by Hilsdorf and Brameshuber (1987), the recommended minimum specimen depth (RILEM, 1985) is 300 mm. Apart from the obvious requirement that the specimen size shall be significantly larger than the aggregate, a more fundamental gauge of the size is given by the non-dimensional size, $\bar{W} = W/l_{\mathrm{ch}}$. The mortar results seem to indicate that the maximum value of G_{If} is obtained for a relative size of 0.65 or less, but the fracture energy for the concrete does not appear to have reached its limiting value until $\bar{W} > 0.65$. A summary of a wide range of tests shows similar dependence on size (Hillerborg, 1985a).

Elices, Planas and Guinea have made a study of the errors involved in the RILEM fracture energy test to determine whether the size effect in the fracture energy is real. The sources of error examined were: (a) hysteresis in the testing equipment and energy dissipated in the lateral supports (Guinea et al., 1992); (b) energy dissipated in the bulk of the specimen away from the fracture plane (Planas et al., 1992); (c) underestimation of the work in the tail of the load-deflection curve. The first error (a) is very minor contributing an

Table 3.1 Size effect on fracture energy and characteristic length (data from Hilsdorf and Brameshuber, 1984)

(i) Tensile strength and Young's modulus

	f_t (MPa)	E_c (GPa)
Concrete	2.7	32.3
Mortar	3.0	25.7

(ii) Fracture energy and characteristic length

Beam depth W (mm)	G_{If} (J/m²)	$K_{If} = (G_{If}E)^{1/2}$ (N/mm$^{-3/2}$)	l_{ch} (mm)	$\bar{W} = W/l_{ch}$
(a) Concrete				
100	141	67.5	625	0.16
400	141	67.3	621	0.64
800	176	75.3	777	1.03
(b) Mortar				
100	53.8	37.2	154	0.65
400	48.4	35.2	138	2.9
800	(34.5)	(29.8)	(98)	8.61

error of less than 1% in the measurement of the fracture energy. The second error (b) is not negligible. The local compressive deformation under the centre load can absorb up to 10% of the energy and increases with size. The bulk energy absorbed away from the fracture plane is only up to 2% of the fracture energy. These errors while contributing to a size effect are far too small to account for the size effect. The energy absorbed under the centre load could be eliminated if the beam was indented by the loading head up to the maximum load expected in the fracture energy test while the beam was supported over its entire length prior to the fracture energy test. Elices *et al.* (1992) claim it is the third error (c) that is the largest error and one which, if allowances are made for it, gives a fracture energy that is independent of size.

Elices *et al.* (1992) used weight compensation in their tests rather than make a correction for the weight. For the smaller sizes they accomplished the weight compensation by testing beams that were twice as long as the span, while prestressed springs were used to take the weight of the beam for the larger specimens. They argue that it is impossible to fracture the specimen right through stably, which is Petersson's (1985) objection to this method of testing. Hence, they argue that there is a tail to the load-deflection curve that is neglected. The angle of bend, θ, is given by $4\Delta/S$, where S is the loading span. Hence during the final stages of fracture the load, P, is given by

$$P = \frac{4\xi B}{S\theta^2} = \frac{\xi BS}{4\Delta^2} \tag{3.30}$$

where ξ is the first moment of the stress-displacement curve given by

$$\xi = \int_0^{\delta_f} \sigma(\delta)\delta \, d\delta \tag{3.31}$$

Elices et al. (1992) assume that the deflection, Δ_f, recorded as that at final fracture occurs at the same angle of beam rotation, θ_f, independent of the size. The work done, $\Delta W_{f(tail)}$, in what would be the tail of the load-deflection curve if the fracture were completely stable is therefore given by

$$\Delta W_{f(tail)} = \int_{\Delta_f}^{\infty} P \, d\Delta = \frac{\xi BS}{4\Delta} = \frac{\xi B}{\theta_f} \tag{3.32}$$

This correction to the work done in fracturing the beam appears reasonable. However, Elices et al. (1992) also assume that, since the final fracture occurs at a small value of the load, P_f, there is an additional work term, $P_f\Delta_f$, which is equal to $\Delta W_{f(tail)}$. This additional work term is unreasonable since it would have already been included in the area under the load-deflection curve, W_f, obtained during the test. Thus the measured work of fracture is

$$W_f = G_{If}Bb - \Delta W_{f(tail)} = G_{If}B\left(b - \frac{\xi}{G_{If}\theta_f}\right) \tag{3.33}$$

in which the correction term differs from that of Elices et al. (1992) by a factor of two. The problem is how to estimate ξ. There are two alternative methods: ξ can be estimated from an assumed form of the stress-displacement relationship for the FPZ, or it can be obtained from the relationship given in eqn 3.31. It is not too clear how Elices et al. (1992) obtain ξ and seem in fact to adjust it so that they show that the true fracture energy is independent of size. The adjustment amounts to only about 10% so that is not too important, but if the correction to the work of fracture is only $\Delta W_{f(tail)}$, and not twice that value as argued by Elices et al. (1992), the correction cannot explain the apparent size effect in the fracture energy as measured by the RILEM method. Hence we believe that Elices et al. (1992) have not demonstrated that the size effect is an artifact of the test method. The size effect on the fracture energy is real. Hu and Wittmann (1992b) have given an explanation of the cause of the size effect.

The fracture energy increases with size because the width of the FPZ increases with ligament size. Since the deformation of cementitious material is relatively independent of the strain path, the specific fracture energy decreases during crack propagation in a notch bend test and also decreases with notch to width (a_0/W) ratio for the same beam size. In the RILEM (1985) method the fracture energy is an average of the specific fracture energy, g_{If}, necessary to completely fracture a unit area of the specimen at a particular position on the notched ligament. Hu and Wittmann (1992b) have used multiple specimens to obtain the distribution of specific fracture energy across a compact tension specimen and have shown for mortar that

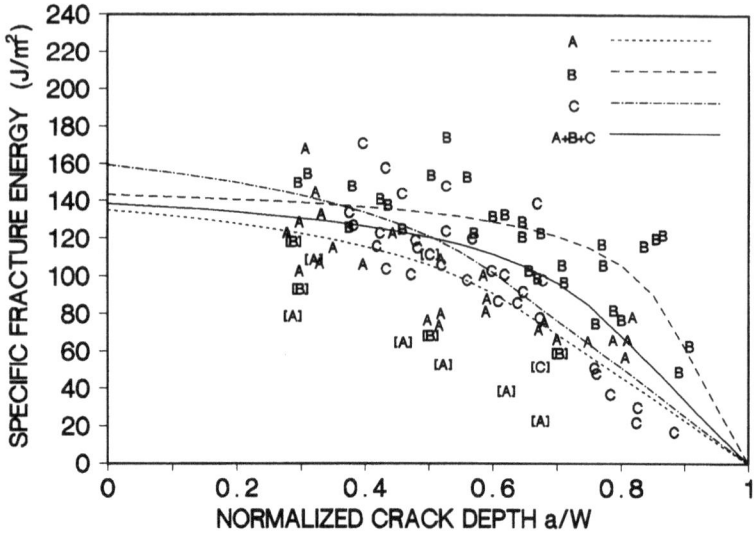

Figure 3.13 Specific fracture energy as a function of a/W (A, B and C are precracked specimens with $W = 102$, 203 and 305 mm, respectively; [A], [B] and [C] are notched specimens with same Ws. (Data from Swartz and Refai, 1987.)

g_{If} is approximately given by

$$g_{If} = g_{Ifm} \qquad \text{for } a_0/W < \alpha$$

$$g_{If} = \frac{g_{Ifm}(1 - a_0/W)}{1 - \alpha} \quad \text{for } a_0/W > \alpha \qquad (3.34)$$

where $\alpha \approx 0.7$. Swartz and Refai (1987) give the fracture energies for a large range of a_0/W ratios for three sizes of beams mainly precracked rather than notched (see Figure 3.13). The scatter in the results for the precracked specimens is perhaps considerably induced by the irregular crack front, but it appears that the specific energy is larger for the precracked specimens. The increase in fracture energy is perhaps caused by the FPZ being wider with a precrack. Curves have been fitted to the results shown in Figure 3.13, using eqs 3.34; the best values of the parameters are shown in Table 3.2.

Table 3.2 Specific fracture energy as a function of position obtained from curves of best fit, using eqns 3.34, from the data of Swartz and Refai (1987) shown in Figure 3.13

Series	W (mm)	g_{Ifm} (J/m^2)	α	g_{If} ($a/W = 0$) (J/m^2)	g_{If} ($a/W = 0.5$) (J/m^2)
A	102	165	0.64	135	106
B	203	153	0.88	143	134
C	305	198	0.61	160	124
A + B + C	—	157	0.77	139	121

There does seem to be an increase in fracture energy with specimen size, but the curve obtained by pooling all the results is reasonable.

The standard specimen size in the RILEM (1985) draft standard for the measurement of fracture energy is 100 mm, with an increase to 200 mm if the aggregate is 16–32 mm and a further increase to 300 mm for larger aggregates. This simple rule may not be adequate to ensure that the fracture energy reaches its maximum value. Practically the effect of testing a specimen that is too small to achieve the maximum fracture energy is probably not that significant. A small specimen will give conservative predictions of the maximum sustainable loads, but the difference may not be significant. A 20% difference in the fracture energy affects the maximum load predictions by less than 7% (Hillerborg, 1985b). Hence for practical design purposes the mode I fracture energy, G_{If}, can be considered a material constant. If the fracture energy is assumed to be a constant, that presupposes that the stress-displacement relationship in the FPZ is also an autonomous material property.

3.6 Modelling the mixed mode strain-softening relationship for cementitious materials

As discussed in section 1.7.1, fractures propagate under mode I conditions unless asymmetry in the FPZ forces the fracture to propagate under mixed mode conditions. Hence, mixed mode strain-softening is not as important as mode I. The mixed mode constitutive equations can be discussed either in terms of a stress-displacement relationship at the edge of the FPZ or a stress-strain relationship within the FPZ. There is much less consensus on the appropriate constitutive equations for a mixed mode FPZ than for pure mode I. The discussion here is limited to two-dimensional problems.

3.6.1 The mixed mode stress-strain relationship

In mixed mode analysis it has been more common to represent the FPZ as a zone of finite width and to specify stress-strain constitutive equations. In the FPZ an increment in strain referred to global coordinates can be decomposed into that due to the elasticity of the uncracked material $d\epsilon^{co}$, and that due to the opening up of microcracks in the FPZ $d\epsilon^{cr}$ so that

$$d\epsilon = d\epsilon^{co} + d\epsilon^{cr} \tag{3.35}$$

Even if the FPZ extends to maintain local symmetry at its tip, shear stress can develop behind the crack tip as the principal stresses rotate. If it is assumed that once a crack is formed it remains fixed in direction, then the matrix that transforms the local strains into global strains is fixed. In symmetrical problems there can be no rotation of the principal axes in a FPZ modelled by a single element and only slight rotation if it is modelled by a number of elements using a non-local crack strain. However, in mixed mode fracture, there can be a significant rotation of the principal axes. In this case the fixed smeared crack concept can be modified by allowing new microcracks to form according to the new principal stress directions. The decomposition of the strain into cracked and uncracked strain then has the advantage that the cracked strain can be further decomposed to give separate contributions from a number of multi-directional cracks (Bažant and Gambarova, 1983; Rots and de Borst, 1987; Rots, 1988). An alternative approach is the so-called rotating crack where the axes of material orthotropy are rotated with the principal directions (Rots and de Borst, 1987; Bažant and Lin, 1988). The rotating crack concept is related to the multi-directional concept and Rots (1988) notes that the two are the same providing that:

(i) The threshold angle for the initiation of new cracks is zero.
(ii) Previous cracks are made inactive.
(iii) The local strain-softening law of the active crack is such that the memory of the previous cracks is maintained and the overall shear modulus ensures that the principal axes of strain and stress coincide.

The crack strain increment referred to the local coordinates (n, t) aligned with the cracks only has two non-zero components, $d\epsilon_{nn}^{cr}$, $d\epsilon_{nt}^{cr}$, since it is assumed that a microcrack does not induce any strain parallel to itself. In general the constitutive relations for the FPZ are given by:

$$\left\{ \begin{array}{c} d\sigma_{nn} \\ d\sigma_{nt} \end{array} \right\} = \left[\begin{array}{cc} D_{nn} & D_{nt} \\ D_{tn} & D_{tt} \end{array} \right] \left\{ \begin{array}{c} d\epsilon_{nn}^{cr} \\ d\epsilon_{nt}^{cr} \end{array} \right\} \tag{3.36}$$

where σ_{nn}, σ_{nt} are the normal and shear stress referred to local coordinates and D is the instantaneous moduli matrix that describes the strain-softening relationship. The instantaneous moduli must satisfy (Bažant and Gambarova, 1980):

$$\frac{\partial D_{nn}}{\partial \epsilon_{nt}^{cr}} = \frac{\partial D_{nt}}{\partial \epsilon_{nn}^{cr}}$$

$$\frac{\partial D_{tn}}{\partial \epsilon_{nt}^{cr}} = \frac{\partial D_{tt}}{\partial \epsilon_{nn}^{cr}} \tag{3.37}$$

It is usually assumed that there is little interaction between shear and normal stress so that the off-diagonal terms in eqn 3.36 are taken as zero (Rots and de Borst, 1987; Bažant and Lin, 1988; Rots, 1988). The mode I strain-softening relationships, discussed in section 3.4, can be used for D_{nn}. If the simplest linear stress-strain relationship is used and the strain is constant over the thickness h of the FPZ, then

$$D_{nn} = -\frac{f_t^2 h}{3G_{If}} \tag{3.38}$$

There is less agreement over the mode II modulus. Bažant and Lin (1988) have suggested that an empirical shear retention factor, β, similar to that proposed for sudden cracking by Schnobrich co-workers (Suidan and Schnobrich, 1973; Yuzugullu and Schnobrich, 1973; Hand et al., 1973) could take into account the reduced shear stiffness in the FPZ by expressing the shear stress as

$$\sigma_{nt} = G\beta\epsilon_{nt}$$

or

$$D_{tt} = \frac{G\beta}{1 - \beta} \tag{3.39}$$

where G is the shear modulus. Such a formulation is probably adequate if the shear strains are moderate, but may lead to unrealistically high shear stresses and mode II fracture energies. If a constant shear retention factor is used, the fracture energy becomes infinite for a pure mode II fracture. It is probable that the mode II fracture energy can be very much larger than the mode I value. Bažant and Pfeiffer (1987), using their size effect law (see section 4.6), obtained values about 25 times that of the mode I fracture energy in tests where the fracture was close to mode II. Also the measurement of the fracture energy by Swartz et al. (1988) from three-point notched beam specimens, where the notches were located at different positions along the beam, indicated a large increase in the fracture energy as the fracture's mode II component was increased.

Rots and de Borst (1987) have suggested that there will be an ultimate shear stress τ_u and that for large strains there will be shear stress-softening which can be approximated by linear behaviour shown in Figure 3.14,

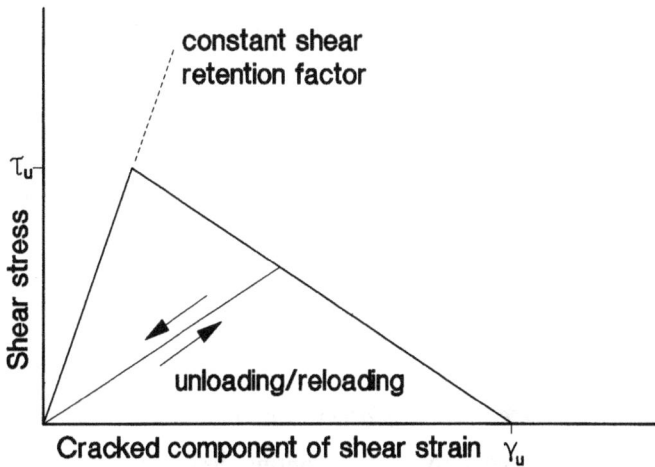

Figure 3.14 Mode II shear-softening relationship. (After Rots and de Borst, 1987.)

where the ultimate cracked shear strain γ_u is related to a mode II fracture energy G_{IIf} by

$$\gamma_u = \frac{2G_{IIf}}{h\tau_u} \qquad (3.40)$$

The main objections to such a proposal are that the ultimate shear stress must surely be related to the mode I damage. Elsewhere, Rots (1988) has given an alternative expression for the instantaneous strain-softening modulus where the shear retention factor, β, depends not on the cracked shear strain, but on the cracked normal strain. The linear version of this proposal is expressed by

$$\beta = \left[1 - \frac{\epsilon_{nn}^{cr}}{\epsilon_u^{cr}} \right] \qquad (3.41)$$

and D_{tt} is given by eqn 3.39. The degradation from full interlock to zero strength corresponds to a degradation of β form 1 to 0. It should be noted that since D_{tt} is infinite when ϵ_{nn}^{cr} is zero that this expression for the mode II modulus can only be used if it is assumed that the FPZ propagates to maintain local symmetry so that at the tip of the FPZ the shear stress τ_u, and hence the cracked shear strain ϵ_{nt}^{cr}, is zero. Provided the assumption of local symmetry is accepted, this proposal seems more reasonable than the other proposal of Rots and de Borst (1987), because surely when a continuous open crack is established, the shear stress in the FPZ must be zero. Continuous cracks can resist shear stress only when there is a normal compressive stress acting between the two surfaces. The mechanism of shear resistance across a continuous crack is one of interlock, which like friction, can only operate if there is a compressive force between the two surfaces. The proposal is open to the objection that in the absence of a large mode I component the shear stress may be very large, but in most situations the mode I component will effectively limit the shear stress. The proposal agrees qualitatively with the results of Swartz et al. (1988) that indicate that the mode II fracture energy is not a material constant, but increases as the mode II component increases. A more serious problem is that since the mode II instantaneous modulus depends on the mode II strain there must be a mixed mode modulus, D_{tn}, which is given by eqn 3.39

$$D_{tn} = -G\frac{\epsilon_u \epsilon_{nt}}{\epsilon_{nn}^2} \qquad (3.42)$$

There is more work needed before the most suitable form of the mixed mode constitutive equations for the FPZ are established.

3.6.2   The mixed mode stress-displacement relationship

Stress can be transmitted across even a continuous crack if the stress normal to the crack is compressive. Studies of the stress-displacement relationships for precracked bodies were undertaken before the constitutive equations for a mixed mode FPZ were considered (Fenwick and Pauley, 1968; Pauley and Loeber, 1974; Fardis and Buyukzturk, 1979; Bažant and Gambarova, 1980; Bažant and Tsubaki, 1980; Divaker et al., 1987). At a fixed crack opening displacement, crack slip will cause both a shear stress and a normal stress to be developed causing a strong relationship between the crack sliding, δ_t, and the normal crack opening, δ_n. Thus

$$\sigma_{nn} = F_n(\delta_n, \delta_t)$$
$$\sigma_{tt} = F_t(\delta_n, \delta_t)$$

(3.43)

where Bažant and Gambarova (1980) have given empirical expressions for the functions on the basis of data obtained by Pauley and Loeber (1974). Although both Ingraffea and his co-workers (Ingraffea and Saouma, 1984; Ingraffea et al., 1984; Ingraffea and Panthaki, 1985) and Liaw et al. (1990) apply an expression for the shear stress obtained from similar data to the propagation of a mixed mode crack, there is no justification and the direct application of such equations to a FPZ can at best be only very approximate. Because of the frictional nature of the shear transfer between crack surfaces, a shear stress can only be developed if there is a corresponding compressive normal stress acting across the crack surfaces. It is possible that behind the FPZ a compressive normal stress could occur and shear be transferred across the continuous crack, but usually the crack faces behind the FPZ will not be compressed together and there will be no shear transfer behind the FPZ.

The stress-strain constitutive equations for the FPZ discussed in section 3.6.1 can be converted into stress-displacement relationships by multiplying by the thickness of the FPZ.

3.7 Experimental techniques for the measurement of the FPZ and FBZ in fibre reinforced cementitious materials

In fibre cements it is necessary to distinguish between the FPZ ahead of the continuous matrix crack and the FBZ behind it. Many of the above-mentioned techniques have been employed to measure the size of these two zones, for example optical and scanning electron microscopy, photography, staining and Moiré interferometry, replicas, electrical potential difference methods, acoustic emissions and compliance measurement. However, difficulties are considerable; Foote (1986) has given a review of these techniques.

The multi-cutting technique is recognized as a more direct method of determining the size of the FPZ in both unreinforced cementitious materials and fibre cements (Mindess, 1991a). It is a variant of the compliance method and it makes use of the compliance difference of an unbridged and a bridged crack of the same crack length (Mai and Hakeem, 1984). The multiple cutting technique described for the unreinforced cement matrices in section 3.2.2 cannot be readily applied to fibre cements to separate and measure the FPZ and the FBZ. An estimate of the size of the FBZ is the more important for the purpose of modelling as in many theoretical analyses the FPZ can be replaced by a critical matrix fracture toughness.

Foote *et al.* (1987) developed a computer-aided method of crack length measurement using a conductive grid whereby cracking can be sensed by the breaking of the lines of the grid. In this method a grid pattern of conductive ink, consisting of fine carbon particles dispersed in a vinyl resin binder and a butyl cellosolve acetate solvent, is screen printed onto the specimen. The grid pattern, used for a compact tension (CT) specimen (see Figure 3.15), has 64 bars in eight blocks of eight allowing the crack to be measured over a distance of about 140 mm. The bars were nominally 1 mm in width and 2.14 mm apart. A schematic diagram of the computerized method is shown in Figure 3.16. The computer scans each bar on the grid serially and tests for continuity. When a bar is broken, indicating cracking at that location, a corresponding voltage is sent to a plotter, and a graphic display of the grid on a monitor shows that the bar is broken. With each scan the latest broken bar detected is plotted and displayed. This technique has made visible the whole crack growth process,

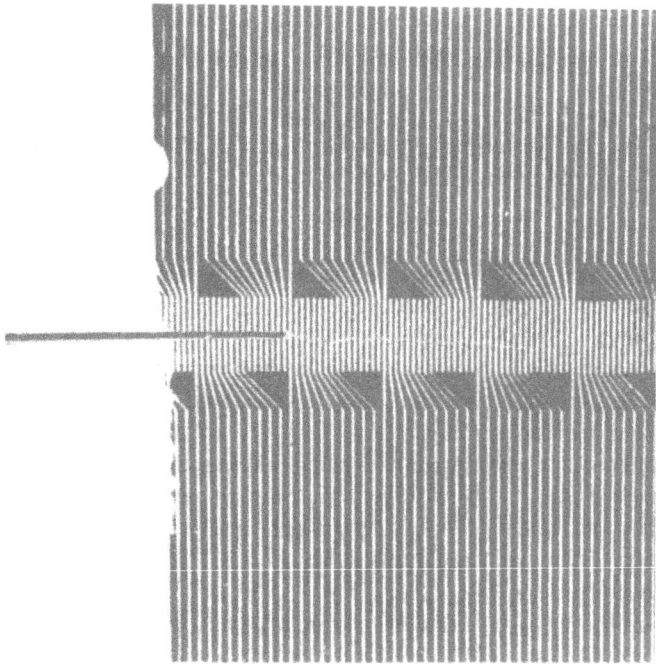

Figure 3.15 Conductive grid pattern printed on CT specimen.

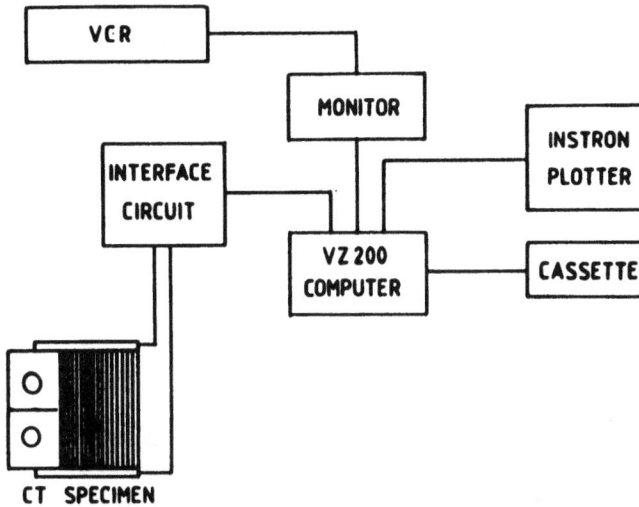

Figure 3.16 Computer-aided crack growth monitoring system.

and both crack extension-time and load-time records can be simultaneously and continuously obtained until final failure. The experimental results for a CT wood fibre cement composite are shown in Figure 3.17. They reveal a complex picture of crack growth in which the bars record breaks and closures followed by second breaks. The region of bars breaking, or the activity zone, corresponds closely with the FPZ and was approximately 20 mm throughout the crack growth. Table 3.3 compares the continuous matrix crack length obtained by optical microscopy and that determined with the present computer-aided method. The leading edge of the activity zone is a good indication of the front of the FPZ and the trailing edge represents the tip of the continuous matrix crack and the leading edge of the FBZ.

An alternative method of locating the extent of the FPZ and the FBZ is to measure the bending stiffness of narrow strips cut from the specimen perpendicular to the crack (Foote *et al.*, 1987). This technique was applied to the CT specimen whose cracking, monitored using the computerized system, is shown in Figure 3.18. The bending stiffness of the strips was measured in pure bending and the location of failure noted. Strips were also cut at the back of the specimen in the relatively unstressed region to represent the undamaged material so that the bending stiffnesses could be normalized by that of the undamaged material. For strips cut within the FBZ the failure sites were the prolongation of the machined notch, but the failure sites for strips cut from the FPZ were scattered randomly about the prolongation of the machined notch. In the FBZ the fracture of the strips takes place close to the prolongation of the machined notch. The bending stiffness in the FBZ is not zero though the matrix is completely cracked because the strips can support a bending moment since their rotation about

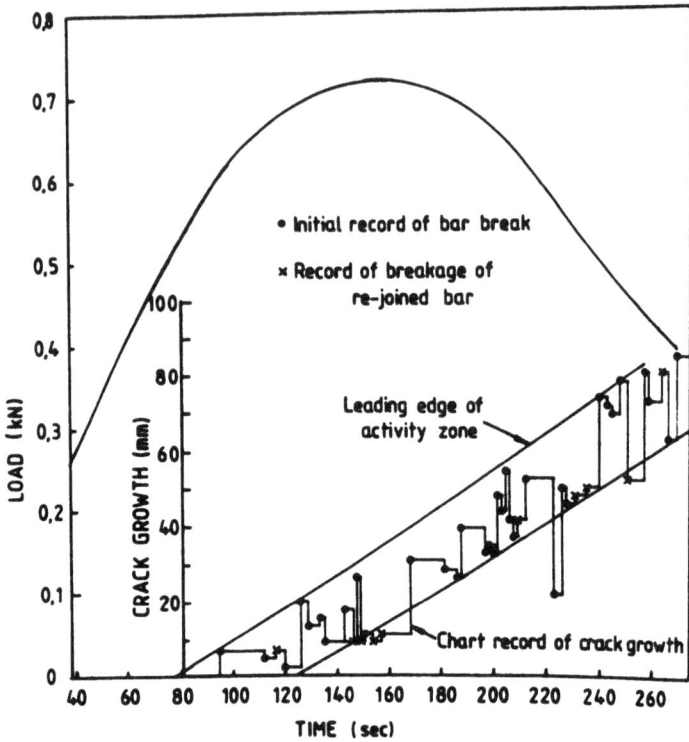

Figure 3.17 Crack growth and load-time traces recorded by the computer-aided system. The leading edge of the activity zone gives a measure of the continuous matrix crack length.

one surface can be resisted by the bridging fibres. Within the FBZ the bending stiffness increases in an orderly fashion from the trailing edge but in the FPZ, though there is an increase in stiffness, there is considerable scatter in the results. Thus from Figure 3.18 the leading edge of the FPZ is at about $a = 80$ mm. Figure 3.18 is poor for evaluating the size of the FPZ because the crack has extended too close to the back face of the CT specimen. In another series of experiments (see Figure 3.19), free of boundary effects, the size of the FPZ can be bounded between 30 and 40 mm.

Table 3.3 Crack growth measurement using the computer aided method, optical microscope and the section and bending stiffness test

Specimen No.	Crack length (mm)		
	Computer-aided	Optical	Bending stiffness
1	62–86	78	77
2	56–88	83	83
3	70–105	99	112

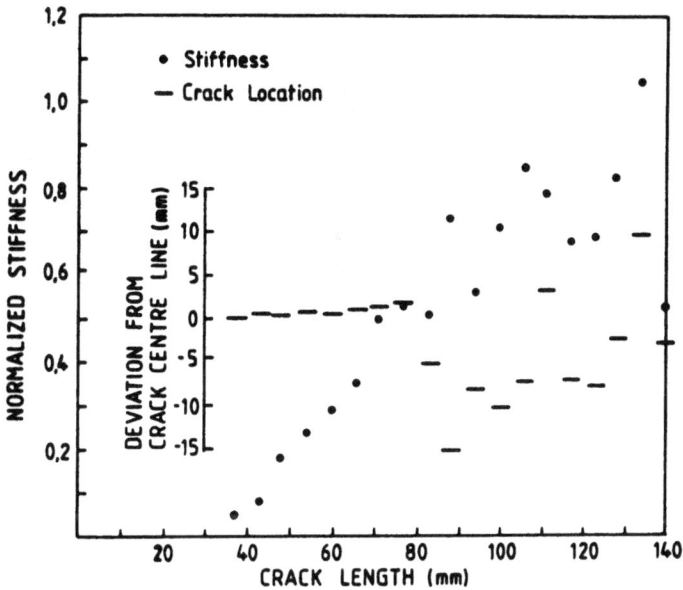

Figure 3.18 A plot of normalized stiffness *versus* crack length. Failure sites of sectioned strips are indicated by horizontal bars.

The estimates of the tip of the continuous crack obtained from optical, computerized grid and stiffness methods are compared in Table 3.3.

3.8   Measurement of the mode I strain-softening relationship for fibre reinforced cementitious materials

Because fibres greatly increase the toughness and crack growth resistance of cementitious materials, fracture is more stable. Thus a direct tension test to determine the stress-displacement relationship is more practical for composites than unreinforced cementitious materials. The other methods of measuring the stress-displacement relationship for unreinforced cementitious materials can be applied to composites, but there are problems with the very large size of the FBZ.

The compliance method of Hu and Mai (1992a) that uses the compliance of a partially developed FPZ or FBZ, described in section 3.3.4, has been applied to wet and dry cellulose cement composites using DCB specimens. The plot of the function, ϕ, against crack extension is shown in Figure 3.20. The size of the FBZ obtained from the crack extension to reach the plateau value of ϕ is 50 mm and 57 mm for the wet and dry composites, respectively. Using the plateau values of the curves in Figure 3.20 and eqn 3.18, the exponent for the stress-displacement relationships, given by eqn 3.17, are

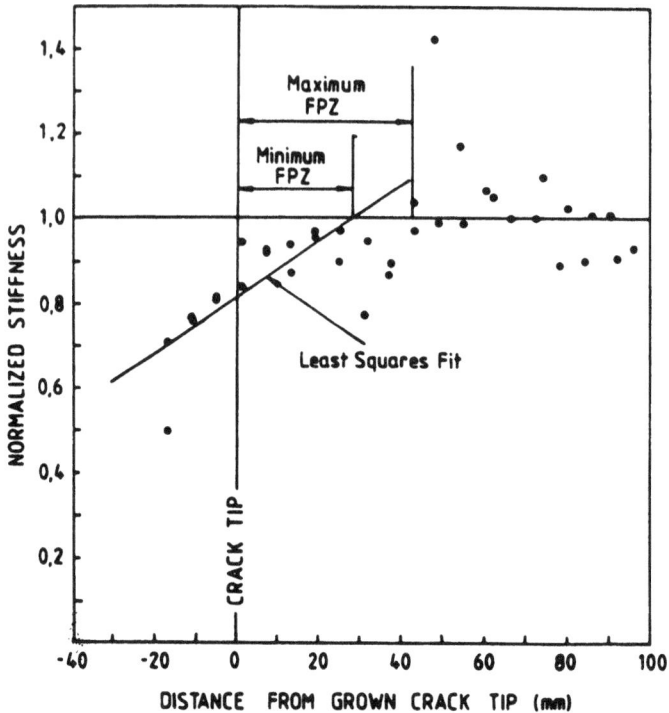

Figure 3.19 Evaluation of FPZ using normalized stiffness-crack tip distance plot.

1.4 and 3.8, respectively. The normalized stress-displacement curves are shown in Figure 3.21.

3.9 The stress-strain relationship for fibre reinforced cementitious materials

3.9.1 Uncracked composites

Prior to matrix cracking a fibre reinforced cementitious material is elastic. For aligned continuous fibres the Young's modulus of the composite, E_c, in the direction of the fibres can be found from the condition of equality of strain, assuming good bonding, and is given by the rule of mixtures

$$E_c = vE_f + (1 - v)E_m \qquad (3.44)$$

where v is the volume fraction of the fibres, and E_f, E_m are the Young's modulus of the fibres and the matrix, respectively.

In discontinuous fibre reinforced composites, the stress in the matrix has to be transferred to the fibre. At low stress levels, this transfer will take place elastically. The stress transfer can be modelled using shear lag theory which was first applied to fibre reinforced composites by Cox (1952). If the

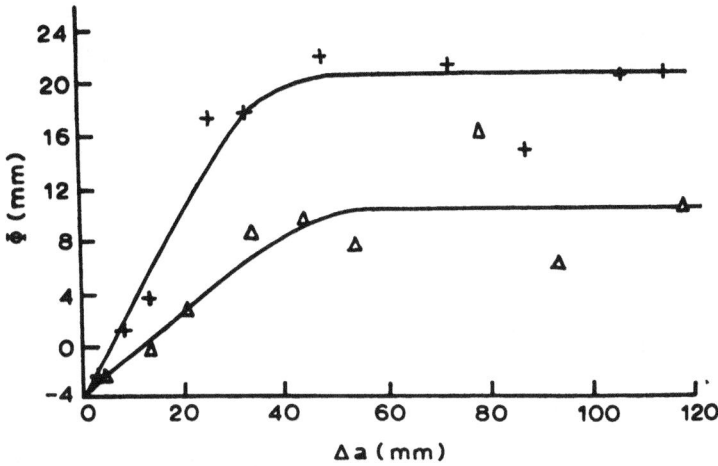

Figure 3.20 $\phi - \Delta a$ curves for wet (+) and dry (Δ) cellulose-fibre reinforced mortar.

fibre is relatively long, the interfacial shear stress, τ, decays exponentially as the stress in the fibre, σ_f, builds up to a plateau value equal to that carried by continuous fibres which is $\sigma_c(E_f/E_c)$, where σ_c is the stress applied to the composite and the stresses[2] are (see Figure 3.22a)

$$\left.\begin{array}{l} \tau(x) = \sigma_c \dfrac{E_f}{E_c} \dfrac{\beta d_f}{4} \exp(-\beta x) \\[2mm] \sigma_f(x) = \sigma_c \dfrac{E_f}{E_c} [1 - \exp(-\beta x)] \end{array}\right\} \tag{3.45}$$

where

$$\beta = \left[\frac{16 G_m}{E_f d_f^2 \ln(2\pi/3v)} \right]^{1/2}$$

for hexagonal packing (Piggott, 1980), where G_m is the shear modulus of the matrix and d_f the fibre diameter. Often fibres in cementitious composites are poorly bonded and debond at low stress levels. Under these circumstances it is usually assumed that the frictional interfacial shear stress, τ_f, is constant and then for long fibres the stresses (see Figure 3.22b) are given by

$$\left.\begin{array}{l} \tau = \tau_f \\[2mm] \sigma_f = \tau_f \dfrac{4x}{d_f} \end{array}\right\} \tag{3.46a}$$

for $x < l^*/2$, and

---

[2] It is usually assumed that the ends of the fibre are stress free, because either they are poorly bonded or, if the modulus of the fibre is much larger than that of the matrix, the stress in the matrix is small compared with the stress in the fibre.

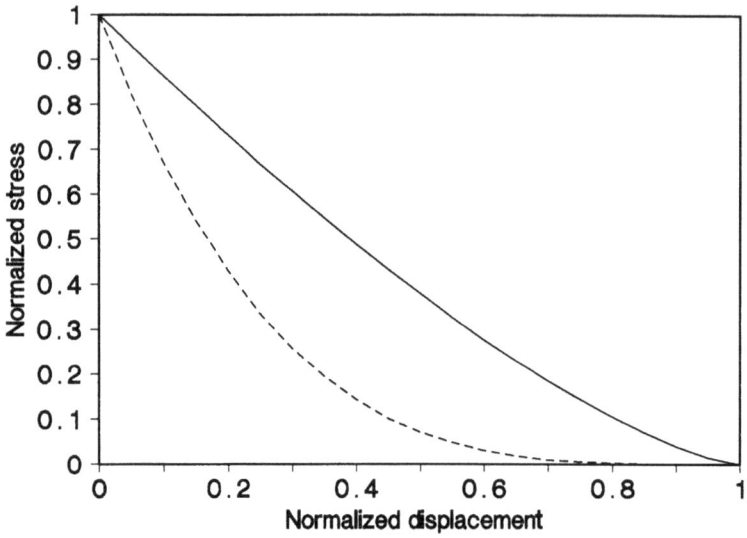

Figure 3.21 Normalized stress-displacement curves for wet (——) and dry (– – –) cellulose-fibre reinforced mortar.

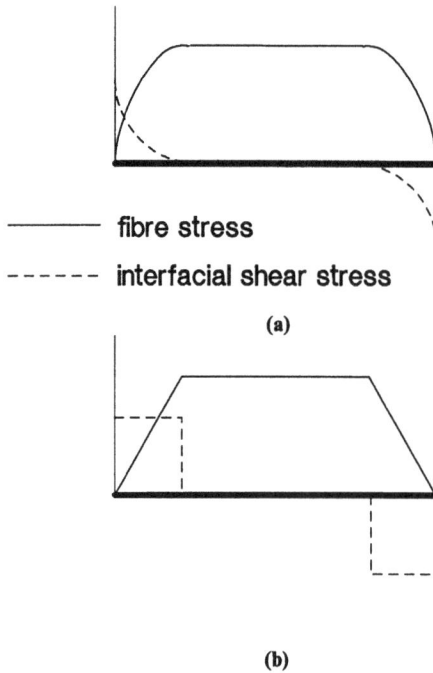

Figure 3.22 Stress transfer to a fibre (a) elastically, (b) constant interfacial shear stress.

$$\left.\begin{array}{c} \tau = 0 \\[2mm] \sigma_f = \sigma_c \dfrac{E_f}{E_c} \end{array}\right\} \qquad (3.46b)$$

for $x > l^*/2$, where

$$l^* = \frac{d}{2}\frac{\sigma_c}{\tau_f}\frac{E_f}{E_c}$$

If the fibres are shorter than a critical length, l_c, defined by

$$l_c = \frac{\sigma_{fc}}{\tau_f}\frac{d}{2} \qquad (3.47)$$

where σ_{fc} is the fibre fracture stress, then the fibres can never break, but must pull out during fracture.

The Young's modulus of composites reinforced with aligned discontinuous fibres is given by

$$E_c = \eta_l v E_f + (1 - v)E_m \qquad (3.48)$$

where η_l is an efficiency factor defined by (for poorly bonded fibres)

$$\begin{array}{ll} \eta_l = \dfrac{l}{2l_c} & l < l_c \\[3mm] = 1 - \dfrac{l_c}{2l} & l > l_c \end{array} \qquad (3.49)$$

If the fibres are not aligned, then the contribution of the fibres to the Young's modulus must be multiplied by a further orientation efficiency factor η_θ, which depends slightly on whether deformation perpendicular to the stress is constrained or not (Krenchel, 1964), and are given in Table 3.4.

3.9.2 Matrix cracking stress and the critical volume fraction

The matrix of cementitious composites usually cracks before the fibres break and the stress in the composite at first matrix cracking, σ_{cmc}, is given by

$$\sigma_{cmc} = E_c \epsilon_{mc} \qquad (3.50)$$

Table 3.4 Orientation efficiency factors for composites; μ is the coefficient of friction (from Cox (1952) and Krenchel (1964))

Fibre orientation factor η_θ	Random 2-D	Random 3-D
Uncracked:		
unconstrained	0.333	0.167
constrained	0.375	0.20
Cracked matrix:		
without snubbing	$2/\pi$	0.50
with snubbing	$2[1 + \mu\exp(\mu\pi/2)]/\pi[1 + \mu^2]$	$[1 + \exp(\mu\pi/2)]/[4 + \mu^2]$

where ϵ_{mc} is the matrix cracking strain. If the volume fraction of the fibres is greater than a critical value, v_c, given by

$$v_c = \frac{E_c \epsilon_{mc}}{\sigma_{fc}} \qquad (3.51)$$

where σ_{fc} is the smaller of the fibre strength or pull-out stress, σ_{fp}, then the fibres alone can withstand the stress at first matrix cracking and the composite is Type I. It should be noted that for randomly orientated fibres, the efficiency factor after matrix cracking is not the same as in the uncracked composite. After the matrix cracks the orientation efficiency factor depends upon the number of fibres per unit area that cross the matrix fracture and there can be an increase in the factor due to snubbing. If the fibres are flexible, the snubbing effect is similar to pulling a rope over a corner. For steel wires of relatively large diameter, there is an additional plastic bending and unbending term. Snubbing increases the pull-out stress for a fibre by a factor $\exp(\mu\theta)$ where θ is the orientation angle of the fibre and μ is the coefficient of friction. For polypropylene and nylon μ is approximately 0.7 and 1.0, respectively (Li $et\ al.$, 1990). The orientation efficiency factors for randomly orientated fibre reinforced composites are given in Table 3.4.

3.9.3   Fibre pull-out

The mechanics of fibre debonding and pull-out in a composite is complex and most studies have focused on a single fibre. There have been two different criteria of debonding: a maximum shear bond strength (Takaku and Arridge, 1973; Lawrence, 1980) or a critical specific work of debonding (Gurney and Hunt, 1967; Atkinson $et\ al.$, 1972; Bowling and Groves, 1979; Stang and Shah, 1986; Gao $et\ al.$, 1988). The simplest fracture mechanics approach is that of Gurney and Hunt (1967). If the fibre volume fraction is small, the change in matrix stress during fibre debonding is small and hence the rate increase in strain energy stored during debonding, which is numerically equal to the decrease in potential energy of the system, is given by

$$\frac{d\Lambda}{dl} = -\frac{d\Pi}{dl} = \frac{1}{2}\frac{\sigma_f^2}{E}\frac{\pi d_f^2}{4} \qquad (3.52)$$

Hence if G_{IIb} is the interfacial debond energy, the debond stress, σ_{fbo}, is given by

$$\sigma_{fbo} = \left[\frac{8 E_f G_{IIb}}{d_f}\right]^{1/2} \qquad (3.53)$$

When a fibre is partially debonded, compressive stresses due to shrinkage of the matrix act on the fibre causing friction between the fibre and the matrix in the debonded zone. Because of Poisson's effect, the pressure between the fibre and the matrix in the debonded zone is reduced and the frictional force is not constant.[3] The Gao–Mai–Cotterell (1988) model assumes a Coulomb friction and takes account of the Poisson effect. If the embedded length of the fibre is long then the initial debond stress is independent of the fibre length and the initial fibre debond stress, σ_{fbi}, is given by

$$\sigma_{fbi} = \sigma_{fbo} \left[\frac{(1+\beta)}{(1-2kv)} \right]^{1/2} \tag{3.54a}$$

$$\text{where } k = \frac{\nu_f \dfrac{E_m}{E_f} + \nu_m \dfrac{v}{1-v}}{(1+\nu_m) + (1-\nu_f)\dfrac{E_m}{E_f} + \dfrac{2v}{1-v}} \tag{3.54b}$$

$$\beta = \frac{(1-2k\nu_m)}{(1-2k\nu_f)} \left(\frac{v}{1-v} \right) \left(\frac{E_f}{E_m} \right) \tag{3.54c}$$

where ν_f and ν_m are the Poisson's ratio of the fibre and matrix, respectively. This model gives a good prediction of the debond stress for long embedded lengths, but overestimates σ_{fbi} for short lengths; Zhou et al. (1992) give the corresponding theory for finite embedded lengths. For fibres with higher moduli than the cementitious matrix, such as steel wires,

$$k \ll 1$$

$$\beta \approx \frac{v}{1-v} \frac{E_f}{E_m} \tag{3.55}$$

$$\sigma_{fbi} \approx \sigma_{fbo} \left[1 + \frac{v}{1-v} \frac{E_f}{E_m} \right]^{1/2}$$

and since the maximum practical fibre volume fraction is of the order of 10%, the initial fibre debond stress is less than 5% greater than σ_{fbo}. For low modulus fibres like polypropylene, $\beta \ll 1$, and the initial fibre debond stress is less than 6% smaller than σ_{fbo}. Hence, the initial fibre debond stress for cementitious composites is reasonably well given by eqn 3.53.

If under load the interfacial clamping pressure is reduced to zero by contraction of the fibres before the initiation of debonding, then the debonding is unstable and complete debonding takes place at σ_{fbi}. The critical interfacial threshold clamping pressure, q_{th}, at which debonding takes place unstably is given by

$$q_{th} = \frac{k\sigma_{fbi}}{1 + \dfrac{E_f \nu_m}{E_m \nu_f} \dfrac{v}{1-v}} \tag{3.56}$$

[3] The interfacial contact pressure decreases if there is significant slip due to compaction of the cement near the wire surface (Pinchin and Tabor, 1978).

and if the initial clamping pressure, q_0, is greater than q_{th}, then the critical pull-out stress for long fibres is $(q_0/q_{th})\sigma_{fbi}$. For a steel wire ($d_f = 0.5$ mm; $v \approx 10\%$) reinforced mortar the approximate threshold clamping pressure $q_{th} = 10$ MPa. The shrinkage of mortar is strongly dependent on the curing details and also on the volume fraction of the steel wires (Malmberg and Skarendahl, 1978) and is of the order of 0.1% which, assuming $E_m = 25$ GPa and $\nu_m = 0.2$, implies an interfacial pressure of some 30 MPa. Thus in theory such steel wires should debond stably. However, in practice the debond strength is not so simple. Pinchin and Tabor (1978) found that the debonding load for steel wires embedded in water-cured specimens was two to four times that for specimens sealed during curing, despite the fact that water-cured mortar expands by about 40 microstrain, whereas sealed mortar shrinks by 400 microstrain.

Once debonded the fibres pull out against friction and for a single fibre the pull-out stress, σ, for a fibre embedded over a length x is given by

$$\sigma = \sigma_{fbi} \frac{q_0}{q_{th}} [1 - \exp -\lambda x] \tag{3.57}$$

where

$$\lambda = \frac{4\mu}{kd_f}$$

Kim et al. (1993) have analysed the pull-out tests of Li et al. (1991) on glass fibres embedded in cement paste using both a normal 3 day cure and an accelerated cure of 1 day at 50°C in lime water using the theory for fibres embedded by a finite length (Zhou et al., 1992). The maximum debond stress and the initial frictional pull-out stress immediately after debonding are compared in Figure 3.23 with the theoretical predictions using the parameters in Table 3.5 that gave the best fit. The theoretical predictions are reasonably convincing. Note that the accelerated cure produces a higher clamping stress, q_0, confirming the comments made in section 2.4.3. Kim et al. (1993) also analysed the results from a wide range of steel–fibre–mortar composites. However, with steel fibres a good fit for both the debond and the frictional pull-out stress could not be obtained using the same clamping stress, q_0. For the debond stress, the value of the clamping stress that gave the best fit was 50 MPa, but this had to be reduced to 24 MPa to obtain a good fit to the frictional pull-out stress. The reduction in the clamping stress can be due to the compaction of the mortar (Pinchin and Tabor, 1978) or by the densification of the porous CSH transition zone between the fibre and the bulk mortar (Bentur et al., 1985).

In practice the friction between a debonded fibre and the matrix is not necessarily controlled by Coulomb friction. If the fibres are not perfectly

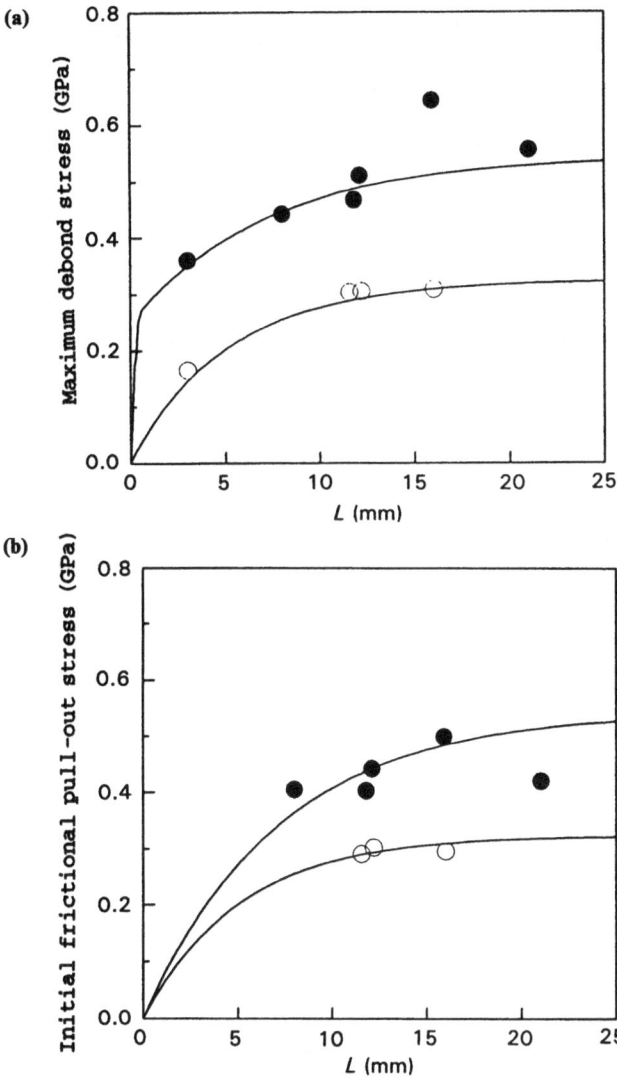

Figure 3.23 Glass-fibre cement paste pull-out tests, comparison of experiment with theory: (a) maximum pull-out stress; (b) initial frictional pull-out stress: O, normal cure, ●, accelerated cure. (After Kim *et al.*, 1993.)

straight and smooth the effects of fibre curvature and roughness dominate and the simpler assumption of a constant frictional shear stress can be more appropriate. The interfacial frictional stress can fall during pull-out not because of Poisson's contraction, but because of compaction of the less dense interphase layer between the fibre and bulk matrix. With the assumption of a constant frictional interfacial shear stress, the pull-out

Table 3.5 Fibre pull-out properties for glass-fibre–cement-paste

	Normal cure	Accelerated cure
Initial debond stress σ_{fbi} (MPa)	10	274
Frictional decay factor λ (mm^{-1})	0.193	0.138
Critical debond stress for very long fibres		
$\quad (q_0/q_{th})\,\sigma_{fbi}$ (MPa)	325	545
Interfacial energy G_{IIb} (J/m^2)	0.03	22.8
Coefficient of friction μ	0.14	0.1
Initial clamping pressure q_0 (MPa)	20.1	33.8

stress, σ, on a fibre embedded to a depth x is given by

$$\sigma = \frac{4\tau_f}{d_f}x \qquad (3.58)$$

3.9.4 Stress-strain curve for Type I composites

Multiple cracking Type I composites have a fibre volume fraction greater than the critical value given by eqn 3.51. The matrix reaches its ultimate strength, σ_{mc}, at a strain, ϵ_{mc}, which can be greater than the fracture strain in the unreinforced matrix. Using energy arguments (the ACK model), Aveston et al. (1971) have estimated the fracture strain for a reinforced matrix to be

$$\epsilon_{mc} = \left[\frac{24\tau_b G_{Im} E_f v^2}{E_c E_m^2 d_f(1-v)} \right]^{1/3} \qquad (3.59)$$

where τ_b is the shear bond strength and G_{Im} is the specific work of fracture of the matrix. This prediction of the fracture strain of the reinforced matrix has been refined by considering the spacing of the fibres (Hannant et al., 1983). In this refinement, which is not expressible analytically, it is assumed that the matrix fracture strain is only enhanced if the fibre spacing is less than the size of the inherent flaw. Thus if the fibre spacing is large, as in the case of steel fibre reinforced cementitious materials, there is no enhancement of the fracture strain. However, both the ACK model and its refinement assume that the specific fracture work is a material constant, whereas there will undoubtedly be an R-curve effect so that the inherent flaw size will not directly control the fracture strain of the reinforced matrix. Aveston et al. (1971) assume that multiple cracks form at an essentially constant stress, $E_c\epsilon_{mc}$, with a spacing of between x and $2x$ where x, the distance over which the matrix strain builds up to its critical value ϵ_{mc}, is given by

$$x = \left(\frac{1-v}{v} \right) \frac{\sigma_{mc} d_f}{4\tau_f} \qquad (3.60)$$

The matrix breaks down into blocks of length between x and $2x$. When this process is complete, the strain in the composite, ϵ_c, is between the limits

(Aveston *et al.*, 1971)

$$\left[1 + \frac{E_m}{2E_f}\left(\frac{1-v}{v}\right)\right] < \frac{\epsilon_c}{\epsilon_{mc}} < \left[1 + \frac{3E_m}{4E_f}\left(\frac{1-v}{v}\right)\right] \tag{3.61}$$

On further straining the fibres stretch until they break or slip through the matrix blocks. In the latter case the modulus of the composite becomes $E_f v$.

The theory of multiple cracking can be extended to cover short fibres (Aveston *et al.*, 1974). Most short fibres pull out, except weak fibres such as aged glass fibres (Hannant *et al.*, 1983) or bleached cellulose fibres (Mai *et al.*, 1983), rather than fracture. If the average stress sustainable on first matrix cracking is greater than $E_c\epsilon_{mc}$ then multiple cracking will occur. If the fibres are well bonded, the stress sustainable by the fibres alone once the matrix cracks can drop suddenly as some of the fibres debond, so that there is only frictional shear between the fibre and the matrix. On further straining the fibres stretch and, provided sufficient fibres remain well bonded, the stress can again rise to the first cracking stress causing multiple cracking.

Consider the crack spacing for aligned short fibres. Let x_d be the distance from the crack face that the strain in the matrix can build up to the matrix cracking strain, ϵ_{mc}, then $N(1 - 2x_d/l)$ fibres per unit area have both ends at a greater distance from the crack plane than x_d and therefore fully transfer their portion of the load carried by the matrix prior to failure. The remaining $2Nx_d/l$ fibres only transfer half of the load. Therefore, the total load per unit area, P, transferred is (Aveston *et al.*, 1974)

$$P = \pi d_f \tau_f x_d N\left[1 - \frac{x_d}{l}\right] = \sigma_{mc}(1 - v) \tag{3.62}$$

and substituting from eqn 3.60, x_d is given by the smaller root of

$$\frac{x_d}{l} = \frac{1 - [1 - 4(x/l)^2]^{1/2}}{2} \tag{3.63}$$

Multiple cracks form with a spacing of between x_d and $2x_d$ at an essentially constant stress. After multiple cracking is complete, the stress will increase until it reaches the maximum pull-out stress, σ_{cp}. The deformation will then localize on a single matrix crack and the stress will decrease as the fibres pull out. This phase in the stress-strain behaviour will be identical to the pull-out behaviour of Type II composites.

3.9.5 Stress-strain curve for Type II composites

With Type II composites the first cracking stress represents the maximum stress sustainable. For Type IIA composites with long fibres it is sufficient to assume that composite is elastic up to the first matrix cracking strain at which the stress drops to a constant pull-out stress, σ_{tu}, which for straight fibres is given by eqn 3.58 (Lim *et al.*, 1987). The stress-displacement curve for Type IIB composites is similar to that for unreinforced cementitious materials. The COD at which a continuous matrix crack occurs will be larger than the value for an unreinforced material, and the COD at which the last fibre pulls out is half the fibre length.

The average embedded length is $l/4$ of the fibres, hence the maximum fibre pull-out stress is obtained by $x = l/4$ in eqns 3.57 and 3.58. After the matrix crack has opened by δ, a fraction $2\delta/l$ will have pulled out and the average embedded length, of the fibres that remain embedded, is $(l/2 - \delta)/2$. The number of fibres bridging the crack when the COD is δ is therefore $(4\eta_\theta v/\pi d_f^2)(1 - 2\delta/l)$. Thus, under the assumptions of Coulomb friction the pull-out stress, σ_{cp}, based on the area of the composite is given by

$$\sigma_{cp} = \eta_\theta v \sigma_{fbi} \frac{q_0}{q_{th}} \left[1 - \frac{2\delta}{l}\right]\left[1 - \exp{-\frac{\lambda l}{4}\left(1 - \frac{2\delta}{l}\right)}\right] \qquad (3.64)$$

and under the assumption of a constant interfacial friction the pull-out stress is given by

$$\sigma_{cp} = \eta_\theta v \tau_f \left(\frac{l}{d_f}\right)\left[1 - \frac{2\delta}{l}\right]^2 \qquad (3.65)$$

The limits to the stress-displacement relationship obtained if Coulomb friction is assumed are linear if λ is very large (i.e. the pull-out stress on each fibre remains constant); parabolic if λ is very small (i.e. a constant interfacial friction stress). In practice the stress-displacement curve for fibre reinforced composites can be parabolic, such as the results obtained by Wecharatana and Shah (1983) for smooth steel wires; can agree with eqn 3.64 as is the case for wet cellulose fibres; or to have an exponent to the power law given in eqn 3.17 greater than 2 which does not conform to either the model using Coulomb friction or a constant interfacial frictional stress, such as dry cellulose fibres (Hu and Mai, 1992).

There is not much loss in accuracy of load-deflection predictions if the stress-displacement curve is assumed to be linear. As with unreinforced cementitious materials, the two most important fracture parameters are the fracture energy and the fracture strength, the actual form of the curve being comparatively unimportant. The strain-softening of the matrix before it cracks completely can also be assumed to be linear with little loss of accuracy provided the fracture work is the same. Hence, the bilinear curve used for unreinforced cementitious materials (see Figure 3.9) is also appropriate for composites, but now the break points are quite different. If the fibres are well bonded the stress at the formation of a continuous matrix crack can be greater than the stress at the initiation of a FPZ. An even simpler strain-softening relationship can be obtained if the FPZ is modelled by the fracture toughness of the reinforced matrix, K_{Ic}, and the fibre pull-out by a linear relationship. Although K_{Ic} is not exactly a material constant, because Barenblatt's hypotheses are not satisfied (see section 1.5), it can in practice be considered constant.

3.10 The toughness of fibre reinforced composites

The concept of fracture energy based upon an area of fracture surface is not appropriate for Type I composites that do not fail on a single fracture plane. However, with Type II composites, apart from minor energies associated with dispersed microcracking activity before the development of a FPZ, the fracture work is concentrated in a thin layer on either side of the single fracture. The major contribution to the total fracture energy comes from the work of fibre pull-out. The fracture of the matrix obviously contributes to the fracture energy, but the work necessary to fracture any fibres[4] and the work of debonding are usually negligible. The fracture energy of the reinforced matrix, G_{Im}, which is approximately given by K_{Ic}^2/E^*, can be significantly larger than the fracture energy of the unreinforced matrix. The specific fracture work of pull-out, w_p, can be obtained from the expression for the stress–displacement relationship given in eqn 3.65 and is

$$w_p = \int_0^{1/2} \sigma_{cp} \, d\delta = \frac{\eta_\theta v}{6} \frac{\tau_f l^2}{d_f} \qquad (3.66)$$

This equation, derived by another method, is given by Kelly and Macmillan (1986).

Because of the well-developed crack growth resistance in fibre reinforced cementitious materials fracture in tension is much more stable than in unreinforced materials. Hence the direct tension method can be used to measure the fracture energy as well as the fracture strength. It is difficult to use the RILEM (1985) method for cementitious composites, because the fibres in a bend specimen cannot be easily pulled out completely. Because of the difficulties in measuring a fundamental toughness for Type I composites, a variety of practical toughness indices have been proposed. All these indices are obtained by taking the ratio of the work of deformation in a beam test up to some specified deflection greater than that necessary to cause first cracking. Barr and Hasso (1985) have reviewed the various toughness indices that have been proposed. The ACI Committee 544 method relies on a standard size ($100 \times 100 \times 300$ mm) bend test and defines the toughness index as the ratio of the work done up to a deflection of 1.9 mm to the work done up to first cracking. The toughness index is dependent on the size and geometry of the specimen. The ASTM Standard specifically warns that the toughness index can be size dependent (Rokugo *et al.*, 1989). The ACI method with its arbitrary deflection is one of the least attractive methods. The ASTM method (Rukugo

---

[4] The fracture of some more ductile fibres such as Kevlar and polypropylene can contribute significantly to the fracture energy.

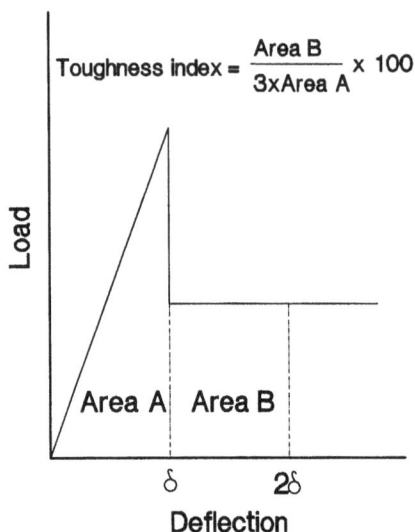

Figure 3.24 Definition of toughness index. (After Barr and Hasso, 1985.)

et al., 1989) and the proposals of Barr and Hasso (1985) all use multiples of the deflection to cause first cracking. Barr and Hasso (1985) show that the same toughness index can be obtained from notched beams of the same size but different notch depths. This result is explained by the fact that K_R curves are not very dependent of the notch depth in beams (see section 5.4.2). The toughness index proposed by Barr and Hasso (1985) seems the best. In their definition (see Figure 3.24) the toughness index, I, is defined as

$$I = \frac{\text{Area } B}{3 \times \text{Area } A} \times 100 \qquad (3.67)$$

where Area A is the area under the load-deflection curve up to first matrix cracking and Area B is the area from first matrix cracking to a deflection equal to twice the first matrix cracking deflection. Thus, if the matrix cracking had no effect, the toughness index would be 100% and if after the matrix cracked the load remained constant, giving a pseudo elastic-plastic behaviour, the toughness index would be 67%.

The toughness index can provide a means of judging the effectiveness of a fibre reinforcement, providing specimens of the same size and geometry are used, but it is of little use in predicting the behaviour of fibre reinforced structures.

3.11   Summary

The most important feature of cementitious materials is their large FPZ and FBZ. Consequently, many methods have been devised to measure the sizes of these zones and their properties. The most basic property is the stress-strain or the stress-displacement relationship. An inherent characteristic of materials that strain-soften is the formation of a single fracture. Strain-softening leads to localization of the deformation which makes it difficult to determine the stress-strain relationship. It is easier to determine the displacement across the FPZ than the strain within it. The analysis of Type I composites that do not strain-soften is analogous to the ductile deformation of metals and it is plastic collapse or large deformations that are the limits rather than fracture. Consequently, Type I composites are outside the scope of this book.

The two most important parameters in the strain-displacement relationship either for the FPZ or the FBZ are the fracture strength, f_t, and the fracture energy, G_{If}. The particular form of the model used to describe the stress-displacement relationship is relatively unimportant provided these two parameters are correct. Consequently, complicated stress-displacement relationships are not necessary and simple linear or bilinear curves are sufficient. The mode I properties are far more important than mixed mode because only in a few cases does the FPZ cause the fracture to propagate as a mixed mode fracture.

The fracture energy, G_{If}, is easier to measure than the fracture strength, f_t. The RILEM (1985) standard method is suitable for measurement of the fracture energy; there is some size effect but for practical purposes the fracture energy can be considered a constant. The fracture strength cannot be measured directly in a bend test and is difficult to measure in a direct tension test. Fibre reinforced composites are more easily tested in tension because of their large crack growth resistance.

4 Theoretical models for fracture in cementitious materials

4.1 Introduction

Cementitious materials are only quasi-brittle. A load-deflection curve obtained from a notch bend test on a small cementitious beam looks very similar to that of a ductile metal. This possibility of ductile behaviour in an apparently brittle material is paradoxical. Not only are small beams ductile but they also have a larger modulus of rupture. Some of the apparent increase in strength can be explained by statistics (see Chapter 6), but the difference is too great for this explanation to be the complete answer. To understand these paradoxes we must model the fracture process.

The flexural behaviour of beams can be modelled approximately using simple beam theory and assuming that plane sections remain plane, even if the beam is notched, and just satisfying global equilibrium (Chuang and Mai, 1989). However, fracture mechanics is needed for accurate modelling. Classic Griffith fracture mechanics cannot generally be used to model fracture in cementitious materials because the FPZs are too large. Even in hardened cement paste where the FPZ is relatively small (of the order 1–4 mm) notch bend specimens need to be deeper than 100 mm before Griffith theory can be used (Cotterell and Mai, 1987). Equivalent crack concepts have been applied to the fracture of cementitious materials where the real crack is replaced by a larger stress free crack (Jenq and Shah, 1985; Nallathambi and Karihaloo, 1986; Shah, 1988; Karihaloo and Nallathambi, 1988, 1989a,b; Alvaredo et al., 1989; RILEM, 1990), but fracture does not occur at the same critical fracture toughness independent of geometry and size as is assumed in these concepts. The two physically sound approaches to modelling the FPZ are the fictitious crack model (Hillerborg et al., 1976; Petersson, 1985) and the crack band model (Bažant and Cedolin, 1979; Bažant and Oh, 1983; Bažant and Lin, 1988; Bažant et al., 1988; Bažant and Ožbolt, 1990). Since the FPZ in cementitious materials is narrow, it can be approximately modelled by a fictitious crack extension to the true stress free crack that carries a stress. The fictitious crack model assumes that there is a unique stress-displacement softening relationship for the FPZ. In the crack band model the FPZ is assumed to have a finite width. Within the FPZ the material is strain softening. The width was, in the original formulation, assigned a somewhat arbitrary value dependent on the aggregate size; but in a subsequent development of the theory, where the fundamental instability of a strain-softened zone is inhibited by using a non-local approach (Bažant and Lin, 1988), the width of the FPZ is determined theoretically. However, unless one is specifically interested in very deep notches or in the last stages of crack propagation, the added preciseness of the crack band model over the fictitious crack model is largely illusory because the heterogeneity of cementitious materials gives rise to a large scatter in results from even supposedly identical specimens. As Bažant and Cedolin (1979) remark "the choice of either [the line crack or the crack band] is basically a question of computational effectiveness".

The approximate method based on the engineers' theory of bending is presented first. The effective crack length models are then reviewed before treating the crack band and the fictitious crack models. Comparison of the various theoretical models with experimental data is difficult for a number of reasons. Firstly, it is almost impossible to produce specimens of differing size that have the same material properties. Secondly, most research workers have concentrated on the three-point notch bend geometry and so there are very few experiments that enable the geometric independence of fracture parameters to be assessed. Where data exist they are often not sufficient to enable other models to be examined because experimenters are usually interested in justifying their own models.

4.2 The flexure of strain-softening materials

Chuang and Mai (1989) have discussed how the engineers' theory of bending can be applied to strain-softening materials. Lim *et al.* (1987) present a similar method for long fibre reinforced cementitious materials for a restricted class of strain-softening materials (Type IIA). The method does not depend on the particular strain-softening relationship chosen. Chuang and Mai (1989) used the power law

$$\sigma = f_t \left[1 - \left(\frac{\epsilon - \epsilon_t}{\epsilon_f - \epsilon_t} \right)^n \right] \tag{4.1}$$

where f_t is the tensile strength of the material, ϵ_t is the strain at the onset of strain softening ($\epsilon_t = f_t/E$), ϵ_f is the strain at which the material fails completely, and $0 < n < 1$. The width of the FPZ, $2w$, is assumed to be constant across the beam. Plane sections are assumed to remain plane and there are three regimes. In regime I the strain ϵ_s at the surface of the beam is less than ϵ_t so that the entire beam is elastic. If $\epsilon_t < \epsilon_s < \epsilon_f$ (regime II), an FPZ develops spreading from the surface of the beam towards the centre. When the surface strain ϵ_s exceeds the maximum strain that can be sustained without complete failure, ϵ_f, a crack develops. The stress distributions in the three regimes are shown schematically in Figure 4.1. The position of the neutral axis, the size of the FPZ and the depth of the crack can be determined from force and moment equilibrium. The curvature $1/R$ within the FPZ is

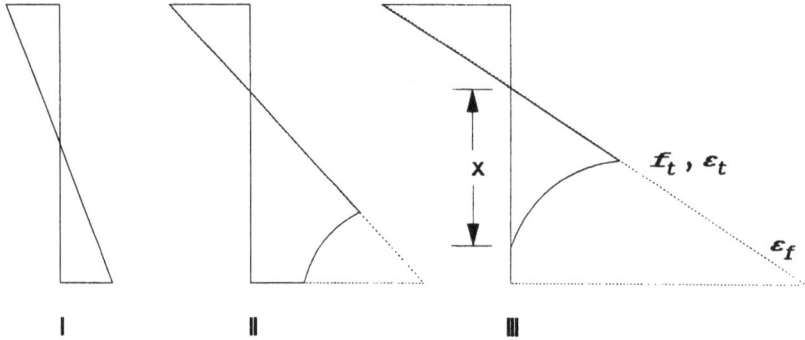

Figure 4.1 Schematic diagram of the strain (- - - -) and stress (———) distribution across the beam for the three regimes. (Chuang and Mai, 1989.)

given by

$$\frac{1}{R} = \frac{\epsilon_t}{x} \qquad (4.2)$$

where x is the distance from the neutral axis to the tip of the FPZ. The contribution of the FPZ to the deflection of a beam under three-point load is calculated from the assumption that the beam has constant curvature over the width $2w$ of the FPZ, i.e. $2w$ is much smaller than the span of the beam. The normalized load-deflection curves for a beam whose properties were deduced from the experiments of Krause and Fuller (1984) are shown in Figure 4.2. The shape of the curve depends on the width of the FPZ. Since the width of the FPZ determines the fracture work within the FPZ the correct width can be determined from the condition that the total work of fracture is equal to the fracture energy, G_{If}, multiplied by the area of the fracture surface. A narrow FPZ implies that the fracture energy is small and hence the material is brittle which leads to the 'snap back' instability shown by curve (a) in Figure 4.2. Snap back will be discussed further in section 4.6. Chuang and Mai (1989) suggest that this approximate method of analysis can be used to solve the inverse problem of knowing the load-deflection curve for a beam to infer the strain-softening behaviour. However, the fictitious crack line model discussed in section 4.5 is more accurate since local rather than global equilibrium and compatibility are satisfied.

4.3 Equivalent crack models

Provided the FPZ is small relative to the size of the specimen or structure, the actual crack and FPZ can be replaced by an equivalent elastic crack as is done in the fracture of metals. To determine the size of the equivalent crack, the

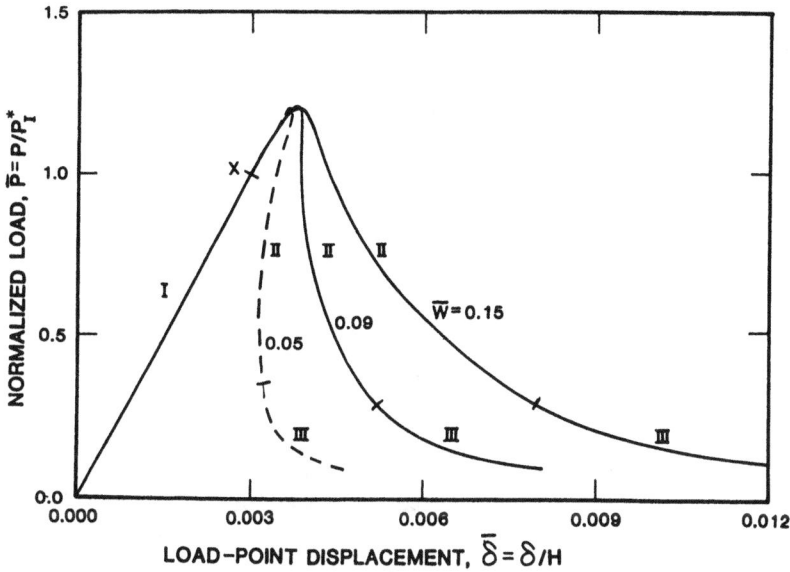

Figure 4.2 Load-deflection for three different FPZ widths; the load is normalized by the maximum elastic load and the deflection and FPZ width by the beam depth. (Chuang and Mai, 1989.)

actual crack and the FPZ can be replaced by a fictitious crack that extends to the end of the FPZ. That portion of the fictitious crack that models the FPZ is subjected to closure forces equal to the forces that exist at the edge of the FPZ. In section 4.5.3 it will be shown that the fictitious crack faces remain almost straight over the FPZ so that when the FPZ is fully developed

$$\frac{x}{d_p} = \frac{\delta}{\delta_f} \tag{4.3}$$

where x is measured from the tip of the fictitious crack and d_p is the size of the FPZ. If the stress field is K-dominated, then equating the equivalent elastic stress at the edge of the FPZ to the ultimate strength of the material yields

$$f_t = \frac{K_{If}}{\sqrt{2\pi d_p (1 - \alpha)}} \tag{4.4}$$

where α is the fraction of the FPZ that is included in the equivalent crack. As in Irwin's (1958) analysis to ensure that the same load is carried by the equivalent elastic system as in the actual one

$$\int_0^{d_p(1-\alpha)} \frac{K_{If}}{\sqrt{2\pi r}} \, dr = -\int_0^{d_p} \sigma \, dx \tag{4.5}$$

where r is measured from the tip of the equivalent crack and σ is the actual stress in the FPZ. Assuming the crack faces remain straight and a bilinear stress-displacement relationship for the FPZ, eqns 4.4 and 4.5 can be solved to give the non-dimensional size of the FPZ, $\bar{d}_p$:

$$\bar{d}_p = \frac{d_p}{l_{ch}} = \frac{2}{\pi(s+v)} \tag{4.6a}$$

and

$$\alpha = 1 - (s+v)/4 \tag{4.6b}$$

Hence

$$\Delta\bar{a}_e = \alpha\bar{d}_p = \frac{1}{2\pi}\left(\frac{4}{s+v} - 1\right) \tag{4.6c}$$

and the equivalent crack length, a_e, is given by

$$a_e = a + \Delta a_e \tag{4.7}$$

Taking typical values of s and v for concrete as 0.16 and 0.19, respectively, the size of a fully developed FPZ in concrete is about 1.8 l_{ch}. Thus, for concrete the size of a fully developed FPZ is huge. Typically, the characteristic lengths of concrete with aggregate sizes of 8 and 32 mm are 250 and 800 mm, respectively; hence, d_p ranges from 450 to 1400 mm and about 90% of the FPZ is included in the equivalent crack length. The distance from the tip of the equivalent elastic crack to the edge of the FPZ, r_p, is given by

$$r_p = (1 - \alpha)d_p = \frac{l_{ch}}{2\pi} \tag{4.8}$$

and is independent of the stress-displacement relationship. As discussed in section 1.5.2, a K-dominated stress field is ensured if the crack and ligament lengths are at least $50r_p$. Hence, for cementitious materials, the crack and remaining ligament lengths should be greater than about eight times the characteristic length, l_{ch}, which implies immense specimens.

The above calculations give the size of a fully developed FPZ. However, the FPZ can only develop to its full extent in a very large specimen. An alternative approach to the estimation of the equivalent elastic crack is to determine the size of the elastic crack that has the same compliance as the secant compliance of the actual crack. This is the method used by the equivalent crack approaches for the analysis of notch bend tests (Jenq and Shah, 1985; Nallathambi and Karihaloo, 1986; Shah, 1988; Karihaloo and Nallathambi, 1988, 1989a,b; Alvaredo et al., 1989; RILEM, 1990).

In the effective crack model (Nallathambi and Karihaloo, 1986; Karihaloo and Nallathambi, 1988, 1989a,b) the Young's modulus is obtained from the linear portion of the load-deflection curve using LEFM expressions with the initial crack length, a_i. Using the same expressions an effective crack length, a_e, is found by iteration that gives the measured deflection at maximum load. It is then postulated that the stress intensity factor, calculated using the effective crack length, attains a critical value K_{Ic}^e at the maximum load that is a material constant independent of geometry and size, with the proviso that the ligament must be more than five times the aggregate size (Karihaloo and Nallathambi, 1988). The approach given in the draft RILEM (1990a) recommendation is somewhat different. In this approach the unloading compliance of the load-crack mouth opening displacement (CMOD) is used. Young's modulus is obtained from the initial linear part of the curve and an effective crack length calculated from the unloading compliance just after the attainment of maximum load (95% of maximum). If the deformation within the FPZ was purely due to the elastic opening of microcracks, the two approaches would give the same effective crack length. However, if there is significant non-elastic deformation such as aggregate slippage within the FPZ or if debris falls into the cracks decreasing the unloading compliance, the RILEM (1990a) method underestimates the effective crack length. In the RILEM (1990a) method not only is a critical stress intensity factor calculated (denoted by K_{Ic}^s) but also the CTOD at maximum load (denoted by $CTOD_c$) is calculated from LEFM analyses using the effective crack length. It is not at all clear why two critical parameters are needed in the RILEM (1990a) method. It is assumed that in the three-point notched bend test, and in those test geometries where the stress intensity factor increases monotonically with crack length, both K and CTOD attain their critical values at maximum load (Jenq and Shah, 1985) which can only be interpreted as meaning that true crack growth starts when the load reaches a maximum. However, Karihaloo and Nallathambi (1989) allow that some true crack growth may occur before the maximum load is reached. In fact, the point at which true crack growth starts is dependent on the size of the specimen and the FPZ is not necessarily fully developed at maximum load. In test geometries where the rate of change in the stress intensity factor with crack length reaches a minimum at some particular crack length for constant load, Jenq and Shah (1985) recognize that the critical values of K and CTOD will be attained prior to the maximum load. In experiments on a panel with a central notch with the load applied at the centre of the notch (K at constant load decreases monotonically with increase in crack length) the critical value, K_{Ic}^s, is taken as the plateau value of a plot of K against effective crack growth and the CTOD at that point is taken as $CTOD_c$ (Alvaredo et al., 1989). The main criticism of these equivalent crack models is that the maximum load is not generally attained when the fracture toughness reaches a critical value. Only in very large specimens will the maximum load occur when the stress intensity factor is equal to K_{If}.

Karihaloo and Nallathambi (1988) have given a comprehensive review of the equivalent crack models. They conclude that the two equivalent crack models are in good agreement and state that the critical stress intensity factor at the tip of the equivalent crack is essentially independent of beam depth apart from the size proviso already mentioned, but their conclusions

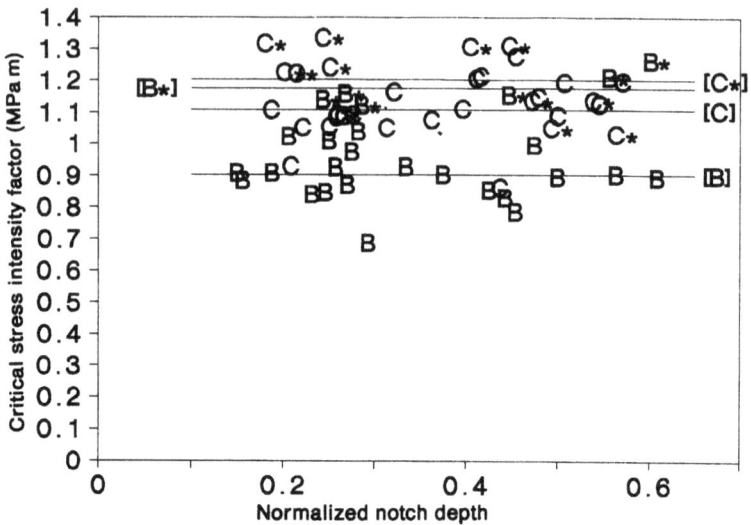

Figure 4.3 The variation of critical stress intensity factor with normalized notch depth, based on results of Swartz and Refai (1987) (K_{Ic}^e: B, $W = 203$ mm and C, $W = 305$ mm; K_{Ic}^s: B*, $W = 203$ mm and C*, $W = 305$ mm).

are worth re-examining. From the limited data available it appears that the critical stress intensity factor is independent of the relative notch depth, a/W. The results of Swartz and Refai (1987) who tested a wide range of a/W ratios, albeit using pre-cracked rather than notched beams, are presented in Figure 4.3 (except for results that fell outside the size limitation suggested by Karihaloo and Nallathambi (1988)). Although there is considerable scatter the critical stress intensity factors are essentially independent of a/W. There does seem a trend for the critical stress intensity factor to increase with beam depth, but since the difference in size is not great, one could not be certain that this difference is significant. There are two sets of data that cover a much wider range of beam size: those of Alexander reported by Karihaloo and Nallathambi (1988) and those of Hilsdorf and Brameshuber (1984). Although the data of Alexander come from three different series of beams, the concrete has the same mix and there does not seem to be any consistent difference between the series and it is assumed that they have identical material properties. An approximation to the characteristic length l_{ch} has been calculated from the mean value of K_{Ic}^e for the largest beams ($W = 800$ mm) and an estimate of the tensile strength f_t of the concrete obtained from the compressive strength of the concrete. The critical stress intensity factor K_{Ic}^e is plotted against the non-dimensional beam depth ($\bar{W} = W/l_{ch}$) in Figure 4.4; the a/W ratios in these tests varied from 0.2 to 0.4, but it is assumed that this variation does not have a significant effect. Although there is some scatter in the results, there is a definite trend for K_{Ic}^e to increase with the non-dimensional beam depth.

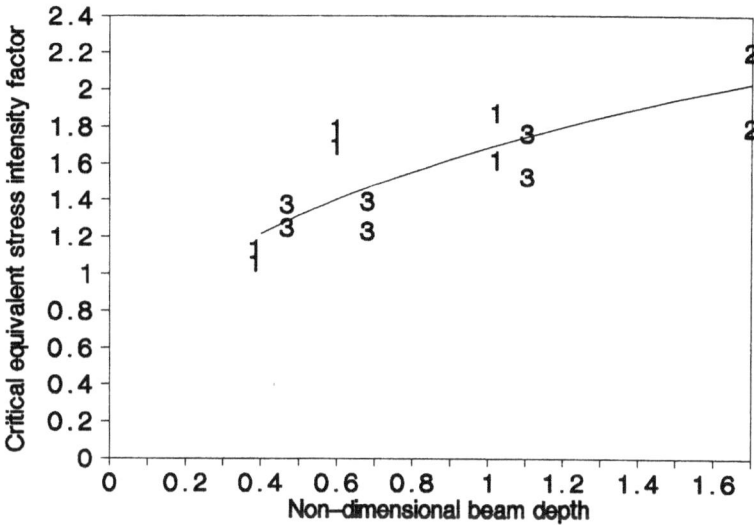

Figure 4.4 The variation of the critical stress intensity factor K_{Ic}^e (in $MPa\sqrt{m}$) with non-dimensional beam depth ($\bar{W}$), based on results of Alexander as reported by Karihaloo and Nallathambi (1988) (numbers represent the series and give a standard deviation on either side of the mean).

A power law

$$K_{Ic}^e = 1.68\bar{W}^{0.354} \qquad (4.9)$$

gives a good fit to the experimental data. The results of Hilsdorf and Brameshuber (1984) encompass an even greater range in beam depths for concrete and mortar beams. When these results are plotted against the non-dimensional depth, power laws with exponents of 0.202 and 0.227 fit the results for concrete and mortar, respectively. Since these exponents are so similar and not that very much different to that obtained for the results of Alexander, the results for both cementitious materials have been pooled in Figure 4.5 by plotting the critical stress intensity factor, K_{Ic}^e, relative to its value for a beam whose non-dimensional depth is 1. These relative critical stress intensity factors lie on a single curve that is given empirically by

$$\frac{K_{Ic}^e}{[K_{Ic}^e]_{\bar{W}=1}} = \bar{W}^{0.214} \qquad (4.10)$$

By combining the results for concrete and mortar specimens, a 20-fold range in $\bar{W}$ is obtained where the relative critical stress intensity factor nearly doubles and does not seem to have reached a plateau value even for the 800 mm deep mortar beams. The reason for the increase in the critical stress intensity factor with beam depth is that the maximum load only approaches the plateau value of the K_R-curve for extremely large specimens. The equivalent crack length models only give an approximately constant

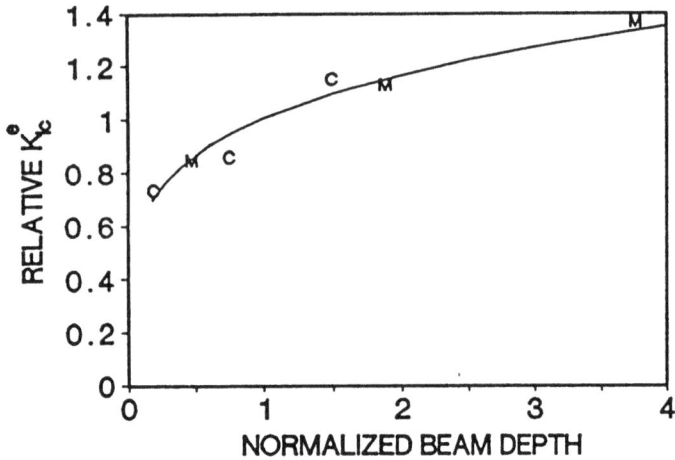

Figure 4.5 The variation of relative critical stress intensity factor with non-dimensional beam depth (based on results by Hilsdorf and Brameshuber (1984); C and M are results for concrete and mortar specimens, respectively).

value for the critical stress intensity over a very limited size range for notch bend specimens and the results cannot be applied to other specimen geometries.

Apart from the fact that the critical stress intensity factor is not a material constant, the equivalent crack models lack a real predictive capability. Karihaloo and Nallathambi (1988) do give an empirical expression for the equivalent crack length as a function of the nominal stress, notch depth, and aggregate size for notch bend specimens, but such a relationship cannot be relied upon outside of the range of the data for which it was obtained and cannot be used for any other geometries. Even if sufficient empirical data are accumulated to enable the equivalent crack length to be estimated, the equivalent crack models are only capable of calculating the maximum load and cannot predict whether a specimen or structure will fail in a comparatively gentle ductile manner with further external work necessary to complete the fracture after the maximum load has been reached, or whether it will fail in a brittle manner with all the energy coming from stored elastic energy without the need for additional external work.

4.4 The crack band model

Bažant and his co-workers (Bažant and Cedolin, 1979; Bažant and Oh, 1983; Bažant et al., 1984; Bažant and Lin, 1988; Bažant and Ožbolt, 1990; Jirásek and Bažant, 1994) have taken an early smeared crack concept used in finite element studies of reinforced concrete for simulating the microcracking and consequent strain softening behaviour of concrete, and made it objective for notched or cracked specimens so that it is not sensitive to the mesh size. Here the theory is presented for specimens where the fracture forms along the line of symmetry, though the method can be used for asymmetrical situations.

In the simplest form of the crack band theory, the FPZ is modelled by a band of width, $w = nd_a$, where d_a is the aggregate size and n a constant greater than 1, which is equal to or less than the element mesh size h. It is assumed that when the maximum principal stress σ_z exceeds the tensile strength f_t of the cementitious material, micro-cracks form which induce a strain-softened FPZ. In the original presentation (Bažant and Oh, 1983) the strain-softening was assumed to be linear and given by:

$$\epsilon^{cr} = f(\sigma) = \frac{(f_t - \sigma)}{D} \tag{4.11}$$

where ϵ^{cr} is the additional strain caused by the microcracks and D is the unloading stiffness. Although in principle any strain-softening relationship could be used, in a later paper (Bažant and Lin, 1988) it is suggested that a truncated exponential relationship is more realistic. A local definition of strain cannot be used in the FPZ and ϵ^{cr} is interpreted as the average strain over the thickness w of the FPZ. Within the FPZ the strain is given by:

$$\begin{Bmatrix} \epsilon_x \\ \epsilon_y \\ \epsilon_z \end{Bmatrix} = \begin{bmatrix} E^{-1} & -\nu E^{-1} & -\nu E^{-1} \\ -\nu E^{-1} & E^{-1} & -\nu E^{-1} \\ -\nu E^{-1} & -\nu E^{-1} & E_t^{-1} \end{bmatrix} \begin{Bmatrix} \sigma_x \\ \sigma_y \\ \sigma_z \end{Bmatrix} + \begin{Bmatrix} 0 \\ 0 \\ \epsilon_0 \end{Bmatrix} \tag{4.12}$$

where $1/E_t = 1/E - 1/D < 0$, E_t is the tangent softening modulus, and $\epsilon_f = f_t/D$ is the strain at which a complete fracture occurs and the stress in the FPZ falls to zero. The use of a total strain relationship implies that path independence is assumed; provided the loading is near proportional this assumption is justified. Objectivity is assured by requiring the specific work performed in separating the faces of the FPZ to be equal to the fracture energy, G_{If}, whether the finite element mesh size h is equal to or larger than the width w of the FPZ. Hence, the tangent softening modulus E_t must be negative, which limits the mesh size to

$$h = \frac{2G_{If}E}{f_t^2} \tag{4.13}$$

In practice the mesh size should be less than about half this limit.

Fracture of a notched or cracked body is solved by an incremental step by step increase in deflection or the crack opening displacement in the case of very large specimens (Rots, 1988). When the principal strain in a certain element exceeds the strain, ϵ_t, for the initiation of microcracking, the compliance switches to eqn 4.12. When the strain exceeds ϵ_f, the material in the element is completely fractured. Bažant and Oh (1983) analysed test data from many sources considering G_{If}, f_t, n and $h = w = nd_\mathrm{a}$, as variables to find the values that gave the optimum fits. In these analyses n varied from 1.5 to 4. However, there is little difference in the solutions if n is always taken as 3.

The crack band model described above is quite suitable in many cases, but it does suffer from a number of limitations. Refinement of the mesh so that h is less than w, which may be necessary to model crack growth accurately under asymmetrical conditions, is impossible and the variations in width along the FPZ which occur in practice cannot be modelled. The stress-strain relationship given in eqn 4.12 can be rewritten as (Bažant and Lin, 1988)

$$\begin{Bmatrix} \epsilon_x \\ \epsilon_y \\ \epsilon_z \end{Bmatrix} = \begin{bmatrix} E^{-1} & -\nu E^{-1} & -\nu E^{-1} \\ -\nu E^{-1} & E^{-1} & -\nu E^{-1} \\ -\nu E^{-1} & -\nu E^{-1} & [E(1-\hat{w})]^{-1} \end{bmatrix} \begin{Bmatrix} \sigma_x \\ \sigma_y \\ \sigma_z \end{Bmatrix} \tag{4.14}$$

where $\hat{w}$ represents the damage and may be regarded as the cracked area fraction; for no damage $\hat{w} = 0$, and for complete fracture $\hat{w} = 1$. Under monotonic loading if $\epsilon_z > \epsilon_\mathrm{t}$ and $\Delta\epsilon_z > 0$,

$$\frac{1}{(1-\hat{w})} = F(\epsilon_z) \tag{4.15}$$

With a linear strain-softening relationship, the damage function is given by

$$\left.\begin{aligned} F(\epsilon_z) &= 1, & \epsilon_z < \epsilon_\mathrm{t} \\ F(\epsilon_z) &= \frac{E\epsilon_z}{E_\mathrm{t}(\epsilon_\mathrm{f} - \epsilon_z)}, & \epsilon_\mathrm{t} < \epsilon_z < \epsilon_\mathrm{f} \\ F(\epsilon_z) &= 0 & \epsilon_z > \epsilon_\mathrm{f} \end{aligned}\right\} \tag{4.16}$$

In a subsequent paper an exponential damage function is suggested (Bažant and Prat, 1988).

The variable that determines the strain-softening must be non-local and this condition is indicated by a chara over the variable. In the earliest versions of the non-local model, the non-local variable is the average strain across the FPZ, but a non-local variable can be obtained by spatial averaging that can be applied to whole continuum rather than just to the FPZ (Bažant and Lin, 1988; Bažant and Ožbolt, 1990). The spatial average strain normal to the FPZ, ϵ_z, is defined by:

$$\hat{\epsilon}_{ij}(\mathbf{x}) = \int_V \alpha(\mathbf{x}, \boldsymbol{\xi}) \langle \epsilon_{ij}(\boldsymbol{\xi}) \rangle \mathrm{d}\boldsymbol{\xi} \tag{4.17}$$

in which

$$\int_V \alpha(\mathbf{x}, \boldsymbol{\xi}) \, \mathrm{d}V = 1 \tag{4.18}$$

the integral is taken over the volume V of the specimen, $\mathbf{x}$ and $\boldsymbol{\xi}$ are general coordinate vectors, $\alpha(\mathbf{x}, \boldsymbol{\xi})$ is a weighting function assumed to be a material property, and the angle brackets, $\langle \ \rangle$, indicate that only the positive values of ϵ are considered. In the original presentation the weighting function is effectively taken as 1 for over the crack band and 0 elsewhere. Bažant and Lin (1988) proposed a Gaussian weighting function. The spatial average strain is used to determine the damage function.

In the latest work from Bažant and co-workers (Bažant, 1994; Bažant and Jirásek, 1994; Jirásek and Bažant, 1994) it is suggested that a non-local inelastic stress increment vector, $\Delta \hat{\mathbf{S}}$, leads to a simpler finite element approximation. The increments in the strain tensor, $\Delta \epsilon$, are decomposed into an elastic component, $\Delta \epsilon'$, and an inelastic component due to strain-softening, $\Delta \epsilon''$, so that the stress increment vector, $\Delta \sigma$, is given by the inelastic increment in stress vector defined by

$$\Delta \sigma = \mathbf{E}:(\Delta \epsilon - \Delta \epsilon'') = \mathbf{E}:\Delta \epsilon - \Delta \mathbf{S} \qquad (4.19)$$

where $\mathbf{E}$ is the elastic modulus tensor of the uncracked material. During unloading $\Delta \mathbf{S}$ is zero and the unloading modulus is the initial modulus, $\mathbf{E}$. In their latest non-local continuum formulation the inelastic stress increment in eqn 4.19 is replaced by a non-local inelastic stress increment, $\Delta \hat{\mathbf{S}}$, which can be defined by an equation similar to eqn 4.17. However, Bažant (1994) has generalized the non-local formulation based on the interaction between microcracks. If only the interactions between dominant microcracks forming in planes perpendicular to the maximum principal stress are considered, the increment in the non-local principal stress component $\Delta S^{(1)}$ is in the generalized formulation given by (Jirásek and Bažant, 1994)

$$\Delta \hat{S}^{(11)}(\mathbf{x}) = \int_V \alpha(\mathbf{x}, \boldsymbol{\xi}) \Delta S^{(1)}(\xi) \, d\xi + \int_V \Lambda^{(11)}(\mathbf{x}, \boldsymbol{\xi}) \Delta \hat{S}^{(1)}(\xi) \, d\xi \qquad (4.20)$$

where $\Lambda^{(11)}$ is a crack influence function.

A problem using a spatially averaged strain to calculate the damage function has to be solved incrementally with iterations at each step to determine the size of the FPZ. A global stiffness matrix is assembled using, for every finite element, the non-local material compliance from the previous iteration (at the start of the iteration procedure for each load step, the compliances from the previous step are used) for the current load step. With the prescribed load and displacement increments the nodal displacement increments are calculated. An estimate of the strains at the end of the load increment is then found. Each integration point is checked to see if cracking has initiated; if it has not, the principal strain directions are calculated but for cracked elements the principal directions are kept the same. The strains in the new principal directions are then calculated. The spatial average strain and the damage function are found from the principal strains. The

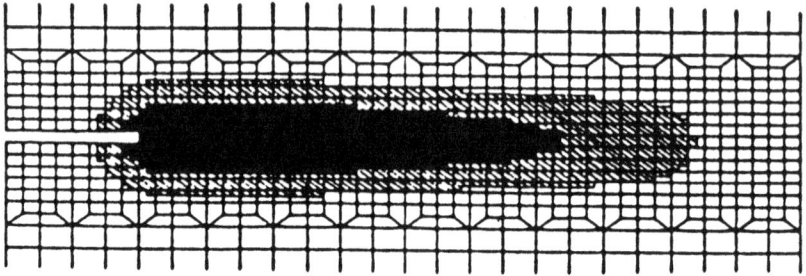

Figure 4.6 The FPZ for a notch bend specimen obtained from the non-local smeared crack model: the softened region is cross-hatched and the completely cracked zone is black. (After Bažant and Lin, 1988.)

non-local compliance matrices for each element are then updated and the nodal forces recalculated. The iteration process is continued until the difference between successive iterations is small. The FPZ for a notch bend specimen obtained by this method is shown in Figure 4.6.

4.5 The fictitious crack model

The fictitious crack has been modelled by finite elements by many researchers (Hillerborg *et al.*, 1976; Ingraffea and Gerstle, 1984; Ingraffea and Saouma, 1984; Petersson, 1985; Rots, 1986; Roelfstra and Wittmann, 1986; Carpinteri *et al.*, 1986, 1987; Wittmann *et al.*, 1988; Liaw *et al.*, 1990; Alfaite *et al.*, 1994) and boundary elements (Harde, 1991; Liang and Li, 1991; Cen and Maier, 1992; Salih and Aliabadi, 1994). However, it can be modelled more simply, if a specimen has a standard geometry, by use of known stress intensity factors and the K-superposition principle (Foote *et al.*, 1986; Cotterell and Mai, 1991; Cotterell *et al.*, 1992).

4.5.1   The finite element method

In the finite element method applied to symmetrical specimens with symmetrical loading, the crack and its fictitious prolongation are represented by pairs of finite element nodes. If the true crack tip is represented by the kth pair of nodes and the end of the fictitious prolongation, that represents the FPZ, by the nth pair of nodes, then the loads p at nodes numbers less than k must be zero and for numbers greater than n the crack opening displacement δ must be zero (see Figure 4.7). Along the fictitious prolongation the displacement is given in terms of the compliance $C(i)$ due to the external load P and the influence coefficients $K(i, j)$ that give the displacement at the ith node due to a unit load at the jth node. Hence, the boundary

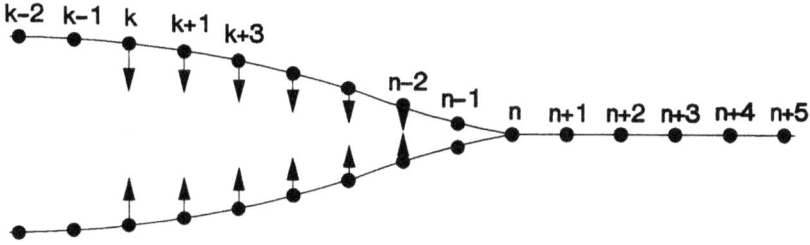

Figure 4.7 The finite element idealization of the fictitious crack in a symmetrical case.

conditions along the line of symmetry are:

$$
\left.
\begin{array}{ll}
p(i) = 0 & i < k \\
\delta(i) = \displaystyle\sum_{j=k}^{n} [K(i, j)p(j) + C(i)]P & k < i < n \\
\delta(i) = 0 & i > n
\end{array}
\right\}
\qquad (4.21)
$$

The compliances $C(i)$ and the influence coefficients $K(i, j)$ can be determined by finite element calculations. The closing forces $p(i)$ can be expressed in terms of the nodal displacements if the stress-displacement law is known. For a given load, P, the n displacements can then be found by solving the system of $2n$ equations defined by eqn 4.21.

In the original finite element analyses of the fictitious crack model a linear strain-softening relationship was used (Hillerborg *et al.*, 1976). However, with a linear relationship it is not possible to duplicate the deep belly that is obtained in the load-deflection curve for notch bend specimens and a variety of stress-displacement relationships have been used to obtain a better description of the load-deflection curve. All the analyses assume that the stress-displacement relationship is a unique material property. The bilinear stress-displacement relationship (see Figure 3.9) is the simplest curve that can accurately model the load-deflection curve of notch bend specimens. Some methods of determining the strain-softening relationship have been discussed in section 3.3, but the most straightforward method is that used by Wittmann and his co-workers who determine the bilinear curve that gives the best fit to the load-deflection curve (Roelfstra and Wittmann, 1986; Wittmann *et al.*, 1987, 1988). In addition to determining the strain softening parameters by optimization, Wittmann and his co-workers also allow the Young's modulus E to be a variable so that in all five parameters E, G_{If}, f_t, v and s are optimized. Experimental and theoretical load-deflection curves for a concrete notch bend specimen are compared in Figure 4.8. The fit between experiment and theory is good but not surprising, since it was obtained from five independent parameters. The fracture energy obtained by curve fitting was $90.5 \, \text{J/m}^2$ as compared with the RILEM (1985) method which gave $112.5 \, \text{J/m}^2$. Wittmann *et al.* (1987) argue that the

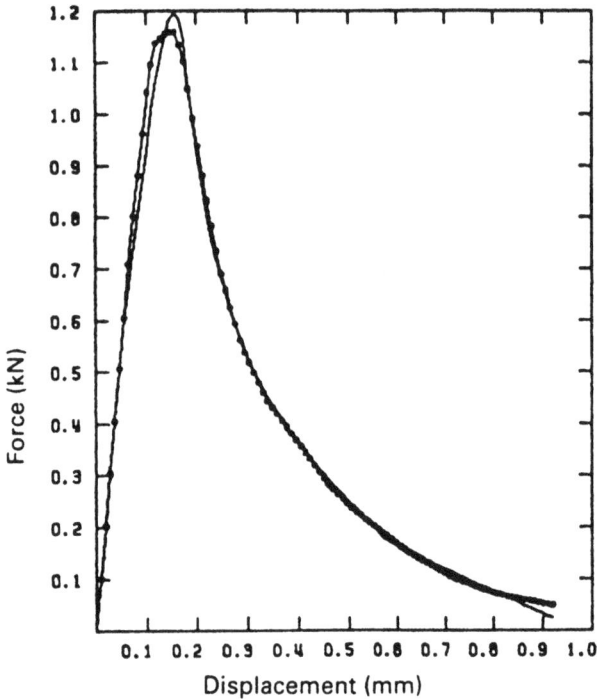

Figure 4.8 The comparison of experimental (curve with circular dots) and theoretical load-deflection curves for a notch bend specimen (After Wittmann *et al.*, 1987).

RILEM method is inaccurate, because it depends on the accurate determination of the deflection at which final instability occurs in the notch bend test. Although it is difficult to determine G_{If} accurately by the RILEM method we are not convinced that Wittmann's method gives any better accuracy because a curve fitting exercise does not ensure that the total work of fracture is exactly the area under the load-deflection curve. What is apparent is that the size restrictions imposed by the proposed RILEM (1985) method are not restrictive enough. Wittmann and his co-workers (1988) have performed a series of tests on compact tension specimens of different size (ligaments 150, 300 and 600 mm). Using their curve fitting technique they found that whereas there was little difference in any of the optimum fracture parameters between the three sizes except that G_{If} for the smallest specimen was significantly less than the values for the two larger specimens. According to the RILEM (1985) method a notch beam only 100 mm deep could be used to obtain a valid fracture energy. However, the load-deflection curve for the largest specimen could only be predicted from the specimen with a 300 mm ligament, as presumably could the load-deflection curves for larger specimens or specimens of different geometry providing they were big enough.

4.5.2 The boundary element method

The use of the boundary element method to the analysis of the fracture of cementitious materials is comparatively new (Harde, 1991; Liang and Li, 1991; Cen and Maier, 1992; Salih and Aliabadi, 1994). The simplest way to use boundary elements is to use the dual boundary element method (DBEM) (Portela et al., 1992). The dual equations on which the DBEM is based are the displacement and the traction boundary integral equations (Salih and Aliabadi, 1994). The boundary integral representation of the displacement components, u_i, in terms of a boundary point is

$$c_{ij}(x')u_i(x') + P\int_\Gamma T_{ij}(x',x)u_j(x)\,d\Gamma(x) = \int_\Gamma U_{ij}(x',x)t_j(x)\,d\Gamma(x) \quad (4.22)$$

where $P\int$ is the Cauchy principal value, $T_{ij}(x',x)$ and $U_{ij}(x',x)$ are the Kelvin traction and displacement fundamental solutions at a boundary point x, and the coefficient $c_{ij}(x')$ is given by the Kronecker delta for a smooth boundary at x'. The boundary integral representation of the traction components is

$$\frac{1}{2}t_j(x') + n_i(x')H\int_\Gamma S_{ijk}(x',x)u_k(x)\,d\Gamma(x)$$

$$= n_i(x')P\int D_{ijk}(x',x)t_k(x)\,d\Gamma(x) \quad (4.23)$$

where $H\int$ is the Hadamard principal value integral, n_i is the component of the outward normal to the boundary, and S_{ijk} and D_{ijk} are linear combinations of the derivatives of $T_{ij}(x',x)$ and $U_{ij}(x',x)$. On the fictitious crack continuity in the tangential displacement and self-equilibrium of the forces acting across it are enforced. If a linear stress-displacement relationship is assumed, then a direct solution of the boundary integral equations can be obtained since the equations can be simplified to (Salih and Aliabadi, 1994)

$$\begin{bmatrix} A & [H_f] & [G_f] \\ 0 & [C_f] & [D_f] \end{bmatrix} \begin{Bmatrix} X \\ \{u_f\} \\ \{t_f\} \end{Bmatrix} = \begin{Bmatrix} F \\ \{S_f\} \end{Bmatrix} \quad (4.24)$$

where $[H_f]$ and $[G_f]$ are coefficients at the nodes on the fictitious crack, $[C_f]$ and $[D_f]$ are the fictitious crack boundary conditions corresponding to the vectors on the fictitious crack $\{u_f\}$ and $\{t_f\}$, and $\{S_f\}$ is a vector of the material parameters. The efficiency of the DBEM as compared with an FEM method is not discussed. It should be noted that the simplified method cannot be used with a more general stress-displacement relationship for the FPZ.

4.5.3 The K-superposition method

The simplest method of analysing specimens with standard symmetrical geometries is by the K-superposition method (Foote et al., 1986; Cotterell *et al.*, 1988; Cotterell and Mai, 1991). This method was originally proposed by Lenain and Bunsell (1979) for asbestos reinforced cement sheet and developed for fibre reinforced cementitious materials (Foote *et al.*, 1980, 1986; Cotterell and Mai, 1988; Cotterell *et al.*, 1988). As in the finite element method, the FPZ is replaced by a fictitious prolongation of the true crack over which bridging stresses exist, but the problem is solved by superposition of the stress system due to the bridging stresses on the stress system caused by the applied loads. If the stresses at the tip of the fictitious crack are to be finite and continuous then the total stress intensity factor (see eqn 1.22) must be zero. The stress intensity factor, K_a, due to the applied loads can be found for standard specimen geometries from handbooks of stress intensity factors (Tada *et al.*, 1973; Rooke and Cartwright, 1976) and the stress intensity factor, K_r, due to the closing bridging stresses can be found by integration of standard K expressions for point loads on a crack face. The crack opening displacements and specimen deflections can be found from Castigliano's theorem (see section 1.6.1). The problem is non-linear because K_r depends on the crack opening displacement which is not known explicitly. However, the problem can be solved quite quickly by iteration because under load the faces of the fictitious prolongation to the crack remain reasonably straight (Foote *et al.*, 1986; Cotterell and Mai, 1988; Cotterell *et al.*, 1988, 1992). The fictitious crack tip must meet in a cusp, but this cusp is very sharp. The crack faces show most deviation from a linear profile when the FPZ is small, but then the stress in the FPZ is close to the ultimate strength of the material and the exact form of the crack opening is comparatively unimportant. Thus, the load-deflection curve can be determined to an accuracy of about 5%, if it is assumed that the fictitious crack faces remain straight. The method of solution is described for a three-point notch bend specimen, but can be applied to any geometry for which expressions for the stress intensity factors are known.

The solution of crack propagation and fracture is driven by the CTOD until the CTOD (δ_t) reaches its critical value δ_f at which the true crack starts to propagate; from this point the solution is driven by the true crack length (a_0). Thus this method can be used even where the behaviour is brittle and the load-deflection curve 'snaps-back' (Biolzi *et al.*, 1989; Carpinteri, 1990). For a particular value of the driving variable, the load P and the normalized length of the fictitious crack a/W can be found by iteration. A flow-chart of the sub-program that carries out this iteration is shown in Figure 4.9. The input into this subroutine is the driving parameter $(\delta_t,$ or $a_0)$ the depth of the beam (W), and an initial estimate of the depth of the fictitious crack (a). The subroutine first calculates the load P that satisfies $K_t = 0$ and then calculates the beam depth (w) that gives the required CTOD. Iteration is performed to find the value of a that gives the required beam depth (W). The subroutine can be used whether the faces of the fictitious crack are assumed to be straight or if the true shape is found by

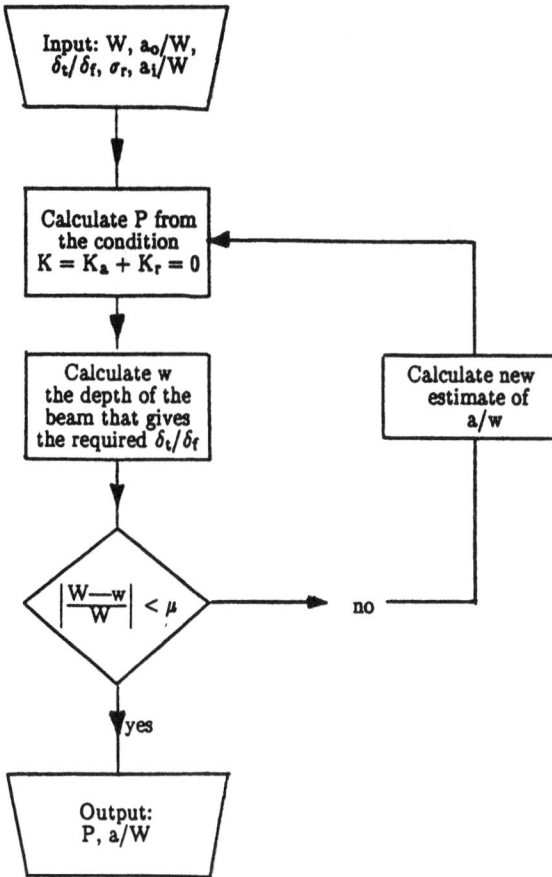

Figure 4.9 Subroutine LOAD: flow chart for determination of P and a/W.

iteration. During the running of subroutine LOAD it is assumed that the relative shape of the crack faces does not change. That is

$$\delta = \delta_t f(\bar{x}) \tag{4.25}$$

where $\bar{x}$ is the distance from the tip of the fictitious crack normalized by the length of the FPZ and the function $f(\bar{x})$ is kept constant. The bridging stress along the fictitious prolongation of the crack is calculated from δ and the strain-softening relationship.

If the actual profile of the fictitious crack is used in calculating the bridging stresses instead of assuming that the faces are straight, subroutine PROFILE is used (see Figure 4.10). The starting point for this iteration routine is a linear profile, where $f(\bar{x}) = \bar{x}$. Subroutine PROFILE uses subroutine LOAD to determine P and a, it then calculates the crack profile determined by the previous estimate of the bridging stresses in the FPZ. The new profile

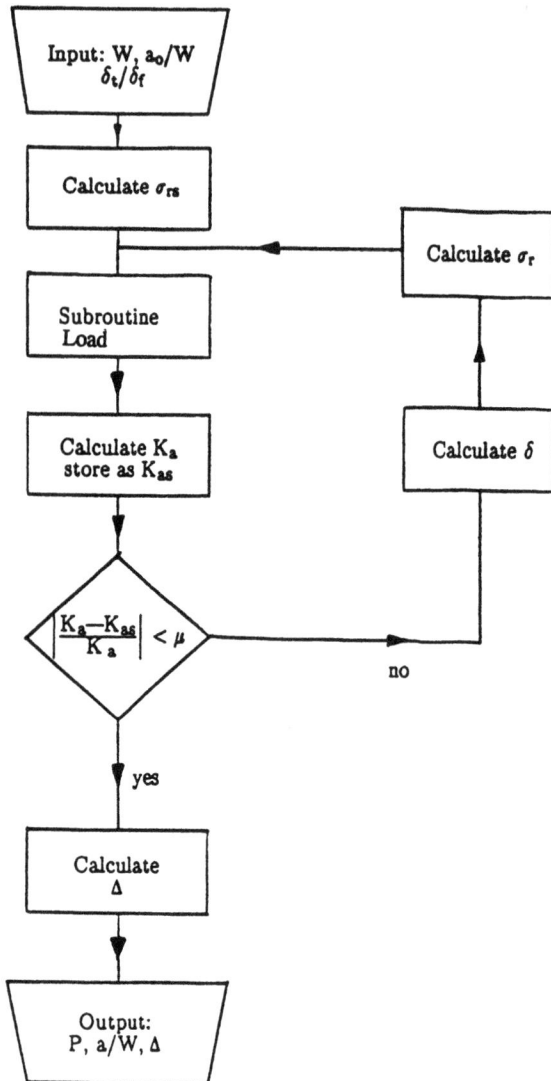

Figure 4.10 Subroutine PROFILE: flow chart for iterative calculation of the fictitious crack profile.

then becomes the input to subroutine LOAD and the process continues until the difference between the current value of K_a and the previous value is small. Since the profile is never very much different to the initial linear one, the convergence is swift. An accuracy of 0.2% is obtained in about three iterations.

To perform the foregoing analysis, assuming a bilinear softening relationship it is necessary to know the parameters G_{If}, f_t, v and s, as well as the Young's modulus E. The load-deflection curve can be modelled with accuracy if the fracture energy is obtained by the RILEM (1985) method so that the number of parameters that need to be optimized is reduced to four. The optimization process makes use of the fact that the maximum load (P_m) of most laboratory sized concrete NB specimens occurs at a relatively small CTOD. Under these conditions the bridging stresses are close to the ultimate strength, f_t, and hence, for a given G_{If}, the maximum load is mainly controlled by f_t. The beam deflection Δ_m at maximum load is largely controlled by the Young's modulus E because the FPZ is only just developing. Thus, for a given break point (v, s) in the bilinear curve, iteration can be used to find the values of f_t and E that give the experimental values of P_m and Δ_m. The break points (v, s) can then be determined by finding values that give the best fit to the rest of the load-deflection curve.

To compare the K-superposition with the finite element method of solving the fictitious crack model the results of Wittmann et al. (1987) shown in Figure 4.8 have been analysed (Cotterell et al., 1992). In this analysis the value of $G_{If} = 112.5\,\text{J/m}^2$ obtained by the RILEM method (1985) has been used and the other material parameters found as described above. The values of these parameters obtained from the K-superposition method are compared with those obtained from the finite element method in Table 4.1, and the theoretical load-deflection curve obtained from the K-superposition method is compared with the experimental results of Wittmann et al. (1987) in Figure 4.11. The material parameters, apart from G_{If}, are very similar and the K-superposition modelled curve fits the experimental data as well as the finite element model (cf. Figure 4.8). Thus, there is little difference in fracture solutions to the fictitious crack model whether the finite element or the K-superposition method is used. The advantage of the K-superposition method is that it is simple and the program can be run on a personal computer. However, it cannot be used for mixed mode fractures.

The profile of the fictitious prolongation to a notch in the NB specimen analysed above is shown in Figure 4.12. Only when δ_t/δ_f is small, is the profile significantly different from a linear profile. The load-deflection curve obtained from the assumption that the fictitious crack faces remain straight is compared with the 'exact' curve in Figure 4.13. The difference between these two curves is much less than the scatter between supposedly identical specimens and there seems little point in finding the 'exact' crack

Table 4.1 Computed parameters for experiments of Wittmann et al. (1987)

Parameter	Wittmann et al. (1987)	Cotterell et al. (1992)
G_{If} (J/m^2)	90.5	112.5
E (GPa)	27.1	32.8
σ_t (MPa)	4.76	4.79
δ_f (mm)	0.132	0.150
s	0.150	0.140
v	0.137	0.174

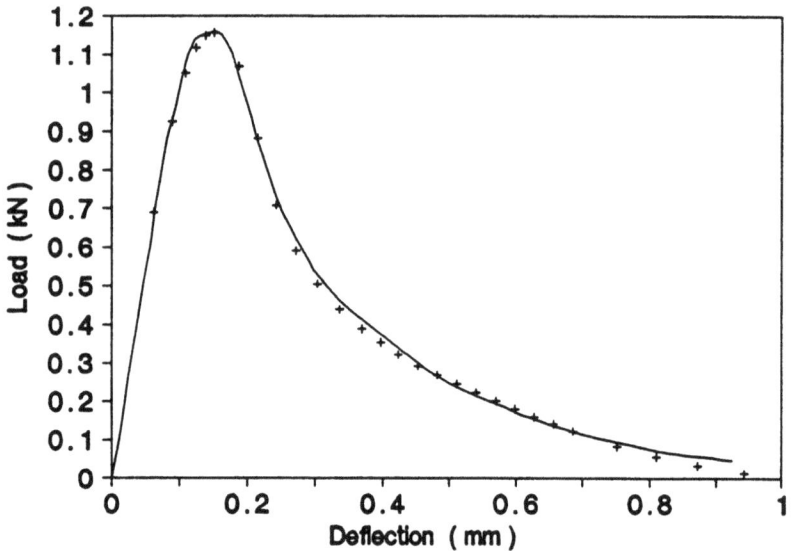

Figure 4.11 Load-deflection curve for a concrete NB test performed by Wittmann *et al.* (1987): (----) experimental data, (+) K-superposition modelling.

profiles. However, the discrepancy between 'exact' load-deflection curves and those obtained from the assumption that the crack faces remain straight is a little more for the double cantilever beam geometry and may be greater for other geometries.

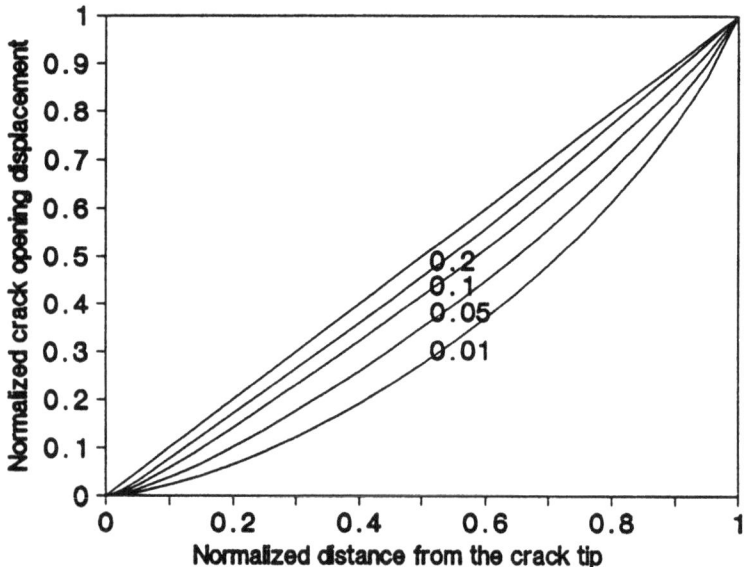

Figure 4.12 Theoretical fictitious crack profiles for the Wittmann *et al.* (1987) NB specimen. The curves for $\delta_t/\delta_f = 0.01, 0.05, 0.1, 0.2$ are compared with a linear profile.

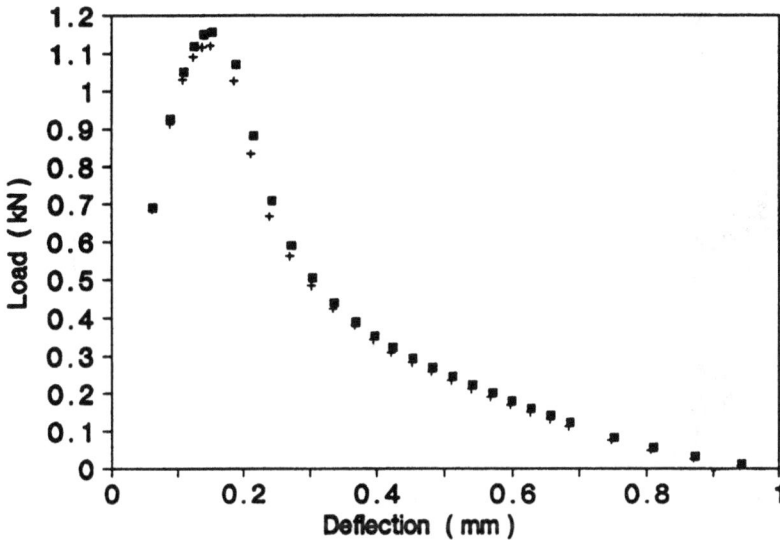

Figure 4.13 Comparison of 'exact' load-deflection curve (■) with that obtained from the assumption that the fictitious crack faces are straight (+).

An essential feature of any fracture model is its predictive capability. The load-deflection curves for two sizes of beam ($W = 100$ and $150\,mm$) are given in Figures 4.14 and 4.15 (Cotterell *et al.*, 1992). There is significant difference in the experimental curves of supposedly identical specimens, but no more

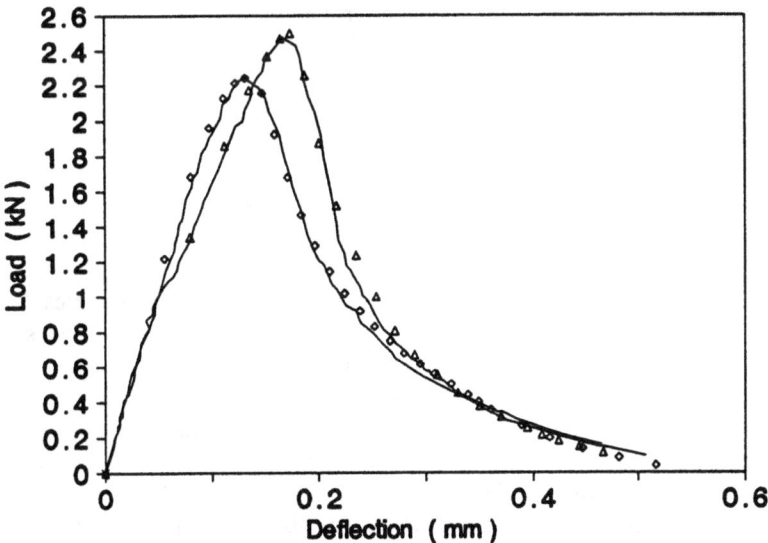

Figure 4.14 Load-deflection curves for mortar NB tests ($W = 150\,mm$): (----) experimental, ($\Diamond, \Delta$) modelled curves for beams 1 and 2, respectively.

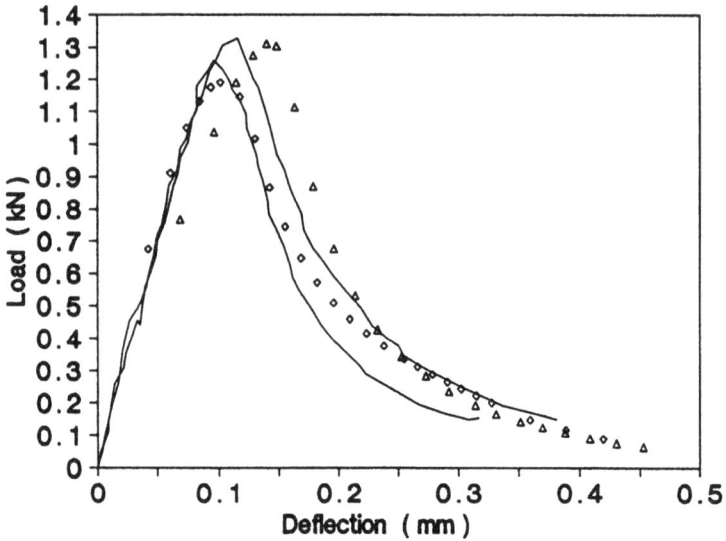

Figure 4.15 Load-deflection curves for mortar NB tests ($W = 100\,\text{mm}$): (----) experimental, ($\Diamond, \triangle$) modelled curves using parameters from beams 1 and 2, respectively.

than that observed by other workers (Hillerborg *et al.*, 1976; Kormeling and Reinhardt, 1981). The fracture energies, obtained by the RILEM (1985) method for the two large beams, are very similar. The theoretical curves given in Figure 4.15 were constructed from the parameters found by the optimization procedure explained above using the load–deflection curves for the larger beams shown in Figure 4.14. The maximum load for the smaller beam predicted from the two sets of material parameters effectively bound the experimental values, but the deflection at maximum load is overestimated somewhat by one set of the parameters. Thus the predictive capabilities shown by these tests are reasonable.

The main characteristics of the stress-displacement relationship that determine the load-deflection curves of cementitious specimens apart from the fracture energy, G_{If}, are the stress, f_{t}, at which strain-softening initiates and the long tail caused by the pulling-out of interlocking aggregates. Smith (1994) has suggested that these characteristics could be modelled by a simple piece-wise softening law (see Figure 3.10) consisting of two constant stress sections. Such a strain-softening relationship is attractive because it is only necessary to determine the COD at the end of each constant stress section. However, an even simpler strain-softening relationship that will give accurate load-deflection predictions is obtained by a combination of the classic LEFM approach and the fictitious crack model. In this simpler model, the FPZ near the tip of the true crack where the stress is close to the tensile strength, f_{t}, is modelled by a critical stress intensity factor, K_{c}, and the remainder of the FPZ by a constant stress region where the bridging stress is $f_{tt} < f_t$. The three parameters defining strain-softening are then K_c, f_{tt} and δ_f. A similar scheme is discussed in section 5.4.1 for fibre reinforced cementitious materials.

4.6 Size and notch effects

If the specimen is large in comparison with the characteristic length l_{ch}, the FPZ will develop to its full extent before the maximum load P_m is reached, and the nominal strength σ_n can be obtained by equating the stress intensity factor at the tip of the crack

$$K = \sigma_n \sqrt{W} F(a/W) \tag{4.26}$$

to the plateau value of the fracture toughness, K_{If}. Thus

$$\frac{\sigma_n}{f_t} = \frac{1}{F(a/W)\sqrt{\bar{W}}} \tag{4.27}$$

where $\bar{W}$ is the non-dimensional size $\bar{W} = W/l_{ch}$ and for a three-point bend specimen σ_n is defined by

$$\sigma_n = \frac{1.5P_m(S/W)}{BW(1 - a/W)^2} \tag{4.28}$$

The nominal strength normalised by f_t for three-point notch bend specimens of different non-dimensional size $(S/W = 8)$ obtained from classic LEFM are compared in Figure 4.16, with the strength obtained from the fictitious

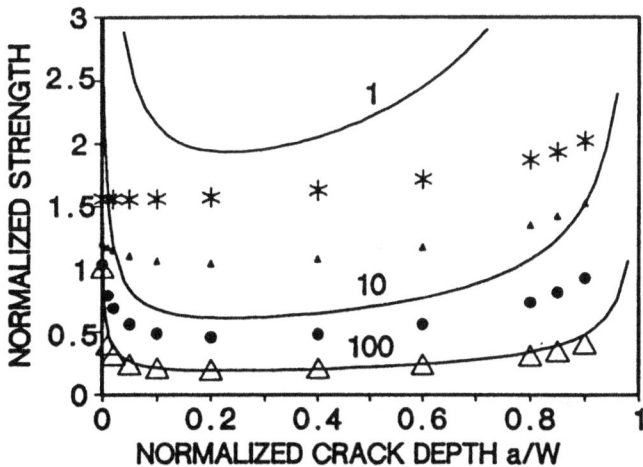

Figure 4.16 Normalised strength of a notch bend specimen—comparison of predictions of classic (--- $\bar{W} = 1, 10, 100$) LEFM and the fictitious crack model ($\bar{W} = 0.1$, *; 1.0, ▲; 10, ●; 100, △).

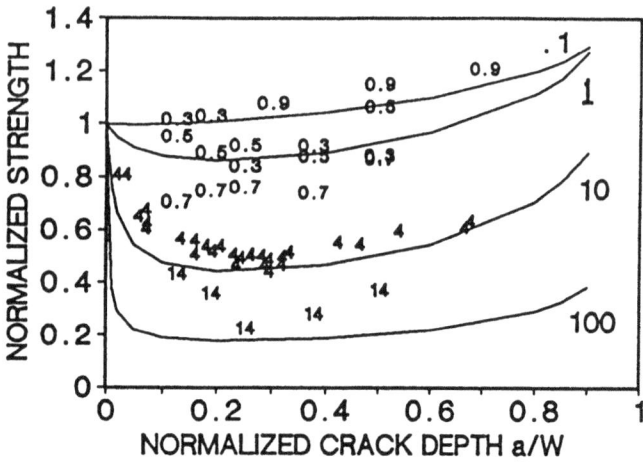

Figure 4.17 Experimental results of notch sensitivity for hardened cement paste, mortar and concrete normalised by the strength of an un-notched beam: the numbers refer to approximate non-dimensional beam depth; the sources of the data are $\bar{W} = 0.3$, 0.5, 0.7, 14 (Shah and McGarry, 1971); $\bar{W} = 4$ (Higgins and Bailey, 1976); $\bar{W} = 0.9$ (Malvar and Warren, 1988).

crack model using the K-superposition method for a bilinear strain-softening material whose properties are given in Table 4.1 (Cotterell and Mai, 1991). A beam whose non-dimensional depth is unity has a physical depth equal to its characteristic length, which for the concrete whose properties are given in Table 4.1 is 160 mm. Hence, it is seen that only for extremely large specimens does classic LEFM give a good approximation to the nominal strength. Over quite a wide range of notch depths the nominal strengths of the notch bend specimens are reasonably constant. Thus, when considering size effect, results in the range $a/W = 0.167–0.5$ can be lumped together. The solution given by the fictitious crack model is not strongly dependent on either the span to depth ratio or the break point (v, s) in the bilinear curve and is therefore compared with experimental results drawn from a variety of sources in Figures 4.17–4.19.

The notch sensitivity of cementitious materials ranging from hardened cement paste to concrete is shown in Figure 4.17, where the strength is normalized by the modulus of rupture for the un-notched specimens. The determination of G_{If} and f_t is not the same in all the references used in Figure 4.17. Often the ultimate tensile strength of the material f_t is not given and it has been necessary to estimate its value from compression tests or to assume a value based on the bend tests. In view of these difficulties in accurately determining the material parameters, the notch sensitivity is reasonably well predicted by the theory.

The size dependence of the nominal strength for plain and notched beams is shown in Figures 4.18 and 4.19. Here, apart from the variety of methods of determining G_{If} and f_t, there is the difficulty of ensuring that specimens of

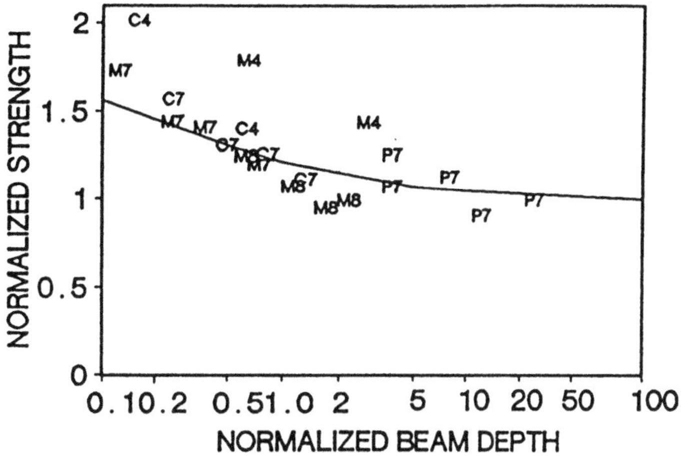

Figure 4.18 Size effect on un-notched beams normalised by f_t: P, M, C refer to cement paste, mortar and concrete, respectively; 4, Hilsdorf and Bramshuber (1984); 7, Strange and Bryant (1979); 8, Ward and Li (1989).

different size have the same material properties, and it is unsurprising that there is considerable scatter in the experimental data. However, the experimental results broadly follow the trends predicted by the theory.

The size effect predicted on the basis of classic LEFM can be modified to take account of the size of the FPZ by using the equivalent crack concept (Cotterell and Mai, 1991), defining a nominal stress σ_N in the same way as

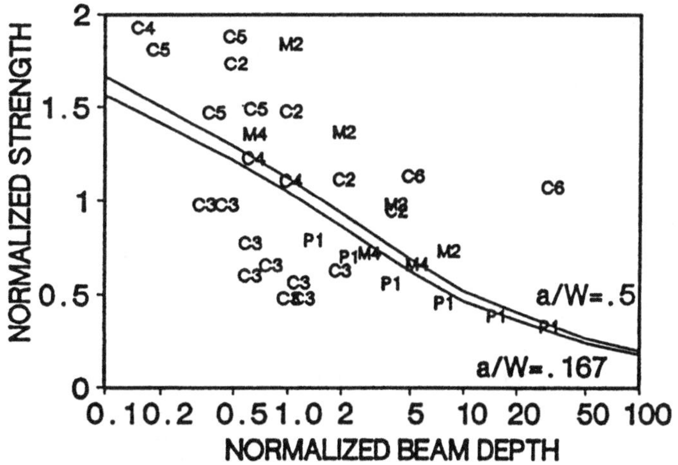

Figure 4.19 Size effect on notched beams normalised by f_t: $a/W = 0.167$–0.5, P, M, C refer to cement paste, mortar and concrete, respectively, data from: 1, Higgins and Bailey (1976); 2, Bažant and Pfeiffer (1987); 3, Mindess (1984); 4, Hilsdorf and Bramshuber (1984); 5, Jenq and Shah (1986); 6, Malvar and Warren (1988).

Bažant et al. (1986) by

$$\sigma_N = \frac{P_m}{BW} \tag{4.29}$$

Since it already has been noted that $F(a/W)$ is not very sensitive to a/W, and provided the FPZ is relatively small, $F(a_e/W) \approx F(a/W)$. Hence

$$\frac{\sigma_N}{f_t} \approx \frac{(1 - a_e/W)^2}{1.5\sqrt{\bar{W}}(S/W)F(a/W)} \tag{4.30}$$

If the proportion of the FPZ that has to be added to the crack length to obtain its equivalent length is small then to a first order

$$\frac{\sigma_N}{f_t} = \frac{(1 - a/W)^2}{1.5(S/W)F(a/W)\sqrt{\bar{W}\left(1 + \dfrac{4\Delta\bar{a}_e/\bar{W}}{1 - a/W}\right)}} \tag{4.31}$$

If the FPZ is fully developed, $\Delta a_e = \alpha d_p$, and is a material property. Hence, for geometrically similar specimens the first order size effect law is given by

$$\frac{\sigma_N}{f_t} = \frac{A}{\sqrt{(1 + \lambda W)}} \tag{4.32}$$

where (A, λ) are constants that depend on the geometry and the material properties. Equation 4.32 is the size effect law (SEL) for geometrically similar specimens originally obtained from dimensional arguments by Bažant (1984). In a later paper (Bažant and Kazemi, 1991), an argument similar to that given above was developed. Bažant et al. (1994) have recently reviewed the application of the SEL to a wide range of specimen geometries. If the SEL is valid, a linear relationship is obtained when $(1/\sigma_N)^2$ is plotted against W with an expected slope m given by

$$m = \frac{[1.5(S/W)F(a/W)]^2}{EG_{If}(1 - a/W)^4} \tag{4.33}$$

The SEL forms the basis of a RILEM draft standard for the determination of the fracture energy (G_{If}) from the above plot (RILEM, 1990b). A SEL plot of data obtained by Bažant and Pfeiffer (1987) is shown in Figure 4.20.

The SEL as originally stated only applies to a single specimen geometry but it can be generalized to include all geometries by a suitable definition of the nominal stress. Following the notation of Bažant and Kazemi (1991), the energy release rate (G) of any notched specimen can be written as

$$G = \frac{P^2 g(\alpha)}{E^* b^2 d} \tag{4.34}$$

where bd^2 is proportional to the volume of the specimen, and $g(\alpha)$ is a function of the crack length normalized by the characteristic dimension d.

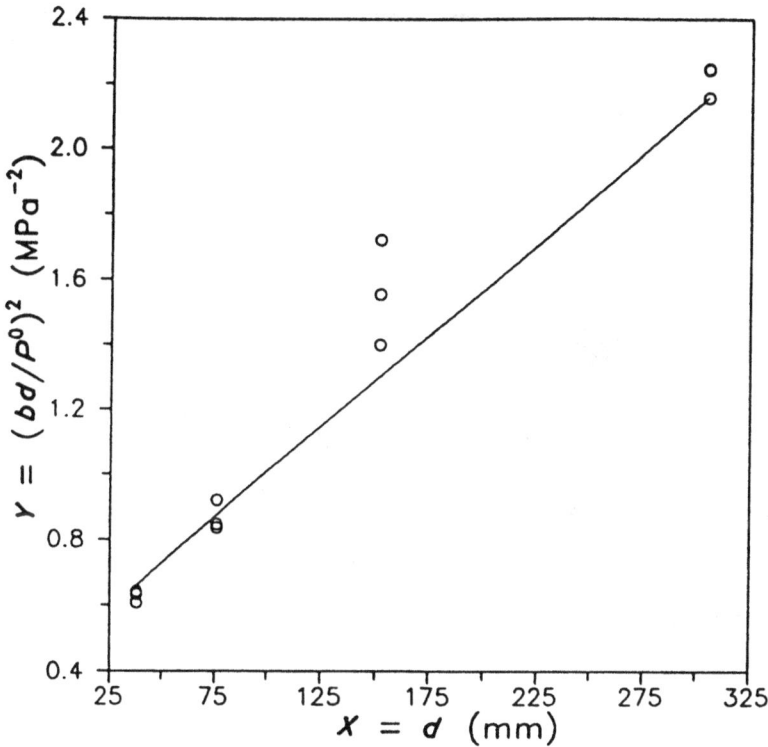

Figure 4.20 A 'Size Effect Law' plot of data from Bažant and Pfeiffer (1987) (After RILEM 1990b).

If it is assumed that the FPZ is fully developed at the maximum load, then the first order approximation to the nominal stress σ_N is given by

$$\sigma_N = \left(\frac{E^*G_{If}}{g'(\alpha)c_f + g(\alpha)d}\right)^{1/2} \qquad (4.35)$$

where $g'(\alpha)$ is the differential of $g(\alpha)$ (it is assumed here that $g'(\alpha) > 0$; Bažant and Kazemi (1991) discuss the implications if $g'(\alpha) < 0$), and c_f is the same as Δa_e. It is tacitly assumed by Bažant and Kazemi (1991) that the extension c_f necessary to the real crack to obtain the equivalent crack is always equal to the limit value for an infinitely large specimen. Redefining the nominal stress as

$$\sigma_N = \frac{\sqrt{g'(\alpha)}P_m}{bd} \qquad (4.36)$$

and introducing a 'brittleness number' (β) defined by

$$\beta = \frac{g(\alpha)d}{g'(\alpha)c_f} \qquad (4.37)$$

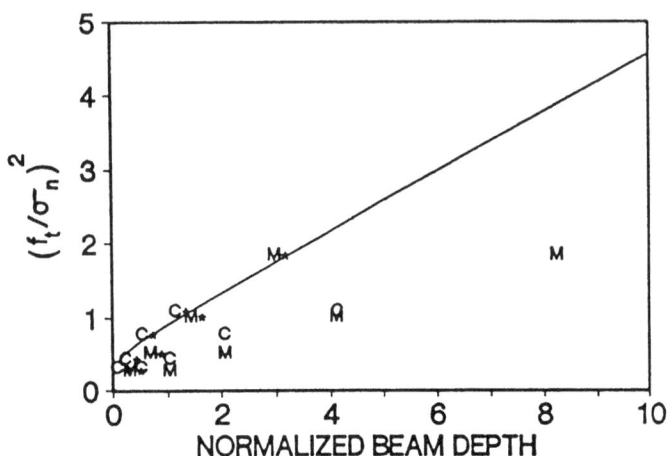

Figure 4.21 A replot of the size effect law of Bažant *et al.* (1986); (---) theoretical curve from the fictitious crack model; C, M experimental data points for concrete and mortar using G_{If} obtained by Bažant's method; C^*, M^* experimental points obtained by choosing the l_{ch} that gives the best fit with the theoretical curve.

Equation 4.35 can be rewritten as

$$\sigma_N = \left(\frac{(E^* G_{If})/c_f}{(1+\beta)} \right)^{1/2} \tag{4.38}$$

This equation looks like a very attractive way of dealing with the vexatious question of size effect. Unfortunately, both eqns 4.31 and 4.38 are only first order approximations and only hold for extremely large specimens.

A plot of $(1/\sigma_N)^2$ against the beam depth W for NB specimens is approximately linear, but the fracture energy obtained from the slope is significantly underestimated. Using the earlier definition of nominal stress, σ_n, the experimental data of Bažant *et al.* (1986) (with G_{If} corrected for the weight of the beams and f_t obtained by extrapolating the notch tension results to zero ligament length) have been replotted in Figure 4.21 in the form of $(f_t/\sigma_N)^2$ against W (Cotterell and Mai, 1991). The theoretical results obtained from the fictitious crack model by the K-superposition method are also given in Figure 4.21. The experimental data based on the values of l_{ch} (74 mm and 37 mm for concrete and mortar, respectively) estimated from the value of G_{If} obtained from the SEL, do not agree well with the theoretical results. The theoretical curve is quite linear except near the origin, but the slope at $\bar{W} = 8$ is about 4, whereas the asymptotic slope obtained from classic LEFM is only 0.275. Thus, unless unrealistically large specimens are tested, G_{If}, obtained by the RILEM (1990b) method is significantly underestimated. If we assume l_{ch} is 200 mm and 100 mm, respectively, for the concrete and mortar specimens, which are far more reasonable values than those given above, then the experimental data fit the theoretical curve (the data

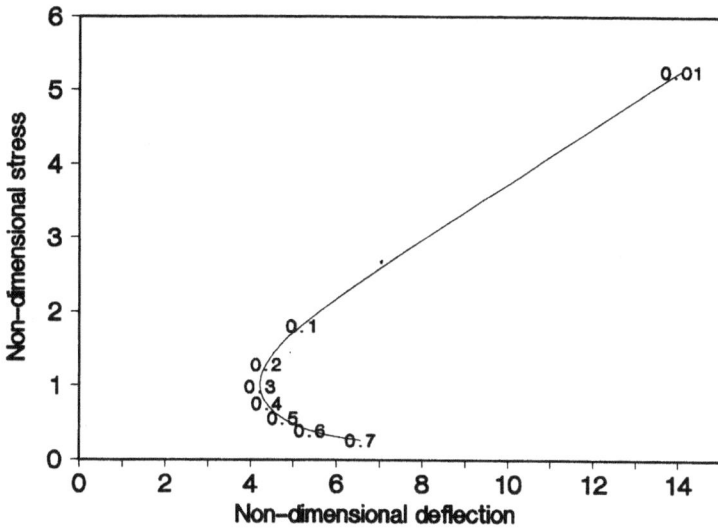

Figure 4.22 The equilibrium fracture stress as a function of the load point deflection (the values of a/W are indicated).

points obtained using these new estimates of l_{ch} are marked with an asterisk in Figure 4.21). This re-analysis of the data of Bažant *et al.* (1986) gives some confirmation to the contention that the RILEM (1990b) draft method for obtaining the fracture energy from the SEL significantly underestimates its value.

The stability during fracture of a flawed structure depends on its material properties, geometry and size of the structure. If the FPZ is small compared with the structure then the stability of the structure can be determined from classic LEFM and its geometry alone. For example, a large tension specimen is unstable whether fracture occurs under force or deflection control, but in other structures may be stable if the loading is controlled by deflection. Beams always fracture unstably under load control; the fracture is stable under deflection control if the notch is deep no matter how brittle the material is. To investigate the inherent stability of a structure we only need to know the stress intensity factor in terms of the notch or crack depth and the load-point deflection in terms of the load and notch depth. Tada *et al.* (1973) give the stress intensity factor and the load-point deflection (Δ) for a three-point notch bend specimen with a span four times the depth of the beam, in terms of the nominal stress

$$\sigma_0 = \frac{1.5PS}{BW^2} \tag{4.39}$$

that would be at the surface of the beam if there were no notch. Equating the stress intensity factor to the fracture toughness K_{Ic} for an ideal brittle material, we obtain an expression for the equilibrium nominal stress σ_0 as a function of the relative crack depth a/W. Knowing the equilibrium nominal stress, the non-dimensional load-point deflection, $E\Delta/(K_{Ic}W^{1/2})$, can be calculated and is shown in Figure 4.22. To maintain equilibrium the load point deflection must decrease as the crack grows if the initial relative crack depth is less than about 0.3. Carpinteri (1989a,b, 1990, 1991) calls this behaviour 'catastrophic'. Hence, notched beams of a brittle material loaded under deflection control fracture initially in an unstable manner if the relative crack depth a/W is less than 0.3, but are stable if the relative crack depth is greater than 0.3. A fracture that is initially unstable may arrest. The possibility of arrest cannot be completely determined from the equilibrium solution because in unstable fracture there is an excess of energy released over and above that needed to create the fracture. Arrest can only be predicted accurately from a dynamic analysis, but arrest is guaranteed if the elastic energy stored is less than the fracture energy required to complete the fracture. In Figure 4.23 the ratio of the energy stored to the energy required to complete the fracture (defined by Carpinteri (1989a) as the 'global brittleness') is shown as a function of the relative crack depth. It can be seen that provided the relative crack depth is greater than about 0.1, a fracture in a three-point notch bend specimen $(S/W = 4)$, though it may be initially unstable, will arrest. As the slenderness ratio (S/W) increases so the behaviour of a beam becomes more brittle. In the above treatment it is assumed that the fracture is not limited by the stress the material is able to sustain, which leads to the impossibility of a plain

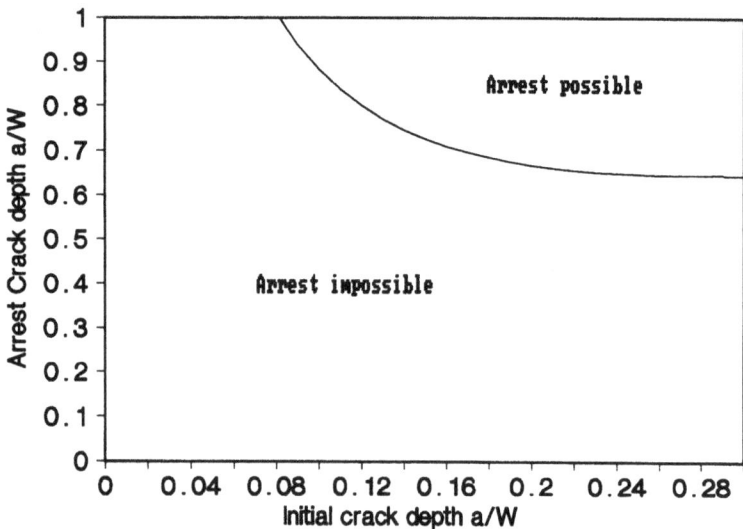

Figure 4.23 The ratio of the elastic energy stored at fracture initiation to the energy required to complete the fracture in a three-point notch bend specimen $(S/W = 4)$.

un-notched specimen being able to sustain infinite load without failure.[1] If the FPZ is small then the global brittleness or 'energy brittleness number' can be expressed as (Carpinteri, 1989a, 1991)

$$\text{Global brittleness} = \frac{\epsilon_t}{18 s_E} \left(\frac{S}{W} \right) \qquad (4.40)$$

where ϵ_t is the elastic strain at the critical stress, f_t, and s_E is the 'stress brittleness number' defined by

$$s_E = \frac{G_{If}}{f_t W} \qquad (4.41)$$

Equation 4.40 under-predicts the slenderness ratio at which catastrophic or snap back fractures occur. Assuming that in the limit situation at complete fracture, the beam effectively rotates about a hinge in the upper face of the beam,[2] an approximate condition for catastrophic or snap back behaviour is given by

$$\frac{s_E}{\epsilon_t} \leq \left(\frac{S}{3W} + \frac{4W}{S} \right) \qquad (4.42)$$

In cementitious materials the FPZ is usually not small compared with the specimen dimensions, and to determine the overall stability one must consider the strain-softening relationship. A finite FPZ implies that a material has a K_R-curve but, if the FPZ is large, it is not independent of the specimen geometry or size. The K_R-curves (normalized by K_{If}) obtained from the concrete parameters given in Table 4.1 for three beams of depth 0.1, 1 and 10 m are shown in Figure 4.24. In the largest beam, the FPZ develops to a length of 730 mm and is almost entirely in a K-dominated stress field. The K_R-curve for this beam has the classic shape and rises to a non-dimensional plateau value of unity. During the very early stages in the development of the FPZ the K_R-curve is unique, but unless the beam is very large it curves concave upwards to non-dimensional values far in excess of unity. The reason why small cementitious bend specimens behave in a ductile fashion is because of the size effect on the K_R-curve. A K_R-curve that curves upwards confers stability, because instability can only occur if the applied stress intensity factor curves upwards even more steeply. It should be noted that the size effect on the K_R-curve is not necessarily always as marked as it is for notched bend specimens. The K_R-curves for DCB and CT specimens have a far smaller size effect

[1] The inability of classic LEFM to describe fracture when a crack is very small is one of its drawbacks. However, if the FPZ is modelled, the fracture of a body with the smallest flaw, or even no flaw at all, can be described. Thus, conventional strength of materials and fracture mechanics can be combined into a single theory.
[2] Rather appropriately this assumption is similar to that made by Carpinteri's fellow Italian, Galileo (1638), in his analysis of the bending stresses in a beam. Although the assumption does appear to give results that agree approximately with more exact theory, the beam is not in global equilibrium which was also Galileo's error, and the agreement must be partly fortuitous.

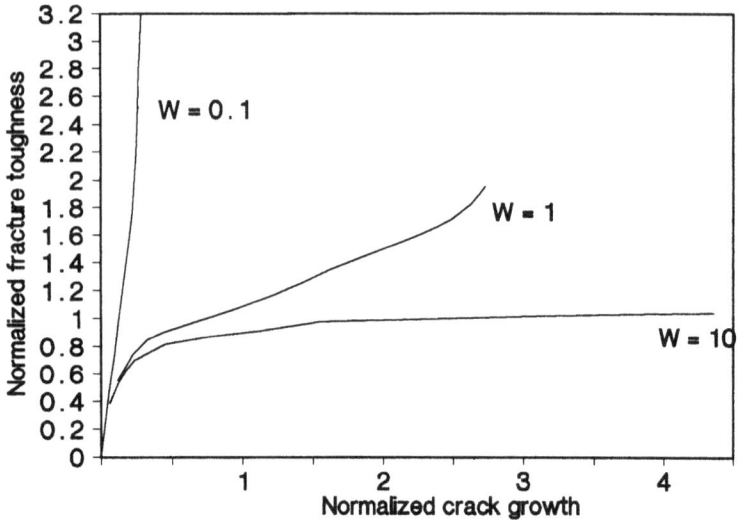

Figure 4.24 K_R-curves for concrete beams of depth $W = 0.1$, 1 and 10 m.

(Cotterell and Mai, 1988). Carpinteri (1991) also discusses the effect of crack growth resistance on the transition from brittle to ductile fracture for a general material. However, the J_R-curve assumed, though appropriate for metals, does not represent the behaviour of cementitious materials.

4.7 Asymmetrical fracture

If the specimen or the load is asymmetrical the crack follows a curved path and the fracture is often referred to as 'mixed-mode'. However, there is debate as to whether such fractures are truly mixed-mode. In isotropic elastic materials with very small FPZ, fractures do not generally propagate under mixed-mode conditions, but propagate so that the fracture is pure mode I as discussed in section 1.7.1.

The Iosipescu geometry shown in Figure 4.25 has been used in a round-robin test series by RILEM TC 89-FMT in an investigation of asymmetrical fracture. In this geometry the bending moment is zero at the notch and a maximum under the central loading points. There are four possible failure modes with this specimen (Ballatore et al., 1990):

(a) Fracture from the notches with the crack path following essentially that predicted by LEFM (referred to as 'mixed mode' crack propagation).
(b) Flexural failure initiated away from the notches opposite the supports.
(c) 'Shear failure' between the central supports.
(d) A Brazilian-like splitting failure in the middle of the specimen.

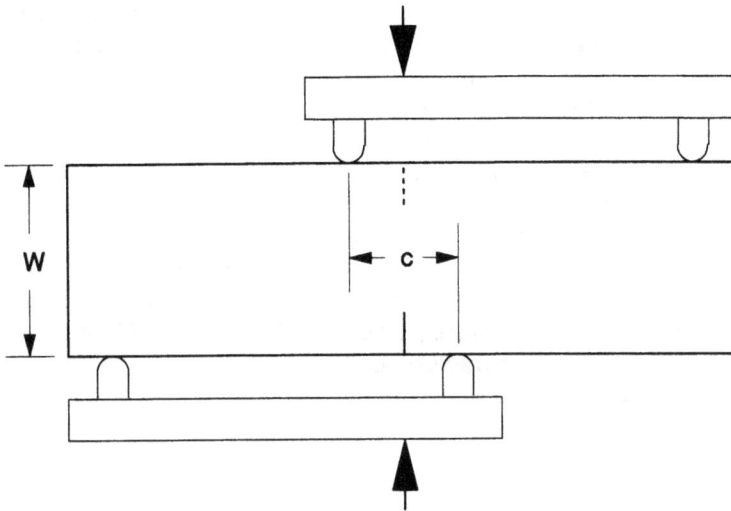

Figure 4.25 The Iosipescu geometry—single or double notches.

The mode of failure is determined by the loading configuration through c/W and is shown in Table 4.2. When the mode of failure is 'mixed mode', the fracture path can be reasonably well predicted by either LEFM (Swartz and Taha, 1990) assuming the crack propagates in the direction of the maximum circumferential stress, or the fictitious crack model (Swartz and Taha, 1990; Ballatore *et al.*, 1990) assuming that the crack grows in the direction of the maximum strain energy density. The fracture energy in these so-called 'mixed-mode' fractures is very similar to G_{If} (Biolzi, 1990; Boca *et al.*, 1990; Ballatore *et al.*, 1990) and one can conclude that the crack propagates so that there is local symmetry at the tip of the FPZ. However, the crack propagation in the specimen where $c/W = 0.167$ (Bažant and Pfeiffer, 1987) is very different (see Figure 1.18). In this case the crack propagates almost vertically along a path that is quite different

Table 4.2 Mode of failure in Iosipescu geometry specimens

Loading configuration c/W	Mode of failure	Reference
0.167	'shear'	Bažant and Pfeiffer (1987)
0.2	'mixed mode'	Ballatore *et al.* (1990)
0.333	'mixed mode'	Bažant and Pfeiffer (1987)
0.4	'mixed mode'	Ballatore *et al.* (1990)
		Rots and de Borst (1987)
0.5	'mixed mode'	Swartz *et al.* (1988)
0.8	11/12 'mixed mode'	Ballatore *et al.* (1990)
	1/12 'flexural'	
1.2	'flexural'	Ballatore *et al.* (1990)
0.24, with axial compression	Transition between splitting and mixed mode	Swartz and Taha (1990)

to the theoretical LEFM path, which predicts that the path deviates sharply away from the vertical, and is the path observed in a brittle polymethyl methacrylate specimen (Melin, 1989). The fracture energy observed in the test with $c/W = 0.167$ is very large, approximately 25 times that of the mode I fracture energy (Bažant and Pfeiffer, 1987). The crack band model was used to analyse specimens with $c/W = 0.167$ and 0.333; in these finite element studies, the crack band was advanced into the element that released the maximum energy, and the simulations indicated that the crack propagated vertically when $c/W = 0.167$ and sideways when $c/W = 0.333$ (Bažant and Pfeiffer, 1987). Although this result for small c/W and large shear indicates a 'shear fracture', Bažant and Pfeiffer (1987) suggest that microscopically, the 'shear fracture' is likely to have formed as a series of isolated inclined tensile microcracks in the region of high stress that stretches between the two notches. When c/W is not small and the shear/normal stress ratio not so high, there is much less difference between the principal stress direction and the direction of high tensile stress. In this case there is little difference between the direction of microcracking and the direction for a mode I fracture and the fracture follows essentially the path predicted by LEFM. Although it is convenient to lump the microcracks into a FPZ, only if the microcracks form essentially the direction in which the FPZ grows will the fracture be able to be predicted from a universal fracture energy. The prime example of the direction of FPZ growth and the direction of microcrack formation being completely different is in a compression fracture. As Hoek and Bieniawski (1965) showed, it is impossible to make a series of small cracks in a homogeneous brittle material link up to form a compression fracture. Hence, the conclusions are that provided the direction of high stress is essentially normal to the maximum principal stress direction, a fracture will propagate so that local symmetry is maintained with a fracture energy of G_{If}; but if the two directions differ significantly, then the fracture will propagate in the direction of high stress and the fracture energy will be greater than G_{If}.

Since there has been some uncertainty in the direction of crack growth in cementitious materials some workers have used a hybrid method of analysis where the crack path has been obtained from experiment and the fracture modelled by the fictitious crack model (Ingraffea and Gerstle, 1984; Ingraffea and Saouma, 1984; Liaw et al., 1990). Provided the crack path can be predicted 'mixed-mode' fractures can be analysed by either smeared or discrete representations of the FPZ. The K-superposition method is inappropriate for the mixed mode problem.

4.7.1 The smeared crack model for mixed mode fracture

Rots and de Borst (1987) have analysed an asymmetrically loaded beam with the Iosipescu geometry used by Arrea and Ingraffea (1982) without imposing any prior conditions on the direction of cracking. The incremental finite element solutions were driven by the crack mouth sliding displacements (CMSD) in a similar manner to the experiments. A non-linear mode I stress-softening relationship proposed by Cornelissen et al. (1986) was used with the crack band width taken as 20.3 mm. The mode I fracture energy was assumed to be $75 \, J/m^2$. Three different assumptions were made about the mode II behaviour. In the first a constant shear retention factor of 0.2 was assumed. The other two assumed a linear shear-softening diagram with the same initial shear retention factor and an ultimate shear strength of 0.5 MPa; two mode II fracture energies were assumed, 10 and $75 \, J/m^2$. There is considerable scatter in the experimental load-CMOD results of Arrea and Ingraffea (1982), but the triangular softening relationship with a mode II fracture energy of $75 \, J/m^2$ fitted the data best. Rots (1988) also analysed the same experimental data of Arrea and Ingraffea (1982), assuming that the fracture is pure mode I and a fracture energy $G_{If} = 100 \, J/m^2$, and obtained very similar results. Thus, it can be concluded that the simpler assumption that the fracture is pure mode I is to be preferred.

One of the problems with the smeared crack approach, that is not encountered by either the crack band or the fictitious crack model, is that geometric discontinuities in the real crack are not modelled. This leads to the problem called 'stress-locking' by Rots (1991). Stress-locking is not important until a true crack has extended significantly. Since the ultimate strength is usually reached before true crack growth, the maximum load can be predicted accurately by the smeared crack model. However, with the smeared crack model the load does not drop to zero, but to a relatively large positive value (Rots, 1991).

4.7.2   The fictitious crack model for mixed mode fracture

The analyses of mixed mode fracture by the fictitious crack model have been made by both the hybrid method of using the experimentally determined crack path (Ingraffea and Gerstle, 1984; Liaw et al., 1990) and without any prior assumptions of the crack path (Ingraffea and Saouma, 1984; Ingraffea et al., 1984; Ingraffea and Panthaki, 1985; Ballatore et al., 1990; Boca et al., 1990; Rots, 1991). Ingraffea and Gerstle (1984) calculated the load-displacement using the experimentally determined crack path and the theoretical mode I path. In the first method they used a shear retention factor that is a function of the CTOD. With this method the maximum load was predicted with a fair accuracy, but the post maximum load-deflection curve was not so well modelled. In the second method and in subsequent papers (Ingraffea and Saouma, 1984; Ingraffea et al., 1984; Ingraffea and Panthaki, 1985) the criterion for the crack path was $K_{II} = 0$ even though at the tip of a fictitious crack both the stress intensity factors must be zero. Starting from the previous iteration step that uses a singular element at the tip, the 'load' is increased which creates small non-zero stress intensity factors at the fictitious crack tip. The position of a new crack tip is now determined on the basis that $K_{II} = 0$ at the new crack tip. Iteration is then performed to determine the value of the 'load' increment that also makes $K_I = 0$. The theoretical crack path agrees well with the experimental one.

The crack paths, in compact tension specimens with a superimposed diagonal compression, are approximately straight, which has enabled their analysis by the hybrid method to be simple (Liaw *et al.*, 1990). Analysis using a trilinear normal stress-displacement softening relationship on its own and combined with an empirical shear-softening relationship showed that there was little difference in the two sets of results (Liaw *et al.*, 1990).

4.8 Summary

Cementitious laboratory specimens are usually too small to enable fracture loads to be predicted from the fracture energy, G_{If} and classic LEFM. The FPZ is also too large to enable the concept of an equivalent stress-free crack to be used. For symmetrical loading the load-deflection curve can be predicted if the stress-displacement softening relationship is known. Because of the inhomogeneous nature of cementitious materials, there is considerable scatter in the stress-displacement curve. The most important fracture parameter is the fracture energy G_{If}, which must be equal to the integral of the stress-displacement curve. The bilinear stress-displacement relationship is sufficiently accurate for all purposes and often even a simpler linear relationship will suffice. The engineers' theory of bending can be used to give a simple approximate prediction of the load-deflection curve for beams once the width of the FPZ is established. An accurate analysis of symmetrically loaded specimens can be made using either the crack band theory or the fictitious crack model. However, the easiest way of modelling the fracture process is with the fictitious crack model and the K-superposition principle.

The analysis of mixed mode fractures can be performed using either crack band models or the fictitious crack model. This latter method requires remeshing at each step. Unless the shear/tensile stress ratio is high, the crack path can be determined from any theory giving local symmetry. The exception is at initiation where there may be a kink in the fracture path. At kinks the criterion of maximum energy release is the most appropriate. More work needs to be done on 'mixed-mode' fracture where the shear/tensile stress ratio is high before it becomes clear how best to treat the problem.

5 Theoretical models for fracture in fibre reinforced cementitious materials

5.1 Introduction

Although high modulus fibres, and to a lesser extent low modulus fibres, can increase the strength of a cementitious material, the main effect of fibres is to increase the toughness by bridging cracks in the matrix. Beams manufactured from Type I composites which form multiple cracks can be analysed by standard elasto-plastic methods (Kalisky, 1989) and are not considered. Type II cementitious composite beams reinforced with long fibres, such as steel wires or polypropylene, where the pull-out length is so large that the maximum load is attained with little loss in fibre bridging stress, can be analysed approximately using the engineers' theory of bending (Nishioka *et al.*, 1975; Babut and Brandt, 1978; Swift and Smith, 1978; Lim *et al.*, 1987a; Naaman *et al.*, 1993a,b). However, for more accurate analysis of all Type II composites fracture mechanics is essential. The analysis of Type II composites using the engineers' theory of bending is considered first.

5.2 Engineers' theory of bending analysis of Type II composites

The analysis using the engineers' theory of bending is simplest for Type IIA composites where, after first cracking, the pull-out stress is almost constant (see Figure 2.11). The tensile elastic modulus, E_{ct}, for strains less than the first cracking strain can be calculated from the rule of mixtures (see section 3.9.1). After the critical stress, σ_{mc}, for first cracking is reached the stress drops to σ_{tu} and, providing the fibres are long, the stress during fibre pull-out is reasonably constant. The compressive stress-strain curve has been idealized by Lim *et al.* (1987a) as elastic up to a critical compressive strain, γ_{cc}, and perfectly plastic for greater strains as shown in Figure 5.1. The composite has a compression elastic modulus, E_{cc}, which is somewhat larger than the tensile modulus, E_{ct}. The effective ultimate compressive strength of the composite is assumed to be $\alpha\sigma_{cu}$, where σ_{cu} is the compressive strength of a standard cylinder test and α is a constant. For steel fibre reinforced concrete, the value of α is not critical, and Lim *et al.* (1987a) suggest that a value of 0.9 is appropriate.

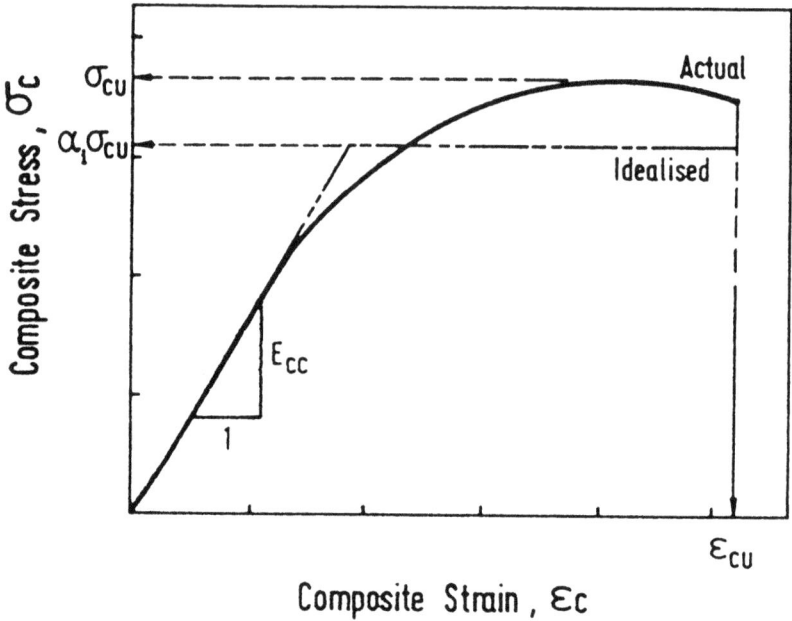

Figure 5.1 Idealized compressive stress-strain relationship for Type IIA composites. (After Lim et al., 1987.)

In the elastic range, the position of the neutral axis can be located by a parameter μ (see Figure 5.2a)

$$\mu = \frac{h_t}{h_c} = \sqrt{\frac{E_{cc}}{E_{ct}}} \tag{5.1}$$

where h_t and h_c are the depths of the tensile and compressive zones. Integration of the moments of the elastic stresses gives the moment curvature relationship

$$M = \frac{E_{ct}h^3}{3R} \left[\frac{\mu}{1+\mu} \right]^2 \tag{5.2}$$

where R is the radius of curvature of the beam. Equation 5.2 holds until the tensile strain reaches the first cracking strain, ϵ_{mc}. The critical curvature is given by

$$\frac{1}{R_c} = \frac{\epsilon_{mc}}{h} \left[\frac{1+\mu}{\mu} \right] \tag{5.3}$$

and the corresponding critical moment, M_c, can be found from eqn 5.2.

The stress distribution after first cracking, shown in Figure 5.2b, can be defined by a parameter, λ, where

$$\lambda = \frac{h_y}{h_t} \tag{5.4}$$

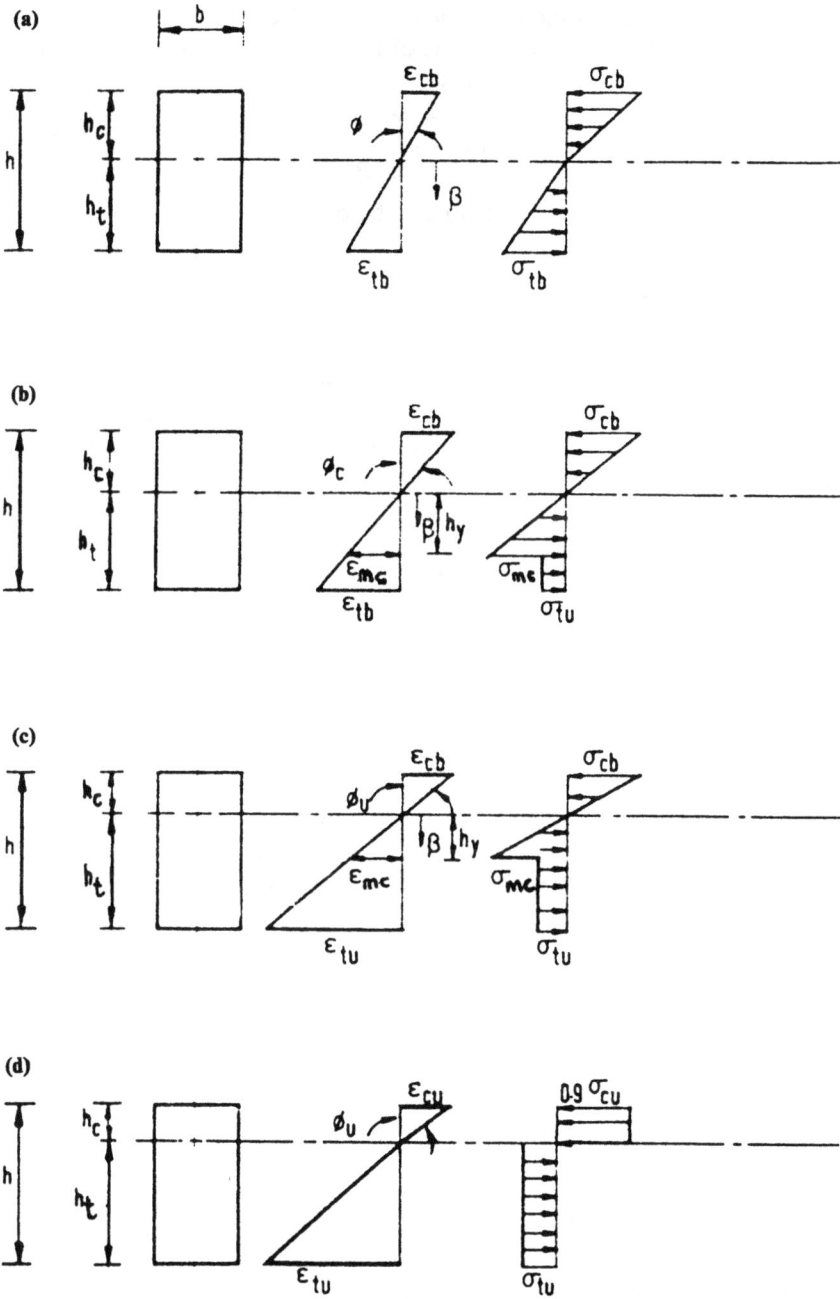

Figure 5.2 Stress-strain relationship in bending (a) elastic range, (b) after first cracking, (c) ultimate condition—case 1, (d) ultimate condition—case 2. (After Lim *et al.*, 1987a.)

and h_y is the distance of the first cracking strain from the neutral axis. The parameter λ can be obtained in terms of the first cracking strain, ϵ_{mc}, from its definition given in eqn 5.4, and is

$$\lambda = \epsilon_{mc}\, \frac{R_c}{h}\left[\frac{1+\mu}{\mu}\right] \tag{5.5}$$

The moment-curvature relationship can be obtained by integration and is given by

$$M = \frac{1}{3R}\left[\frac{h}{1+\mu}\right]^3 [E_{cc} + E_{ct}(\lambda\mu)^3] - \frac{\sigma_{tu}}{2}\left[\frac{h\mu}{1+\mu}\right]^2 [\lambda^2 - 1] \tag{5.6}$$

Ultimate tensile failure can occur either (1) without crushing in the compressive zone (see Figure 5.2c), or (2) after crushing in the compressive zone (see Figure 5.2d). In case (1) the ultimate moment is given by eqn 5.6 with μ_u and λ_u given by

$$\mu_u = \left[\frac{E_{cc}\epsilon_{tu}}{E_{ct}\epsilon_{mc}^2 + 2\sigma_{tu}(\epsilon_{tu} - \epsilon_{mc})}\right]^{1/2}$$

$$\lambda_u = \frac{\epsilon_{mc}}{\epsilon_{tu}} \tag{5.7}$$

where ϵ_{tu} is the ultimate tensile strain. The ultimate curvature is given by

$$\frac{1}{R_u} = \frac{\epsilon_{tu}}{h}\left[\frac{1+\mu}{\mu}\right] \tag{5.8}$$

For case (2) the ultimate moment can be calculated with little loss of accuracy if the stress in the tension and compression sides of the beam are assumed to be constant and equal to σ_{tu} and $\alpha\sigma_{cu}$, respectively (see Figure 5.2d). The ultimate moment is then given by

$$M_u = \tfrac{1}{2}\sigma_{tu}hh_t \tag{5.9}$$

where

$$\frac{h_t}{h} = \frac{\alpha\sigma_{cu}}{\alpha\sigma_{cu} + \sigma_{tu}} \tag{5.10}$$

and the ultimate curvature is given by the smaller of

$$\frac{1}{R_u} = \epsilon_{tu}\frac{1+\mu_u}{h} \quad \text{or} \quad \frac{\epsilon_{cu}}{h - h_t} \tag{5.11}$$

There are two types of moment-curvature relationship for long steel wire reinforced cementitious materials: (a) when $M_u < M_c$ and (b) when $M_u > M_c$. Lim *et al.* (1987a) suggest that the two types of behaviour can be idealized by the straight line relationships shown in Figure 5.3. With type (a) behaviour there will be only a single crack, but with type (b) behaviour,

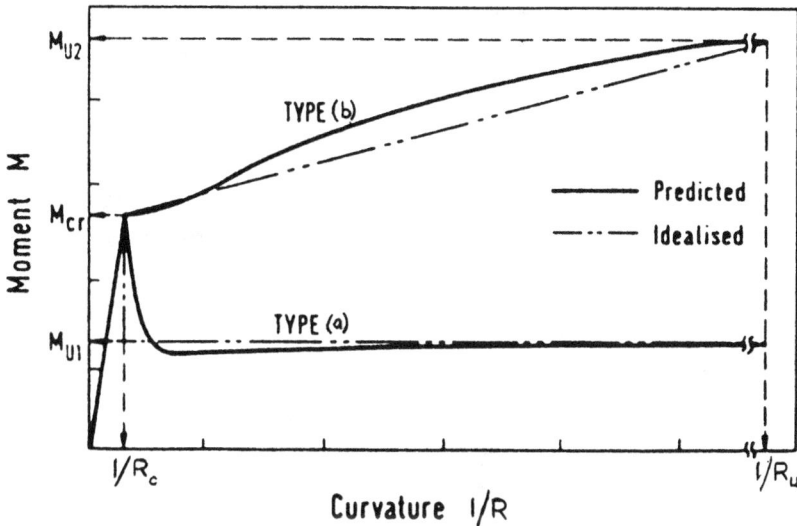

Figure 5.3 Predicted $M - 1/R$ curves and their idealizations. (After Lim *et al.*, 1987a.)

though in tension only a single crack will form, under bending, multiple cracks occur. The load-deflection relationship for type (b) beams loaded by three- or four-point bending can be obtained by double integration of the moment-curvature relationship. The predicted load-deflection curves have been compared with experimental data by Lim *et al.* (1987a). They tested two series of steel fibre reinforced concrete specimens one using hooked and the other straight fibres of diameter 0.5 mm and 30 or 50 mm long. The volume fraction of the fibres varied from 0.5 to 1.5%. All the straight fibre reinforced specimens showed a drop in load at first cracking, whereas those reinforced with hooked fibres showed either type (a) or (b) behaviour depending upon the fibre length and volume fraction. The predictions were generally good and an example of type (a) and type (b) behaviour are shown in Figure 5.4, which also contains predictions from K-superposition which will be discussed in section 5.4.1.

The idealized moment-curvatures were used by Lim *et al.* (1987a) to obtain simple expressions for the toughness index. Using the toughness index, I, defined by eqn 3.67 (Barr and Hasso, 1985), the toughness index for type (a) composites is given by

$$I = \frac{1}{4} + \frac{1}{2}\left(\frac{M_u}{M_c}\right) \tag{5.12}$$

For type (b) composites the toughness index is given by

$$I = \frac{3}{4} + \frac{1}{4}\left(\frac{M_2}{M_c}\right) \tag{5.13}$$

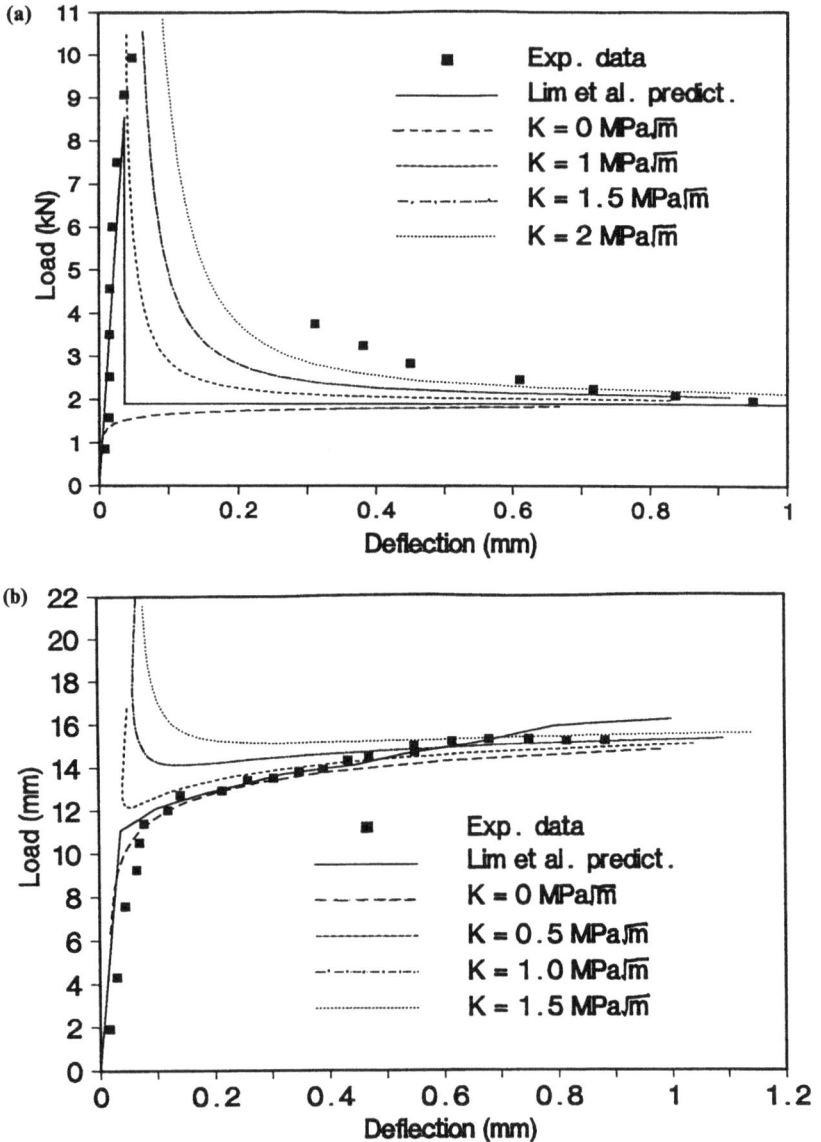

Figure 5.4 Load-deflection relationship for steel fibre reinforced concrete beams, (a) 30 mm straight fibres, 0.5% volume fraction; (b) 30 mm hooked fibres, 1.5% volume fraction. Theoretical curves from Lim *et al.* (1987a) and from application of fracture mechanics.

where M_2 is the moment at a curvature twice the critical curvature, $1/R_c$, for first cracking.

SIFCON, slurry infiltrated fibre concrete, can have a higher fibre fraction ratio than conventional fibre reinforced concrete manufactured by mixing

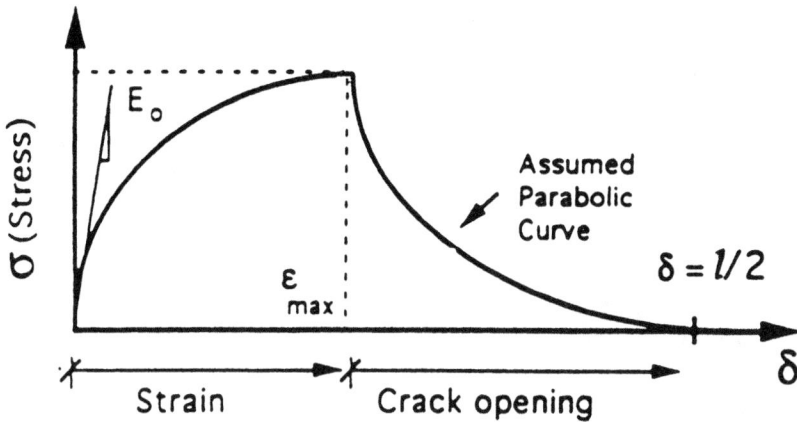

Figure 5.5 Schematic representation of the stress-elongation curve for SIFCON in tension. (After Naaman *et al.*, 1993.)

the fibres into the concrete mix and the stress does not drop suddenly on first cracking. A schematic representation of the stress-strain-displacement relationship for SIFCON in tension is shown in Figure 5.5. Because localization must take place with strain-softening, the descending portion of the curve cannot be related directly to strain and can only be expressed in terms of the crack opening. Localization is, of course, a major problem in the analysis of cementitious materials and prevents a straightforward finite element solution. In section 4.2 a strain-softening solution based on the engineers' theory of bending was presented in which a finite width FPZ was introduced to cope with the problem of localization. Naaman *et al.* (1993) have attempted a somewhat similar analysis for reinforced concrete beams using a SIFCON matrix. The reinforcing bars simplify the problem somewhat because they are a constraint to strain localization. A length scale is still needed to convert displacements in the strain-softening part of the stress-displacement relationship to strains, but the choice of that length scale is not quite so critical. Naaman *et al.* (1993) found that they could get good load-deflection predictions for reinforced SIFCON beams by taking the length scale equal to the depth of the beam. Knowing that in pure bending plane sections must remain plane, Naaman *et al.* (1993) solved the non-linear problem of reinforced SIFCON iteratively. The maximum loads were predicted reasonably accurately, but there was less success in predicting the load-deflection relationship. The load dropped after the maximum load was obtained in their experiments, but the analysis of Naaman *et al.* (1993) failed to predict any load drop. Probably the experimental load drop was not predicted because the tensile localization behaviour was not modelled well. Also, all the beams tested by Naaman *et al.* (1993) had depths in a rather narrow range from 18.75 to 26.25 mm. For much larger beams it is questionable whether the assumption that the length scale for converting displacement to strain would be equal to the beam depth.

Hence, while the engineers' theory of bending approach is suitable for Type IIA cementitious materials, reinforced with a low volume fraction of long fibres, that show a drop in strength at first cracking and little change in stress with further straining, its suitability for general Type II cementitious materials is doubtful.

5.3   Fracture behaviour of short fibre Type II reinforced cementitious composites

Although the work of fibre pull-out provides a major contribution to the fracture energy of short fibre composites, the load carrying capacity of beams and other structures cannot be determined accurately without fracture mechanics consideration of the propagation of cracks in the matrix. When a crack propagates a FBZ develops behind the matrix crack tip. The load taken by the bridging fibres impedes the growth of the crack in a similar way to the bridging grains that impede crack propagation in an unreinforced cementitious material. However, the fibres can span a very much larger crack opening, and hence the FBZ is much bigger than the FPZ and the crack growth resistance of fibre reinforced cementitious materials is much greater than that of unreinforced ones. Because of the large size of the FBZ, the crack growth resistance curves are generally both geometry and size dependent (Cotterell and Mai, 1988a). Only in very large specimens, where the FBZ is small compared with the overall size of the specimen, can the crack growth resistance curve be considered a material constant. Thus the concept of the crack growth resistance curve, which is very useful for materials such as fibre reinforced ceramics that have a relatively small FBZ, is inappropriate for most fibre reinforced cementitious materials except in its asymptotic form for very large specimens.

5.4   Crack growth models for fibre reinforced Type II cementitious composites

Most theoretical models for crack growth in fibre reinforced cementitious materials are based on the fictitious crack model of Hillerborg *et al.* (1976). The main problem in analysing the fictitious crack model for reinforced cementitious materials is the same as those discussed in Chapter 4 for unreinforced materials, namely the non-linearity caused by the crack closure stresses being dependent on the crack opening displacement. Hillerborg (1980, 1983) has solved this problem using the finite element method described in section 4.5.1 and this method will not be discussed further.

Jenq and Shah (1984, 1986a,b) proposed that the two-parameter model they developed for concrete could be applied to fibre reinforced composites. The two parameters are a critical stress intensity factor, K_{Ic}^s, at the tip of the effective crack and a critical crack tip opening displacement, $CTOD_c$, measured at the tip of the original crack (RILEM, 1990a). For what Jenq and Shah (1986a) call type G specimens,[1] it is assumed that K_{Ic}^s and $CTOD_c$ are the values of the applied stress intensity factor at the tip of an effective crack and the crack opening displacement at the tip of the original crack at maximum load. Some crack growth may occur at lower loads, but once the maximum load is achieved, it is assumed that the crack propagates so that the total stress intensity factor remains constant at K_{Ic}^s. In their earlier papers, Jenq and Shah (1984, 1986a,b) attempted to partition the total load P so that

$$P = P^M + P_k^f + P_s^f \qquad (5.14)$$

where P^M is the contribution due to the matrix and is related to the applied stress intensity factor, P_k^f accounts for the effect of the bridging fibres, and P_s^f satisfies global equilibrium due to the fibre bridging forces. However, in later works Shah and his co-workers (Ouyang et al., 1990; Mobasher et al., 1991) appear to have abandoned their earlier method and instead partition the stress intensity factor in a more readily understood fashion that is similar to the original suggestion of Lenain and Bunsell (1979). Their earlier method will not therefore be discussed.

Fracture of Type II composites usually occurs by the formation of a single crack. Multiple cracking can occur in some small un-notched specimens, such as the bend specimen, where there are stress gradients. The discussion here is limited to single cracks. For symmetric geometries, where the expressions for the stress intensity factors are known, the simplest method of analysis is to use the K-superposition principle which was first introduced for fibre reinforced cementitious materials by Lenain and Bunsell (1979). The total stress intensity factor K_t at the tip of a continuous matrix crack is given by

$$K_t = K_a + K_r \qquad (5.15)$$

where K_a is the stress intensity factor due to the applied loads, and K_r is the stress intensity factor due to the bridging fibres. K_r is negative, since the fibre forces tend to close the crack. Lenain and Bunsell (1979) assumed that, for equilibrium crack growth, the total stress intensity factor must be equal to

---

[1] In type G specimens, later identified as 'positive geometry' specimens, the applied stress intensity factor increases monotonically with crack length (Ouyang et al., 1990). The notch bend specimen is of type G and the measurement of K_{Ic}^s and $CTOD_c$ in such specimens is the subject of a draft RILEM recommendation (RILEM, 1990a). In type N, or 'negative geometry' specimens, the second derivative of the applied stress intensity factor is positive, but for small crack lengths the derivative is negative, so that there exists a minimum applied stress intensity factor.

the fracture toughness of the reinforced matrix, K_{Ic}. This value is greater than the fracture toughness of the unreinforced material because the strength of the reinforced matrix in the FPZ and the critical crack tip opening displacement are both larger than the corresponding values for the unreinforced material. For example, Lenain and Bunsell (1979) measured K_{Ic} at crack initiation for an asbestos reinforced mortar to be 1.7 MPa$\sqrt{m}$ as compared to 0.6 MPa$\sqrt{m}$ for the unreinforced mortar. At the initiation of a continuous crack, K_r is zero and the applied stress intensity factor, K_a, is equal to K_{Ic}. As the load is increased the matrix crack extends so that $K_t = K_{Ic}$ and the relationship between the applied stress K_a and the crack extension Δa gives what is known as the K_R-curve. However, as noted previously, crack growth resistance curves obtained from reinforced cementitious materials are geometry and size dependent (Cotterell and Mai, 1988a).

The use of the K-superposition principle does not remove the problem of non-linearity. The closing stress in the FBZ is in general dependent on the crack opening displacement. However, this problem can be solved by a simple iteration routine as discussed in section 4.5.3. What is not so clear is whether the FPZ of a fibre reinforced cementitious material can be modelled by a critical stress intensity factor K_{Ic} for the toughness of the matrix and the effective crack tip taken as the tip of the continuous matrix crack, or whether it is necessary to model the FPZ as a fictitious crack and take the effective crack tip at the tip of the FPZ. Hillerborg (1980, 1983) does not distinguish the FPZ as a separate entity and so effectively uses the latter model. If the FPZ is treated as a separate zone, then the equation for the total stress intensity factor K_t (eqn 5.15) has to be replaced by

$$K_t = K_a + K_r + K_m = 0 \qquad (5.16)$$

where the stress intensity factors are calculated at the tip of the fictitious crack tip, and K_m is the stress intensity factor due to the stress in the FPZ.

5.4.1   The K-superposition method applied to fibre reinforced Type II composites using a critical stress intensity factor

Type IIA composites, where the fibre pull-out stress after first cracking is practically constant, are the simplest to analyse because, since the fibre bridging stress does not depend upon the COD, the problem is linear. The load-deflection curves for the steel wire reinforced concrete three-point bend specimens tested by Lim et al. (1987a) have been calculated from the tensile pull-out data (Lim et al., 1987b). Since Lim et al. (1987a,b) do not give the fracture toughness of the reinforced concrete, the load-deflection curves have been calculated for a range of toughnesses, K_{Ic}, ranging from 0 to 2 MPa$\sqrt{m}$, which cover the expected range in values for concrete (Pak and Trapeznikov, 1981) and are shown in Figure 5.4. Modelling the FPZ by assigning a critical stress intensity factor for first cracking is only valid if the crack is large compared with the FPZ and cannot be used to obtain initiation accurately. For specimen reinforced with smooth wires (Figure 5.4a), the load drops suddenly after the matrix cracks. It is not surprising that the model overestimates the load for small deflections. The K-superposition method should give an accurate load-deflection curve once the matrix crack is deep, but it appears to underestimate the load as does the simpler theory developed by Lim et al. (1987a). There is an increase in the load after first cracking in the specimens reinforced with hooked wires (Figure 5.4b) and in these specimens two closely spaced parallel matrix cracks formed near the centre of the beam. For both the smooth and hooked ended wires the load at large deflections is little affected by the fracture toughness of the matrix, because the crack has already penetrated almost to the back surface of the beam and most of the tensile stresses are being resisted by the fibres with the matrix carrying the compressive stresses. At loads greater than the first cracking load, K_{Ic} needs to be greater than $2 \text{MPa}\sqrt{m}$, which is unrealistically high, to give a good prediction of the load-deflection curve for smooth wires (see Figure 5.4a), but for the hooked ended wires the more reasonable assumption that $K_{Ic} = 0.5 \text{MPa}\sqrt{m}$ gives an accurate prediction of the load-deflection curve apart from the initiation of the matrix crack (see Figure 5.4b).

Mobasher et al. (1991) also suggest that in the absence of knowledge of the actual stress-displacement relationship, the fibre bridging stress should be considered basically constant during the growth of a matrix crack. However, they complicate their analysis by assuming that the bridging stress builds up to the constant value from zero over a distance proportional to the critical crack length. This assumption is unrealistic. After a continuous matrix crack forms in a Type II composite the stress carried drops, not rises. Mobasher et al. (1991) are confused by the fact that in a pull-out test, the stress does first rise with displacement as the fibres stretch elastically before pull-out is initiated, but in a Type II composite, the fibre volume fraction is less than the critical value, and the stress carried drops. At the tip of a crack in a matrix there is considerable strain in the FPZ, which in the fictitious crack model is equivalent to a CTOD, δ_m, and the fibres are already stressed before a continuous matrix crack forms. The initial build-up in fibre bridging stress makes the problem of Mobasher et al. (1991) non-linear.

For a general Type II composite the fibre bridging stress is a function of the COD and the problem is non-linear. Solutions can be found by the use of Muskhelishvili's (1953) method. The method has been applied to a specimen with a semi-infinite crack (Foote et al., 1980). In this solution the closing stress on the crack face was represented by a fifth order polynomial which enabled the displacements of the crack face to be found analytically. Iteration was then used to obtain the fibre bridging stresses that matched the crack face displacements. A similar method was used for the stress intensity factor for a double cantilever beam specimen (Foote *et al.*, 1986b). Although the analysis of the semi-infinite crack by this means was comparatively simple, there were problems in convergence with the DCB specimen. A simpler approximate method gives results that are sufficiently accurate for practical purposes.

In the FBZ, the crack faces are very nearly straight except for a theoretically necessary parabolic opening right at the crack tip (Foote *et al.*, 1986b; Cotterell *et al.*, 1988). The approximate method makes use of the fact that the crack faces in the FBZ remain reasonably straight as they also do in the FPZ of unreinforced cementitious materials. Llorca and Elices (1993) use the same approximation to analyse the fracture of fibre reinforced ceramic matrix composites. If the crack faces are assumed to remain straight and the general stress-displacement curve is represented by

$$\frac{\sigma}{f_t} = F\left(\frac{\delta}{\delta_f}\right) \tag{5.17}$$

where f_t is the maximum stress resisted by the bridging fibres, the stress in a fully developed FBZ is given by

$$\frac{\sigma}{f_t} = F(1 - x/d_f) \tag{5.18}$$

where d_f is the length of the FBZ and x is measured from the tip of the matrix crack. The problem of calculating the load for a fully developed bridging zone knowing K_{Ic}, E, f_t and δ_f, then reduces to finding by iteration the length of the saturated bridging zone d_f while satisfying the condition for equilibrium growth that the total stress intensity factor at the continuous crack tip, K_t, is equal to the fracture toughness of the reinforced matrix, K_{Ic}. It is often sufficient to model the stress-displacement relationship for fibre pull-out by the linear expression

$$\frac{\sigma}{f_t} = \left[1 - \frac{\delta}{\delta_f}\right] \tag{5.19}$$

though the theoretical relationship is given by either eqn 3.64 or 3.65, depending upon whether a Coulomb or a constant interfacial frictional stress is assumed and in practice by eqn 3.17 with the exponent n greater than 1. The crack profile calculated from the external load and the fibre bridging stresses is reasonably linear (see Figure 5.6) which justifies the use of eqn 5.18. A similar iteration procedure can be used if the FBZ is only partially developed. However, a simplified method that requires no further iteration once the saturated length of the FBZ d_f is known is often sufficient (Foote *et al.*, 1986). This method assumes that the fibre bridging stress is given by eqn 5.19, even when the FBZ is only partially developed. The approximate method differs little from an 'exact' method where the actual crack profile is used, as can be seen from the comparison of crack growth

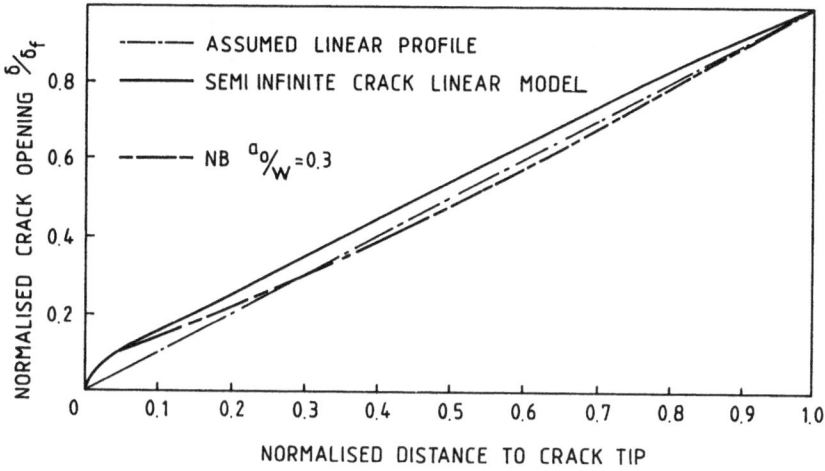

Figure 5.6 Shape of the crack faces in a fully developed FBZ for a semi-infinite and a notch bend specimen.

resistance curves for a DCB specimen calculated by the two methods shown in Figure 5.7. Once the length of the FBZ is known, the load-point displacement can be obtained by the method described in section 1.6.1.

For fibre reinforced cementitious materials it is possible, in theory, to model the debonding and pull-out of the fibres and predict the stress-displacement

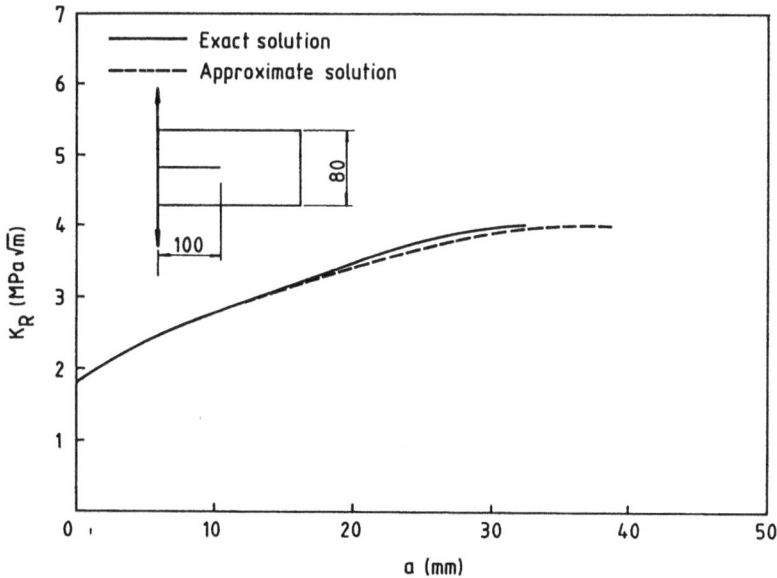

Figure 5.7 Comparison between the 'exact' and approximate crack growth resistance curves for a DCB specimen ($E = 6\,\mathrm{GPa}$, $f_t = 6\,\mathrm{MPa}$, $\delta_f = 0.8\,\mathrm{mm}$, $K_{Ic} = 1.8\,\mathrm{MPa}\sqrt{m}$).

relationship as has been discussed in section 3.9.5. At a less fundamental level one could start from an experimentally determined stress-displacement relationship, but the best method is to use an experimentally determined load-deflection curve and find the parameters that give the best fit to the curve. If a linear relationship is assumed for the stress-displacement curve, the fracture parameters required are: E, K_{Ic}, f_t and δ_f. In this case the post-ultimate load-deflection curve will probably not be modelled precisely but all the other essential features can be accurately modelled. The effective Young's modulus can be found from the initial slope of the load-deflection curve. The fracture toughness of the reinforced matrix, K_{Ic}, can be found from the load to cause first cracking. The maximum pull-out stress, f_t, can be estimated from the maximum load. The final pull-out displacement, δ_f, mainly controls the shape of the post-ultimate load-deflection curve and ensures that the fracture energy, G_{If}, is given by

$$G_{If} \approx \frac{K_{Ic}^2}{E} + \tfrac{1}{2} f_t \delta_f \qquad (5.20)$$

An optimization program to determine the fracture parameters from these initial estimates will converge with little difficulty. Once the fracture parameters are determined the load-deflection curve or the crack growth resistance can be predicted for any geometry or size of specimen.

Mobasher *et al.* (1991) propose a method that is somewhat similar to that outlined above. They obtain K_{Ic}^s and $CTOD_c$ from notch bend tests performed according to the RILEM (1990a) draft recommendation. As with the above method, Mobasher *et al.* (1991) assume that the stress intensity factor at the tip of the matrix crack, calculated according to eqn 5.15, is K_{Ic}^s. It is difficult to understand the other criterion for equilibrium crack growth of Mobasher *et al.* (1991), which is that the CTOD at the tip of the original crack should be equal to $CTOD_c$. Since the value of the $CTOD_c$ is obtained at maximum load, it cannot be the same as the critical COD, δ_f, at which the crack surfaces are stress free. However, the logical use of a critical COD seems to be limited to determining the extent of the bridging fibre free crack after a fully developed FBZ has been formed. Mobasher *et al.* (1991) use an iterative routine to obtain the applied load that results in a bridging stress determined from fibre pull-out tests for a particular matrix crack using the two criteria given above. Instead of calculating the load-deflection curve directly, Mobasher *et al.* (1991) use an energy balance to calculate an R-curve[2] which is then used to obtain the load-deformation response. It is difficult to form an opinion of the validity of their method but it is in any case a very indirect method of predicting the load-deflection curve.

---

[2] Shah and his co-workers recognize that the R-curve is geometry dependent (Ouyang *et al.*, 1990), but it is not clear whether they consider the R-curve to be size dependent.

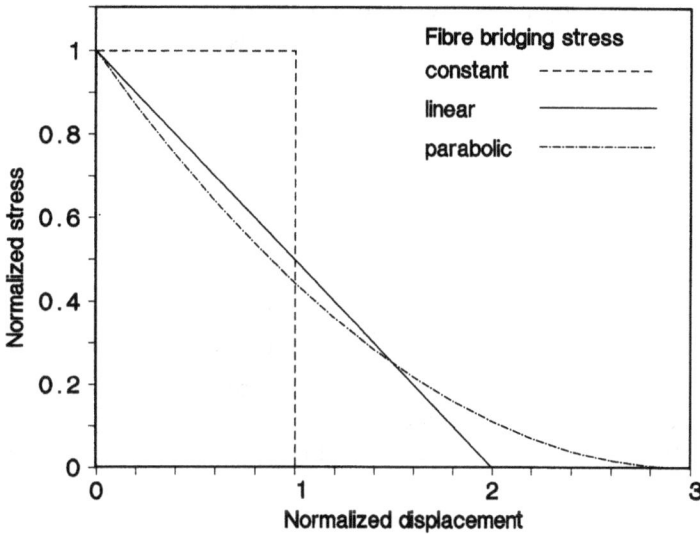

Figure 5.8 Constant, linear, and parabolic stress-displacement curves giving the same work of pull-out.

5.4.2 Crack growth resistance curves

The exact shape of the stress-displacement relationship for the FBZ has only a comparatively small effect on the analysis, providing the fracture energy is the same, as can be judged from a study of the crack growth resistance curves for a semi-infinite crack. These resistance curves are the limiting curves for all geometries as the size increases and are particularly easy to obtain by the approximate method given above because the non-linearity is in the form of a quadratic equation. The theoretical K_R curves have been constructed for three FBZ stress-displacement relationships: (i) a constant stress, (ii) a linear variation, and (iii) a parabolic variation. Thus, the stress-displacement relationship is assumed to be given by

$$\frac{\sigma}{f_t} = \left[1 - \frac{\delta}{\delta_f}\right]^n \tag{5.21}$$

where $n = 0$, 1, or 2. The three different stress-displacement relationships that give the pull-out work are given in Figure 5.8. Assuming that the crack faces remain almost straight, the stress in a fully developed FBZ, of length d_f, is given by

$$\frac{\sigma}{f_t} = [1 - x/d_f]^n \tag{5.22}$$

The stress intensity factor due to a unit point load a distance x from the crack tip is given by

$$K = \sqrt{\frac{2}{\pi x}} \tag{5.23}$$

Hence, for a bridging stress given by eqn 5.22, the stress intensity factor due to the bridging forces in a fully developed FBZ is

$$K_r = f_t \alpha_n \sqrt{\frac{2d_f}{\pi}} \qquad (5.24)$$

where $\alpha_0 = 2$, $\alpha_1 = 1.333$, and $\alpha_2 = 1.067$. For a semi-infinite crack with a fully developed FBZ, both of Barenblatt's (1959, 1962) hypotheses hold (see section 1.5) and hence the plateau value of the fracture toughness, K_{If}, is given by

$$K_{If}^2 = K_{Ic}^2 + \frac{E^* f_t \delta_f}{n+1} \qquad (5.25)$$

The crack opening displacement at the tip of the crack has two components; that due to the applied stress intensity factor, K_a, and that due to the closing stresses exerted by the bridging fibres. The crack opening displacement, δ_a, due to the applied load is given by

$$\delta_a = \frac{4K_a}{E^*} \sqrt{\frac{2x}{\pi}} \qquad (5.26)$$

The crack opening displacement due to the bridging fibres can be found by Castigliano's method (section 1.6.1) and for a fully developed FBZ is

$$\delta_r = -\frac{4\alpha_n f_t d_f}{\pi(n+1)E^*} \qquad (5.27)$$

Hence, the equation that determines the length of the fully developed FBZ, d_f, is given by

$$\delta_f = \delta_a + \delta_r$$

$$= \frac{4}{E^*}\left[K_{Ic}\left(\frac{2d_f}{\pi}\right)^{1/2} + \frac{\alpha_n f_t}{2}\left(\frac{2n+1}{n+1}\right)\left(\frac{2d_f}{\pi}\right) \right] \qquad (5.28)$$

This equation can be best expressed by introducing the non-dimensional terms:

$$\bar{K}_{Ic} = \frac{K_{Ic}}{K_{If}}$$

$$\bar{a} = \frac{a}{l_{ch}} \qquad (5.29)$$

where the characteristic length, l_{ch}, is defined by

$$l_{ch} = \left(\frac{K_{If}}{f_t}\right)^2 \qquad (5.30)$$

Using eqn 5.23 the length d_f of the fully developed FBZ can be found as a function of K_{Ic} from eqn 5.28. Assuming that the closing bridging stress for

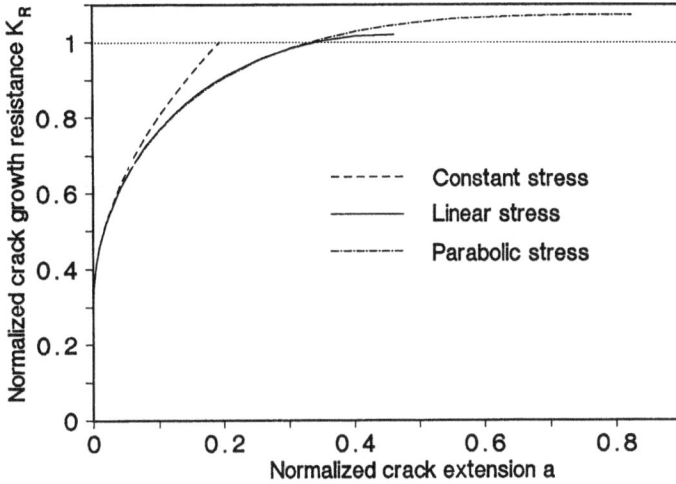

Figure 5.9 Reference K_R-curves for the semi-infinite crack model ($\bar{K}_{Ic} = 0.3$).

a partially developed FBZ is also given by eqn 5.22, enables the K_R-curves, shown in Figure 5.9 for $\bar{K}_{Ic} = 0.3$, to be obtained directly. The K_R-curve for case (i), constant stress, is exact and reaches the exact non-dimensional plateau value of unity. The slight discrepancy between the plateau value, $\bar{K}_{If}$, and unity, for the other stress-displacement relationships, is caused by the assumption of straight crack faces. There is little difference between the K_R-curves for the linear and parabolic stress-displacement relationships, which shows that the shape of the stress-displacement curve is relatively unimportant providing it gives the same fracture energy and justifies the use of a linear relationship.

The crack growth resistance curves for all specimen geometries tend to that for the semi-infinite crack as their size increases and it is only this curve that can be considered as a reference crack growth curve independent of geometry. If the FBZ is not small compared with the specimen's dimensions then the crack growth resistance curve can be very much different to the reference curve. The non-dimensional crack growth resistance curves have been calculated for two different specimen geometries: the double cantilever beam (DCB) and the notch bend (NB) specimen under pure bending (Cotterell and Mai, 1988a). A non-dimensional reinforced fracture toughness $\bar{K}_{Ic} = 0.3$ has been chosen that is representative of asbestos and cellulose reinforced mortars. The non-dimensional crack growth resistance curves for DCB specimens of varying size, but constant notch to height ratio $a_0/H = 0.3$, are shown in Figure 5.10. All the curves reach a plateau value $\bar{K}_{If}$ close to unity, the only difference being the size of the fully developed FBZ. The ratio a_0/H has little effect on the crack growth resistance curve (Foote et al., 1986). The NB specimens behave quite differently to the

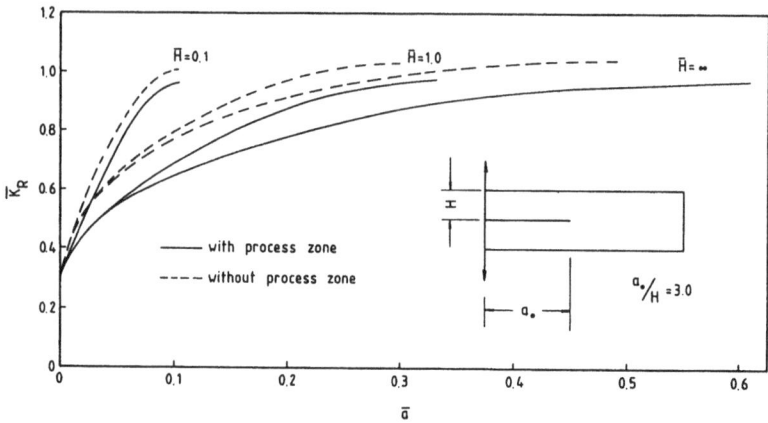

Figure 5.10 Crack growth resistance curves for the DCB geometry, with and without modelling of the FPZ $(a_0/H = 3)$.

DCB specimens, as can be seen from Figure 5.11, which gives the K_R-curves for NB specimens of varying size, but constant notch to depth ratio a_0/W. Only for very large beams does the crack growth resistance curve reach a plateau close to that of the reference crack growth resistance curve of the semi-infinite crack model. For small beams, the K_R-curve increases with crack growth and never attains a plateau. Similar shaped crack growth

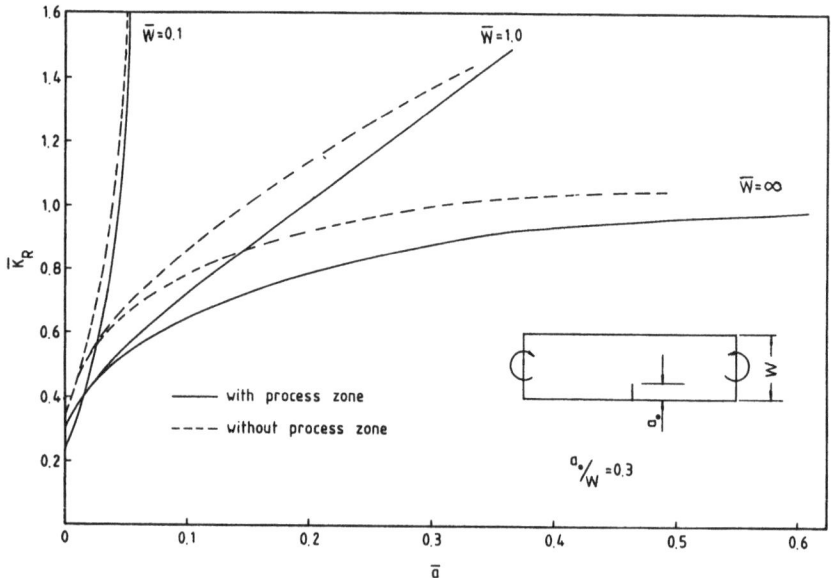

Figure 5.11 Crack growth resistance curves for the NB geometry, with and without modelling of the FPZ $(a_0/W = 0.3)$.

resistance curves have been obtained by Llorca and Elices (1993) for ceramic-matrix composites by the same method, the only difference between the behaviour of the ceramic and the cementitious matrix is due to the very different characteristic lengths. For the LAS glass-ceramic reinforced with 50% SiC fibres the characteristic length is about 4 mm as compared with 250 mm for the asbestos/cellulose reinforced mortar considered here. Llorca and Elices (1993) also show that the centre notch tension, and the single edge notch tension specimens exhibit similar crack growth resistance curves that increase sharply as the crack, in a small specimen, approaches a free edge. The reason for the difference in behaviour of the DCB specimen to the other geometries is that the compliance in the DCB specimen increases relatively gradually, in proportion to the square of the crack length, whereas in the other geometries the compliance increases without limit as a crack approaches the free surface. Hence in the NB geometry, the notch to depth ratio has a large effect on the K_R-curve for small specimens (see Figure 5.12). The crack growth resistance curve for crack extensions greater than that necessary to develop the full FBZ can be obtained by joining the values for the fully developed zones. If the solutions were exact, then all the curves in Figure 5.12 would be smooth. 'Exact' solutions can be obtained

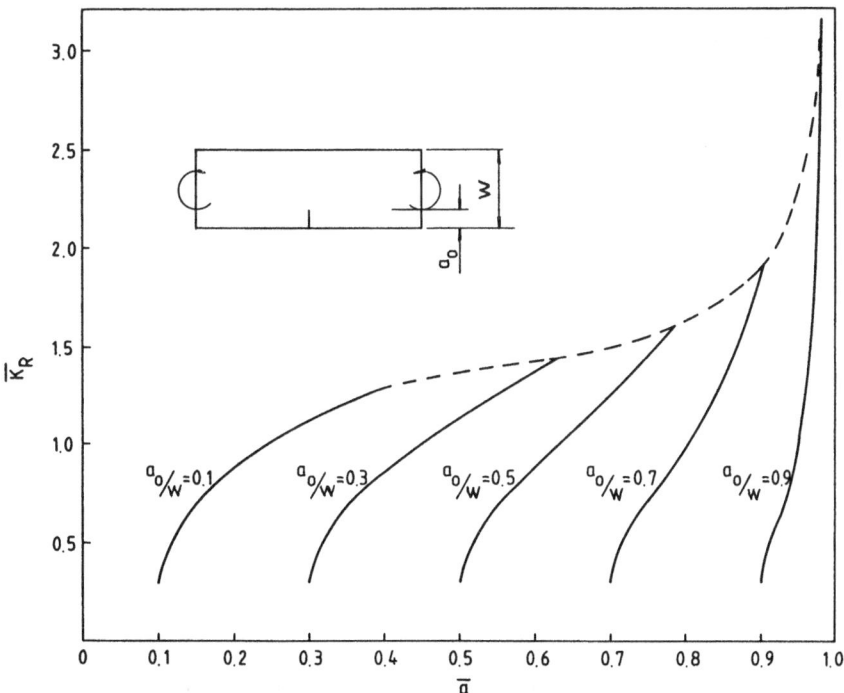

Figure 5.12 Crack growth resistance curves for the NB geometry ($\bar{W} = 1$).

Table 5.1 Properties of asbestos/cellulose reinforced mortar

	Asbestos	Cellulose
(a) *Fibre properties*:		
Fibre length	2 mm	3.5 mm
Aspect ratio	80	135
Volume fraction	8%	7%
Bond strength	0.8 MPa	0.35 MPa
(b) *Composite properties*:		
Young's modulus (E)		6 GPa
Tensile strength (f_t)		10 MPa
Reinforced matrix toughness (K_{Ic})		1.9 MPa$\sqrt{m}$
Plateau value of fracture toughness (K_{If})		5 MPa$\sqrt{m}$

by iteration to obtain the true crack profile, as described for unreinforced materials in section 4.5.3, but the gain in accuracy is slight.

Experimental crack growth resistance curves obtained from asbestos/cellulose/mortar NB specimens (whose properties are shown in Table 5.1) are compared with the predicted crack growth resistance curves in Figures 5.13 and 5.14. The bond strengths shown in Table 5.1 are only estimates and were selected to give reasonable agreement between theoretical and experimental fracture strengths (Mai *et al.*, 1980). The fibres in the mortar sheet, which was manufactured by the Hatschek process, were not randomly aligned. The crack growth resistance curves were obtained for cracks propagating in the weak direction for which the efficiency factor was estimated to be

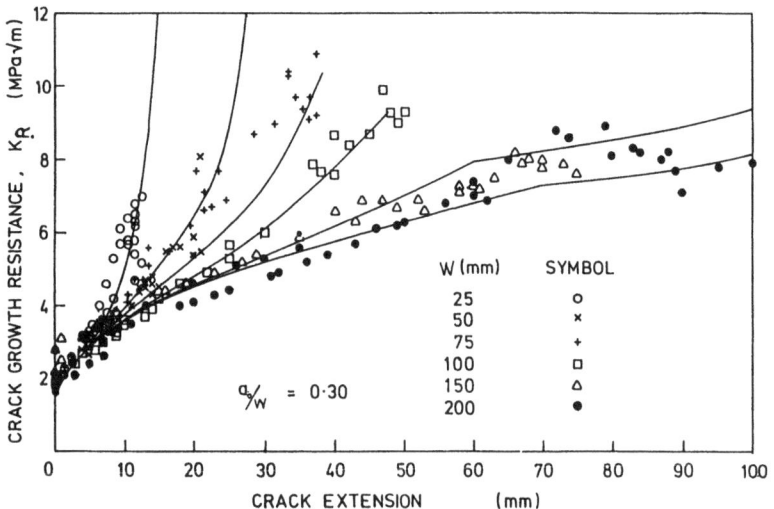

Figure 5.13 Experimental crack growth resistance curves for asbestos/cellulose mortar obtained from NB specimens, $a_0/W = 0.3$. Parameters used to calculate the predicted curves: $f_t = 10$ MPa, $K_{If} = 5$ MPa$\sqrt{m}$, $K_{Ic} = 1.9$ MPa$\sqrt{m}$.

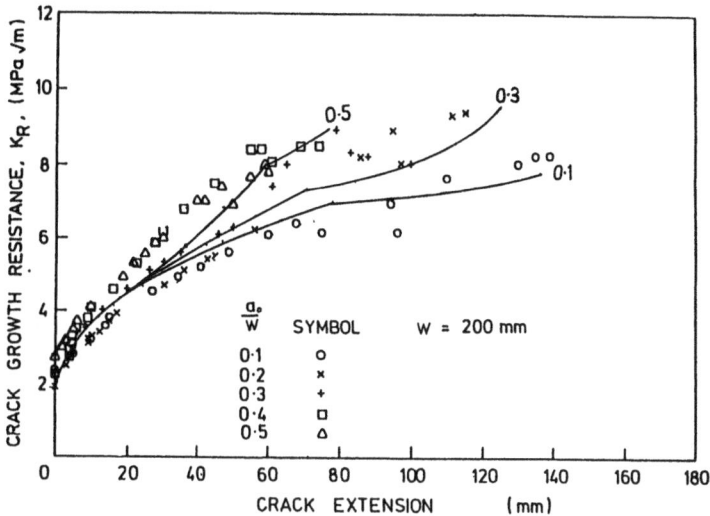

Figure 5.14 Experimental crack growth resistance curves for asbestos/cellulose mortar obtained from NB specimens, $W = 200$ mm. Parameters used to calculate the predicted curves: $f_t = 10$ MPa, $K_{If} = 5$ MPa$\sqrt{m}$, $K_{Ic} = 1.9$ MPa$\sqrt{m}$.

0.31 (Mai *et al.*, 1980). Using the data given in Table 5.1, the plateau value of the reference crack growth resistance curve was estimated to be 8.4 MPa$\sqrt{m}$. This value is very similar to the plateau value for the largest specimen ($W = 200$ mm). However, this agreement is, to some extent, fortuitous because the non-dimensional length of the largest beam is close to unity and for this case the plateau value is significantly larger than that of the reference curve, as Figure 5.11 shows. The parameters for the predictions of the crack growth resistance curves were those which gave the best fit to the experimental data for the largest specimen ($W = 200$ mm, $a_0/W = 0.3$). The agreement with the smaller beams and the other a_0/W ratios is good and gives justification to the approximate model.

5.4.3 The K-superposition method applied to fibre reinforced Type II composites modelling the FPZ as a fictitious crack

The FPZ for fibre reinforced cementitious composites are not generally small and may not always be accurately modelled by assuming that equilibrium crack growth occurs when the stress intensity factor at the tip of the matrix crack is equal to the fracture toughness of the reinforced matrix. The FPZ can be more exactly modelled by replacing the FPZ by a fictitious extension to the continuous matrix crack (Cotterell and Mai, 1988b). It can be difficult to separate that part of the stress-displacement curve that is due to the formation of the FPZ from that due to the development of an FBZ; it is for this reason Hillerborg and his co-workers do not distinguish the FPZ

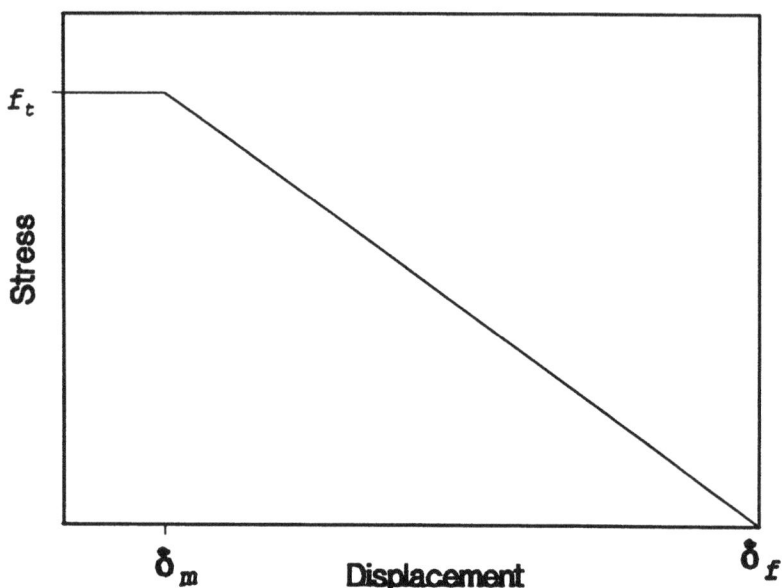

Figure 5.15 Idealized stress-displacement relationship for the FPZ and FBZ.

from the FBZ in their analysis (Hillerborg *et al.*, 1976; Hillerborg, 1980, 1983). However, such a division can be useful especially if there is a large drop in strength when the matrix cracks. In the analysis presented here it is assumed that the strength of the FPZ is constant until the crack opening displacement reaches the critical value, δ_m, at which a continuous matrix crack forms and there is no drop in strength at first cracking. A linear stress-displacement is assumed for fibre pull-out. The idealized stress-displacement relationship is illustrated in Figure 5.15. For a semi-infinite crack with a fully developed FBZ, the reference fracture toughness of the reinforced matrix, K_{Ic}^*, is given by

$$K_{Ic}^* = \left[E \int_0^{\delta_m} \sigma \, d\delta \right]^{1/2} \tag{5.31}$$

Because Barenblatt's two hypotheses are not satisfied, generally the fracture toughness, K_{Ic}, of the reinforced composite is only approximately equal to its reference value—the fracture toughness of a specimen which is large compared with the characteristic length and propagating under self-similar conditions.

In this analysis the crack faces in both the FBZ and the FPZ are assumed to be straight, but the crack opening angle is allowed to be different in each region. The first stage in the solution is to determine the lengths d_f of the FBZ and d_p of the FPZ when they are both fully developed. The lengths of these zones are found from the three conditions: (i) the total stress intensity factor at the tip of the fictitious crack is zero, (ii) the crack opening displacement at the tip of the continuous matrix crack is δ_m, and (iii) the crack opening displacement at the tip of the FBZ is δ_f. The applied force necessary to develop full FPZ and FBZ can be found and the crack growth resistance calculated. For crack growths less than that necessary to produce a fully developed FBZ, it is assumed that the crack opening angle in the FBZ remains the same, but the size of the FPZ changes with crack propagation and its current size is found from the condition that the crack opening displacement at the tip of the continuous matrix crack is δ_m. Three specimen geometries have been analysed: a semi-infinite crack (SIC), a double cantilever beam (DCB), and a notched beam under pure bending (NB). A hypothetical cementitious composite where the non-dimensional reference matrix crack stress intensity factor, $\bar{K}_{Ic}^* = 0.3$, which implies for the idealized behaviour illustrated in Figure 5.15 that $\delta_m/\delta_f = 0.047$, has been used so that the results can be compared with section 5.4.1. The non-dimensional sizes of the FPZ and the FBZ at the initiation of a matrix crack and when the FBZ is fully developed are shown in Figures 5.16 and 5.17 for the DCB and NB specimens. There can be a significant difference between the size of the FPZ at the initiation of a matrix crack and when the FBZ is fully developed,

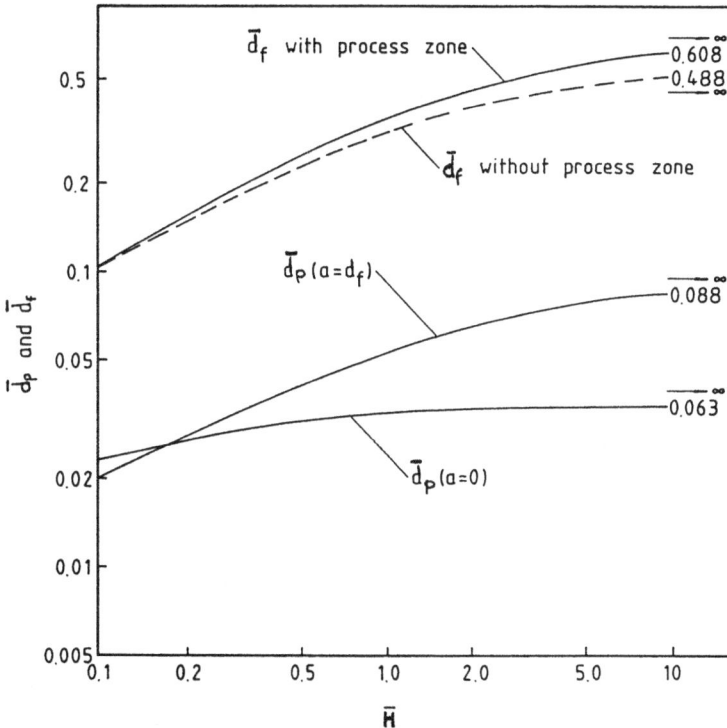

Figure 5.16 Size effect on the FPZ and FBZ for DCB specimens ($a_0/H = 3$).

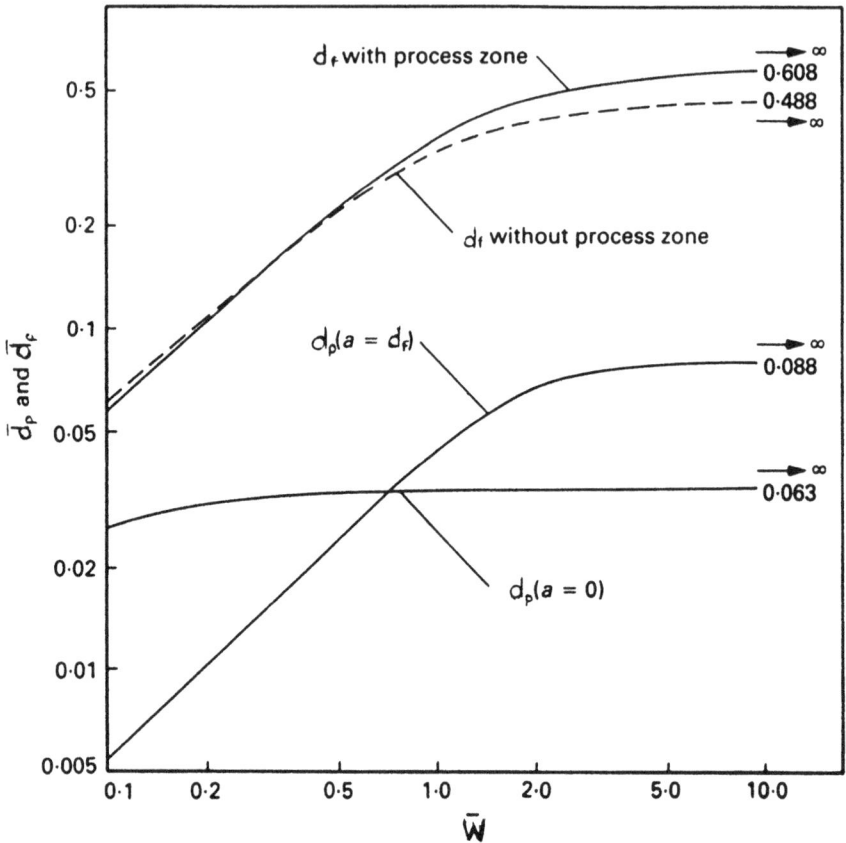

Figure 5.17 Size effect on the FPZ and FBZ for NB specimens ($a_0/W = 0.3$).

especially for small NB specimens, where the crack growth occurs well before the FBZ is fully developed. There is not a large difference between the size of the fully developed FBZ calculated by the alternate methods of modelling the FPZ. The crack growth resistance curves for DCB and NB specimens are compared with the curves given by the method of section 5.4.1 in Figures 5.10 and 5.11. The main difference in the crack growth resistance curves is caused by the difference in the size calculated for the FBZ by the two methods. One feature of the more exact modelling of the FPZ is the prediction that the fracture toughness of the reinforced matrix of small NB specimens is less than that for large ones; this variation in toughness has been observed experimentally (Mai *et al.*, 1980). The same experimental data presented in section 5.4.1 were analysed by this method, but gave no better agreement (Cotterell and Mai, 1988b).

In the practical application of crack growth models the prediction of the maximum load is more important than the crack growth resistance curve.

The nominal bending stress as a function of crack growth is given for three different sizes of NB specimen in Figure 5.18. There is reasonable agreement between the predicted maximum bending stresses for the two larger specimens obtained by either modelling the FPZ or replacing it by a singularity. However, there is a significant difference for the smallest specimen. The difference in the two predictions is mainly caused by the reduction in the effective fracture toughness of the reinforced mortar for small specimens.

5.5   Size effect and the R-curve

The strength of a fibre reinforced cementitious material is size dependent. In the limit for very large notched structures the strength can be determined from LEFM and the plateau value of the K_R-curve or the total fracture energy of the composite, G_{If}. However, the size required before LEFM becomes applicable is much larger than that for unreinforced cementitious materials because the characteristic length of fibre reinforced materials can be very large as can be seen from the characteristic lengths given in Table 5.2 for a range of fibre reinforced mortars. Bažant and his co-workers (Bažant et al., 1986; Bažant and Kazemi, 1991) have suggested that the SEL can be used to obtain an R-curve for unreinforced cementitious materials, since the fracture strength gives the slope of the R-curve at that point. However, the premise upon which the construction of the R-curve is based is that it is independent of size. As has been shown in this chapter, the R-curve is very size dependent

Table 5.2 Fracture energy, Young's modulus, tensile strength, and characteristic length of a number of fibre reinforced mortars (Ong and Ohgishi, 1989)

	Fracture energy G_{If} (J/m^2)	Young's modulus E (GPa)	Tensile strength f_t (MPa)	$l_{ch} = G_{If}E/f_t^2$ (mm)
Plain mortar	44	24	3.4	91
PAN-carbon	860	25.5	6.45	530
Vinylon (RMS182E)	6 840	23.6	4.57	7730
Tyrano (Si-Ti-C-O)	6 890	25.5	5.34	6160
Al- resist. glass	4 310	24.2	6.89	2200
Mild steel	8 890	25.6	7.86	3860
Alumina	5 580	25.6	8.64	1910
Silicon carbide	10 050	25.6	8.64	3450
Aramid (Kevlar 49)	13 370	24.8	10.3	3130
Polyacrylate	22 340	24.2	10.1	5300
Amorphous metal	48 670	24.7	12.7	7453

Fibre fraction 3%, except for amorphous metal fibres which were 2%; fibre length 25 mm; Young's modulus estimated by rule of mixtures assuming a two-dimensional randomness; tensile strength obtained from split tests.

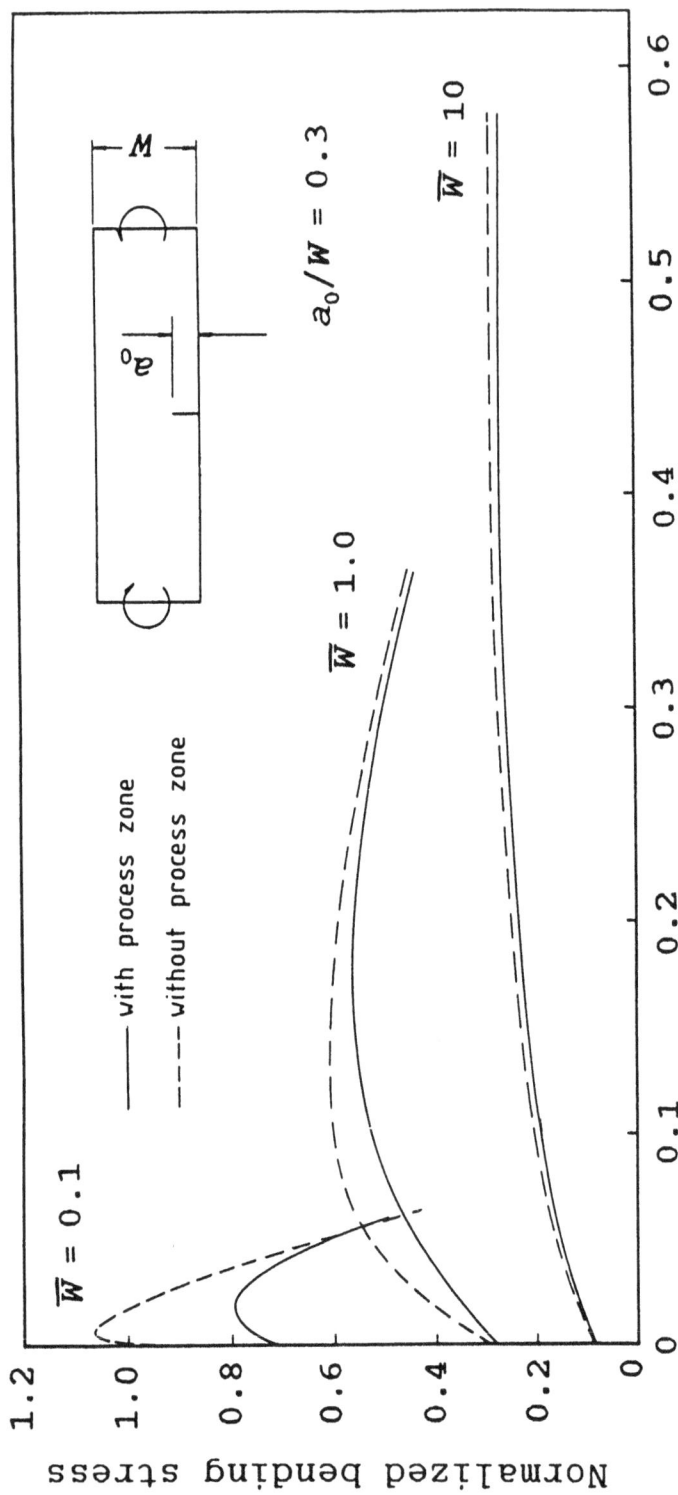

Figure 5.18 Nominal bending stress as a function of crack growth for three sizes of NB specimens.

and only if the size is large compared with the characteristic length does it approach a limiting form. What Bažant and his co-workers obtain is a composite R-curve for the whole range in specimen sizes that approximates to the limiting or reference R-curve. Tarng *et al.* (1991) apply Bažant's method to concrete reinforced with Fibercon 19 mm steel fibres. The expected general trend that the plateau value and the range of the R-curve increase with the fibre volume fraction is obtained, but curiously for small crack extensions, the R-curve for reinforced concrete is less than that for concrete unreinforced with 2% fibre volume fraction. Also the plateau value of the R-curve of $120 \, \text{J/m}^2$ is very low.

5.6 Summary

Fibre reinforced cementitious materials are important as structural materials because of their post-cracking ductility. Typically fibre reinforced cementitious matrix structures develop their ultimate loads after the matrix cracks. The fracture behaviour of such materials can be modelled by the K-superposition principle as outlined in section 5.4, assuming that for crack growth the stress intensity factor at the tip of the matrix crack is equal to the fracture toughness, K_{Ic}. It is not necessary to model the FPZ other than by the fracture toughness of the reinforced matrix unless the specimen is very small ($\bar{W} < 1$). In most cases a linear stress-displacement relationship can be assumed with little loss of accuracy. It is also sufficient to assume that the crack faces in the FPB remain straight. The fracture parameters that are necessary for predicting the ultimate strength of a structure are K_{Ic}, f_t, E, and K_{If}, or G_{If} or δ_f. These are best found by optimization of the fit to an experimentally determined load-deflection curve for a particular specimen, such as a three-point notch bend specimen. Once obtained these fracture parameters can be used to predict the fracture behaviour of any other specimen or structure. Other stress-displacement relationships can easily be used and it is possible to find the exact crack profile by iteration in a similar manner to that described in section 4.5.3 for unreinforced materials, but these refinements do not usually affect the load-deflection curve significantly.

Crack growth resistance curves, though useful for the analysis of the fracture behaviour of high strength metals and fibre reinforced ceramics which have small characteristic lengths, cannot usually be assumed to be material constants for fibre reinforced cementitious materials. In most relatively small specimens with geometries that lead to a very large increase in compliance as the crack approaches a free edge, the crack growth resistance does not tend to a plateau value. Hence, in general the concept of the crack growth resistance curve is not useful for prediction of the ultimate strength. However, the reference crack growth resistance curve for large specimens can be used to compare the toughnesses of different fibre reinforced cementitious materials.

An approximate method that satisfies global equilibrium, but does not consider the fracture mechanics of matrix crack growth can be used with what we have termed Type IIA fibre reinforced cementitious materials where the fibres are long so that the pull-out force is essentially constant for large crack openings. However, the K-superposition principle can also be used in these cases. In fact, since, in these cases, the fibre-bridging stress does not depend on crack opening, the K-superposition method is particularly simple.

6 The statistical nature of fracture in cementitious materials

6.1 Introduction

The strength of brittle solids is governed by the fracture toughness and the flaw size distribution. Under uniform tensile stress, in such a material, catastrophic fracture will be initiated at the largest flaw. This is the concept of the weakest link originally proposed by Weibull (1939, 1951). Cementitious materials are not ideal brittle materials and cracks can be initiated that will grow stably under increasing stress. Fracture is even more stable for fibre reinforced cementitious materials. In the limit case cementitious materials, reinforced by long fibres whose critical strain to failure is greater than the matrix, behave like bundles of fibres which share the load. The statistical theory of bundles was first considered theoretically by Daniels (1945). The weakest link and the bundle theories bound the behaviour of all materials.

6.2 The strength distribution for ideal brittle solids

In Weibull's (1939) analysis it is assumed that a structure is composed of N small 'elements', ΔV. The strength of each of these 'elements' is assumed to be independent of the strength of any adjacent elements. Weibull defined the 'risk of rupture', B, of a component as

$$B = -\ln S(\sigma) \tag{6.1}$$

where $S(\sigma)$ is the probability of the survival of the component. Weibull assumed that the risk of rupture, ΔB, of a single element could be given as

$$\Delta B = -\ln(\Delta S) = (\sigma/\sigma_0)^m \tag{6.2}$$

where σ_0 is a reference stress. The probability of survival of the whole component is the joint probability that each element survives or

$$S(\sigma) = \prod_1^N \exp{-\Delta B} = \exp{-\sum_1^N \Delta B} \tag{6.3}$$

If it is assumed that the 'elements' can be made as small as one likes, the probability of failure $P(V)$ of a specimen of volume V can be expressed as

$$P(\sigma, V) = 1 - S = 1 - \exp\left\{-\int_V (\sigma/\sigma_0)^m \rho \, dV\right\} \tag{6.4}$$

where ρ has the dimension of $1/V$. Since a brittle fracture originates at a flaw, ρ can be interpreted as the density of the flaws. Although in practical terms this interpretation of ρ is satisfactory, it does cause some problems. Another slight problem is that the strength of the 'elements' is not bounded, which implies that the solid is a continuum down to an infinitesimal scale and the Weibull distribution implies that the fracture of a specimen is not certain except at infinite stress. The statistical strength distribution of most brittle materials is reasonably modelled by the Weibull distribution.

In a uniformly stressed tensile specimen of volume V the probability of failure at a stress σ or less is given by

$$P(\sigma, V) = 1 - \exp[-\rho V(\sigma/\sigma_0)^m] \tag{6.5}$$

and the probability density function is given by

$$p(\sigma, V) = \frac{\mathrm{d}P(\sigma)}{\mathrm{d}\sigma} = \frac{m\rho V}{\sigma_0} \left(\frac{\sigma}{\sigma_0}\right)^{m-1} \exp[-\rho V(\sigma/\sigma_0)^m] \tag{6.6}$$

The mean strength of the tensile specimen is given by

$$\bar{\sigma} = \int_0^\infty \sigma p \, \mathrm{d}\sigma = \frac{\sigma_0}{(\rho V)^{1/m}} \Gamma(1 + 1/m) \tag{6.7}$$

There is little difference between the mean strength and the median strength, σ_m, given by

$$\sigma_m = \sigma_0 \left(\frac{0.693}{\rho V}\right)^{1/m} \tag{6.8}$$

for $m > 10$. The coefficient of variation of the strength of the uniform tensile specimens μ is given approximately by

$$\mu = 1.283/m \tag{6.9}$$

for $m \geq 10$ (Coleman, 1958).

Under three-point bending the stress, in a beam of rectangular cross-section with a span S and depth W, is given by the engineers' theory of bending as

$$\sigma = 4\sigma_{max} \frac{xy}{SW} \tag{6.10}$$

where σ_{max} is the maximum stress at the load point and (x, y) are the coordinates measured from the centre of one end of the beam. If it is assumed that failure only occurs on the tension side of the specimen, the probability of failure at a maximum stress of σ_{max} or less is found by integration of eqn 6.4 to be given by

$$P(\sigma_{max}, V) = 1 - \exp\left[-\frac{\rho V}{2(1 + m)^2} (\sigma_{max}/\sigma_0)^m\right] \tag{6.11}$$

Equation 6.11 can be refined by including the Seewald–Karman correction for the stress near the load point (Diaz and Kittl, 1988). If the median strength of a three-point bend specimen is compared to that for a tensile specimen of the same volume the ratio in strengths is given by

$$\sigma_{bend}/\sigma_{tensile} = [2(1+m)^2]^{1/m} \tag{6.12}$$

This ratio is predicted to be independent of the size of the specimen. Other assumptions of risk of rupture do show a decrease in the ratio with an increase in the size as does the theory for two-phase brittle materials (Hu et al., 1985). However, the major contribution to size effect in cementitious beams comes from the crack growth resistance of the material rather than from a statistical distribution of the strength (see section 4.6). There is no deterministic size effect under tension in cementitious materials since once the material reaches its maximum strength, at the onset of strain-softening, a tensile specimen must fail. Hence, for a deterministic cementitious material the ratio of bending to tensile strength must decrease as the size increases, tending to unity for very large beams.

The Weibull distribution as defined by eqn 6.5 is a two-parameter distribution where the Weibull modulus, m, defines the variance of the distribution. The Weibull modulus m is usually found from linear regression applied to a double logarithmic plot

$$\ln\ln\{1/[1-P(\sigma)]\} = m\ln(\sigma/\sigma_0) + \ln(\rho V) \tag{6.13}$$

To find the cumulative probability of failure a number of experiments N are performed on identical specimens. These experiments will give strengths that can be ranked in order of ascending values so that the ith strength is σ_i and the probability that the strength of a specimen is σ_i or less is $i/(N+1)$.[1] A double logarithmic plot of the bending strength of porcelain taken from Weibull's original 1939 paper is shown in Figure 6.1; the results conform to the Weibull distribution except for very high probabilities of failure greater than about 99.5%. However, in practice it is the low probabilities of failure that are important and here the Weibull distribution is excellent. For many other materials the Weibull distribution describes the variation in strength well, but for some materials the double logarithmic plot gives a curve rather than a straight line. To accommodate these materials, Weibull (1939, 1951) introduced a third parameter σ_u in his distribution so that

$$P(\sigma, V) = 1 - \exp{-\rho \int_V \left(\frac{\sigma - \sigma_u}{\sigma_0}\right)^m dV} \tag{6.14}$$

[1] It has been argued that this interpretation of the cumulative probability is based on erroneous assumptions of the weighting function and that a better estimate of the cumulative probability is given by $(i - 0.5)/N$ (Bergman, 1986). However, provided a sufficiently large sample size is tested there is little practical difference in the probability.

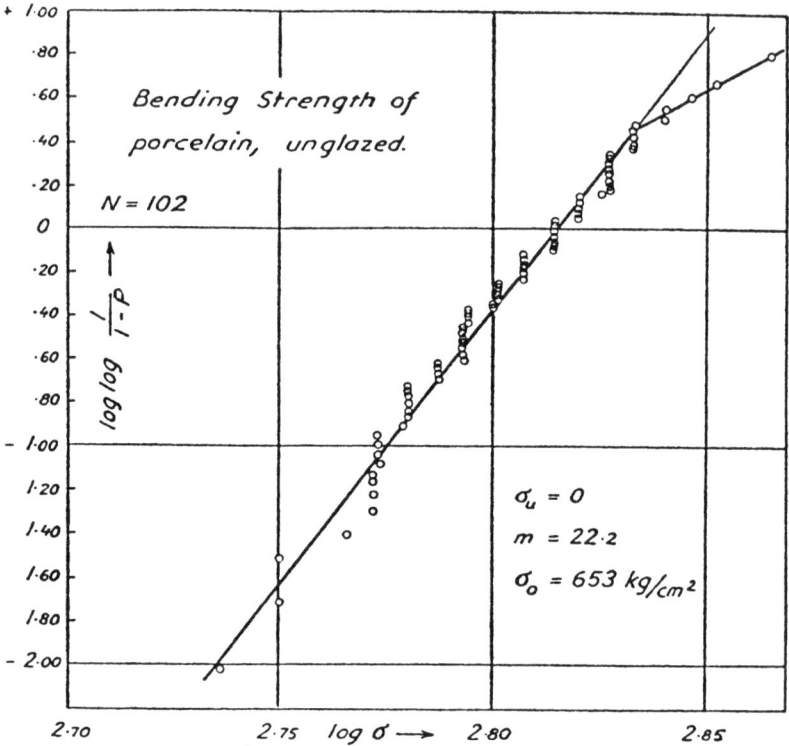

Figure 6.1 The bending strength of unglazed porcelain (Weibull, 1939).

An equation with three parameters that is of the right general form naturally gives a good fit to most data.

6.2.1   The fracture mechanics approach to strength distribution

Although Weibull's formulation of the statistical strength distribution of brittle materials has stood the test of time, it does not give a direct physical insight to the problem. More recent discussions have applied LEFM to the statistics of fracture (Jayatilaka and Trustrum, 1977; Hunt and McCartney, 1979; McCartney, 1979; Trustrum and Jayatilaka, 1983). In these analyses it is assumed that the brittle material possesses defects in the form of Griffith cracks that are sufficiently widely spaced so that there is no interaction between them. These crack-like flaws are not necessarily orientated so that they are normal to the maximum principal stress and some analyses have taken the orientation of the flaws into consideration (Jayatilaka and Trustrum, 1977; McCartney, 1979; Trustrum and Jayatilaka, 1983). However, since the exact distribution in flaw size is difficult to determine, there seems little point in this refinement and it is sufficient to describe a distribution of equivalent Griffith cracks that are orientated normal to the maximum principal stress. These equivalent cracks have the same critical stress for propagation as the real flaws. In the analysis of strength, the fracture toughness is usually considered to be a deterministic material property. McCartney (1979) does consider variations in the fracture toughness of the material, but such refinements seem unjustified because these variations can be absorbed into the equivalent Griffith crack distribution.

Here the argument of Hunt and McCartney (1979) is followed. Assume that the flaws can be replaced by equivalent penny shaped cracks of radius a and that the density of the flaws is sparse so that there is no interaction. If a tensile specimen has a strength σ and its fracture toughness is K_{Ic} its largest equivalent crack is given by

$$a = \left(\frac{K_{Ic}}{2\sigma} \right)^2 \pi \qquad (6.15)$$

Let $q(a)\, da$ be the expected number of flaws per unit volume in the size range a to $a + da$. If $f(a)\, da$ is the probability of a flaw having a size range a to $a + da$, and ρ is the density of the flaws, then

$$q(a) = \rho f(a) \qquad (6.16)$$

In a small element, dV, the probability of finding more than one flaw must decrease to zero as the size of the element becomes infinitesimally small because the integral

$$\int_0^\infty f(a)\, da = 1 \qquad (6.17)$$

is bounded. The probability of finding no cracks in the volume element dV in the size range a to $a + da$ is

$$R(a) = [1 - q(a)\, da\, dV] \qquad (6.18)$$

In the limit as the volume element becomes infinitesimal

$$R(a) = \exp[-q(a)\, da\, dV] \qquad (6.19)$$

The probability that the equivalent cracks are all less than a in the elemental volume is given by

$$T(a, dV) = R(a)R(a + da)R(a + 2da) \ldots$$
$$= \exp\left[-\int_a^\infty q(a)\, da\, dV \right] \qquad (6.20)$$

The probability that the largest equivalent crack in a specimen of volume V is less than a is the joint probability that each crack is less than a

$$T(a, V) = \exp\left[-\int_V \int_a^\infty q(a)\, da\, dV \right] \qquad (6.21)$$

The probability that the strength of a specimen of volume V is less than or equal to σ is

$$P(\sigma, V) = 1 - T(a(\sigma), V) = 1 - \exp\left[-\int_V \int_{a(\sigma)}^{\infty} q(a)\, da\, dV\right] \qquad (6.22)$$

where $a(\sigma)$ is given by eqn 6.15.

Weibull's definition of the 'risk of rupture' interpreted in fracture mechanics terms implies a cumulative flaw size distribution

$$F(a) = \int_0^a f(a)\, da = 1 - \exp-\left(\frac{a_0}{a}\right)^{m/2} \qquad (6.23)$$

where the reference crack size, a_0, is given by

$$a_0 = \left(\frac{K_{Ic}}{2\sigma_0}\right)^2 \pi \qquad (6.24)$$

Hence, the probability that the strength of a tension specimen of volume, V, is less than or equal to σ is given by

$$P(\sigma, V) = 1 - \exp\{-\rho V[1 - \exp-(\sigma/\sigma_0)^m]\} \qquad (6.25)$$

This failure probability is not quite the same as Weibull's. If the expected number of flaws in the specimen, ρV, is large, or the stress small, then eqn 6.25 is the same as Weibull's (eqn 6.5). However, there is a difference if the expected number of flaws is small (Hu *et al.*, 1988). The difference between the expression obtained by fracture mechanics considerations and that of Weibull arises, because if fracture only initiates at a flaw, the probability of there being no flaw in an element must be considered. Weibull in his analysis tacitly assumed that the number of flaws in a specimen is always equal to the expected number.

For tensile specimens under constant stress the expected number of flaws in a specimen has to be about 20 or less to cause a significant difference in the failure probability, but if the stress is not constant then the effective volume of the specimen is much smaller and the difference is more significant. For a three-point bend specimen of volume V with a rectangular cross-section the probability of failure at a stress σ or less is given by (Hu *et al.*, 1988)

$$P(\sigma, V) = 1 - \exp\left\{-\frac{\rho V}{2}\int_0^1 \int_0^1 \{1 - \exp[-(\sigma/\sigma_0)^m]u^m v^m\}\, du\, dv\right\} \qquad (6.26)$$

The integral in eqn 6.26 has to be evaluated numerically unless the expected number of flaws is very large when the usual Weibull strength distribution given by eqn 6.5 is recovered.

The extension to Weibull's analysis presented above implies that if a straight line relationship is obtained for a double logarithmic plot (eqn 6.13), it may be curved for small specimens. If a straight line of best fit is forced through the

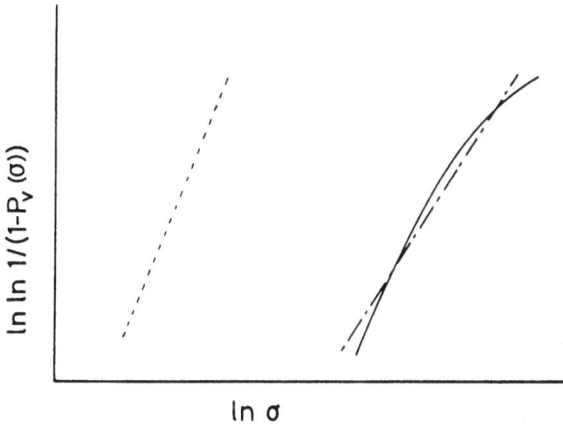

Figure 6.2 Schematic plot of $\ln\ln[1/(1 - P)]$ against $\ln \sigma$ for large and small specimens (small specimen —, best straight line through small specimen – — – —, large specimen – – – –).

data for a small specimen, as shown schematically in Figure 6.2, an effective Weibull modulus m^* that decreases with specimen size will be obtained. The effective modulus m^* partly depends on the sample size but reaches a limit when the sample size is greater than 50 (Hu *et al.*, 1988). The effective Weibull modulus is shown as a function of the number of flaws in Figure 6.3 for tensile, pure-bend, and three-point-bend specimens. The size effect on the Weibull modulus of tensile specimens is small and may not be observed, but the size effect for bend specimens is appreciable.

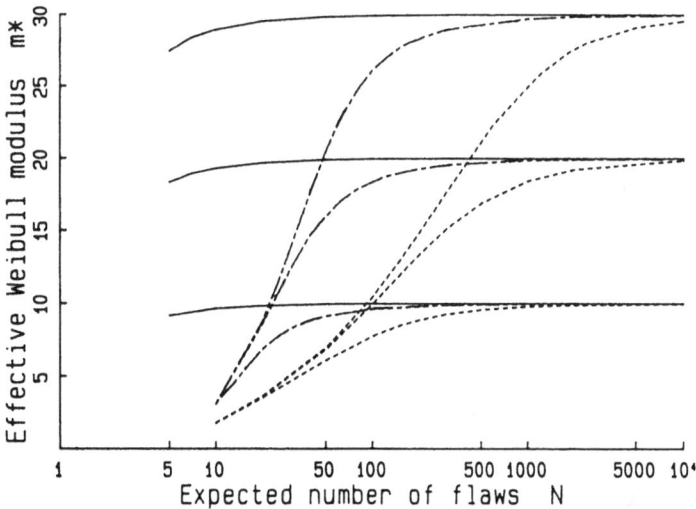

Figure 6.3 The effective Weibull modulus as a function of the expected number of flaws for: a tensile specimen —, a pure-bend specimen – — – —, a three-point-bend specimen – – – –; for three materials with a Weibull modulus of 10, 20 and 30.

Although Weibull's analysis is slightly at variance with fracture mechanics, the discrepancy is unimportant. Weibull's strength distribution can be obtained at all but the highest strengths if the flaw size distribution, $f(a)$, is assumed to be given by the Pareto distribution:

$$f(a) = 0 \qquad\qquad\qquad \text{for } a < a_0$$
$$f(a) = \frac{m}{2a_0}\,(a_0/a)^{(m/2+1)} \quad \text{for } a > a_0 \qquad (6.27)$$

This flaw size distribution gives exactly the same probability of failure as the Weibull probability function, provided the maximum stress is less than the reference stress, σ_0. For a tensile specimen of volume, V, and flaw density, ρ, the probability of fracture increases to $1 - \exp -\rho V$ at the reference stress, σ_0, which is the probability of finding no flaws in the specimen and is very small unless there are very few flaws expected. There will be a greater effect if the stress is not uniform. However, since one is usually more concerned with high survival rates, the probability of fracture at the higher stresses is relatively unimportant. For high survival rates the distribution in failure stress is insensitive to the flaw size distribution (Trustrum and Jayatilaka, 1983). This insensitivity partly explains the success of the Weibull distribution.

If the density of flaws is high, it cannot be assumed that there is no flaw interaction. On average flaw interaction causes a slight increase in the mean stress intensity factor, but the flaws need to be very close together before the increase is appreciable. Two-dimensional analyses (Lam *et al.*, 1991) show that the most probable nearest neighbour flaw must be closer than the average flaw size before the interaction causes an enhancement of more than about 10% in the stress intensity factor. The standard deviation in the stress intensity factor is about three times the mean enhancement in the mean stress intensity factor, so distortion in failure probability because of flaw interaction will only be very slight. Since it is impossible to divorce the flaw size distribution from the density of flaws, it is unnecessary to consider flaw interaction for static strength in an inert atmosphere. However, crack interaction can be significant in time dependent fracture where the rate of crack growth can be highly dependent on the stress intensity factor.

6.2.2   Strength distribution materials that exhibit R-curves

Cementitious materials have considerable crack growth resistance due to the pull-out of grains or aggregate behind a crack tip and do not fail when the first crack initiates. The effective Weibull modulus for such materials is enhanced (Kendall *et al.*, 1986). It is instructive to consider a simple model that demonstrates the enhancement in the Weibull modulus though a more rigorous model has been recently given by Duan *et al.* (1995).

As a crack propagates so generally its resistance to growth increases. The development of crack growth resistance from a flaw may not be the same as that measured with a long crack because the natural flaw is of the same dimension as the inhomogeneities causing the crack growth resistance. For the same reason there may be a variation in the crack growth resistance curve for different flaws. In this simple study of the effect of crack growth resistance it is assumed that the crack growth resistance is the same for all flaws and can be given by the power law

$$K_R = k(a - a_i)^{1/n} \tag{6.28}$$

where a_i is the radius of the equivalent initial crack and a is its value after some crack growth and $n > 2$. The 'knee' in the K_R-curve sharpens as n increases and the limit is an ideally brittle material $(n = \infty)$ that has no crack growth resistance. The general crack growth resistance expression implies that cracks start growing from flaws immediately load is applied. While in practice some stress can be sustained without crack growth the neglect of an initiation phase will not negate the qualitative conclusions. Using the criteria for unstable crack growth given in section 1.8 we obtain

$$2\sigma(a/\pi)^{1/2} = k(a - a_i)^{1/n} \tag{6.29}$$

and

$$\frac{4\sigma^2}{\pi} = \frac{2k^2}{n(a - a_i)^{(n-2)/n}} \tag{6.30}$$

Solving eqns 6.29 and 6.30 simultaneously gives the initial size of the equivalent crack a_i that causes instability at a stress σ

$$a_i = \left(\frac{2}{n}\right)^{2/(n-2)} \left[\frac{\pi}{4}\left(\frac{k}{\sigma}\right)^2\right]^{n/(n-2)} \left[1 - \frac{2}{n}\right] \tag{6.31}$$

Comparing eqn 6.31 with eqn 6.15 it is seen that the strength follows the Weibull distribution with an increased modulus $m^* = mn/(n - 2)$. In the limit when $n = \infty$, $m^* = m$ and we have the Weibull distribution for an ideally brittle material. As n decreases the modulus of the distribution increases. In other words, the greater the range of the crack growth resistance curve the narrower is the range in strengths. This simple analysis demonstrates that crack growth resistance causes the strength of a material to be more reliable. Naturally, this analysis does not imply that all materials with a crack growth resistance are more reliable than those without, since the distribution of initial flaw sizes is independent of the crack growth resistance.

6.3 The statistics of heterogeneous brittle materials

In brittle homogeneous materials, fracture under uniform stress initiates at the largest flaw even if the material possesses a crack growth resistance.

Therefore failure is governed by the weakest link and can be modelled by the Weibull distribution. However, concrete and mortar are highly inhomogeneous. Many microcracks initiate before complete failure and the most severe flaw or defect in the highly stressed region is not necessarily the flaw that leads to final failure. Weibull's theory cannot therefore be applied directly to such materials (Bažant and Xi, 1991; Bažant et al., 1991). Because Weibull's theory is inappropriate, it does not necessarily follow that the probable strength of supposedly identical specimens will not be modelled approximately by the Weibull distribution. Zech and Wittmann (1977) found that the tensile strength of concrete specimens followed a Weibull distribution with a modulus, $m = 12$. However, where Weibull theory fails is in predicting the size effect (Bažant and Xi, 1991; Bažant et al., 1991; Mazars et al., 1991).

6.3.1   The fracture of two-phase brittle materials

Wittmann (1983) has pointed out that in inhomogeneous materials such as concrete, fractures can be arrested at stronger second phase particles. Under these conditions the final fracture does not necessarily initiate at the weakest link. The arrest and reinitiation of a crack in a real three-dimensional material is extremely complex. A micro-fracture nucleated at an internal flaw under tensile stress will grow as a penny-shaped flaw until it meets a second-phase particle. The propagation as a penny-shaped crack will then cease, but it may be able to continue propagation around the particle or in another direction. Here a simple two-dimensional model (Hu et al., 1985) based on an extension of the theory of Hunt and McCartney (1979) is discussed.

Consider a two-phase material in the form of a plate of unit thickness, where the second phase particles are modelled by prisms with square cross-sections with sides of length D and are randomly distributed throughout the matrix with a density, ρ_p. The elastic constants of the two phases are assumed to be similar so that the particles do not affect the stress distribution significantly. The flaws are randomly distributed in the matrix, and their size can be characterized by the length a of the equivalent Griffith crack lying normal to the maximum principal stress. It is assumed that the distribution of second phase particles and flaws are independent. The density of the flaws in the range a to $a + da$ is $q(a) da$. Although the density of the flaws is assumed to be sparse, there is no restriction on the density of second phase particles. Overlapping of particles is permitted. In the model a flaw may intersect a particle. Two models of the intersection have been considered: in one the flaw is bridged by the particle and in the other the particle is intersected by the flaw. Since there is little difference in these two models only the bridging model will be presented in detail.

It is assumed that a crack initiates at a flaw when the stress intensity factor at its tip reaches the fracture toughness of the matrix, K_m. When both tips of a crack meet tougher second phase particles, the crack is arrested if the stress intensity factor is less than the effective fracture toughness, K_p, of the particle. If the distance between particles is s, a crack will be arrested if

$$s < k^2 a \tag{6.32}$$

where $k = K_p / K_m$. Failure occurs when either a micro-fracture initiated at a flaw is not arrested or when an arrested crack is reinitiated. To avoid confusion, the term flaw is used for the original defect and the term crack is used for an arrested micro-fracture spanning two particles. The inequality (6.32) enables the flaws to be separated into two groups. The first group contains flaws that satisfy the inequality and are arrested, whereas the second group are not arrested. As the density of the second phase particles is reduced or if the fracture toughness of the second phase particles approaches that of the matrix, the number of flaws falling into the second group increases and the failure is governed by the weakest link. On the other hand, if the second phase particles are dense or their toughness high most flaws will be arrested and failure will be governed by the largest crack which again implies a weakest link theory. Hence, in a two-phase material, most deviation from weakest link theories will be expected for moderate densities of second-phase particles whose fracture toughness, while being higher than that of the matrix, is of the same order.

Assume that flaws are bridged by any intersecting second-phase particles. Failure can initiate from either a flaw or a crack spanning two particles. Consider first the probability of fracture initiation from a crack. Fracture occurs at the largest crack. The probability that a crack whose length is between s and $s + ds$, and which lies within a strip of width dy, has a particle centred within the small elements shaded in Figure 6.4 is $(\rho_p D \, ds)^2$, and the probability that there are no other particles intersecting the crack is $\exp[-\rho_p D(s + D)]$. Hence, the joint probability $F(s, D) \, ds^2$ that there are

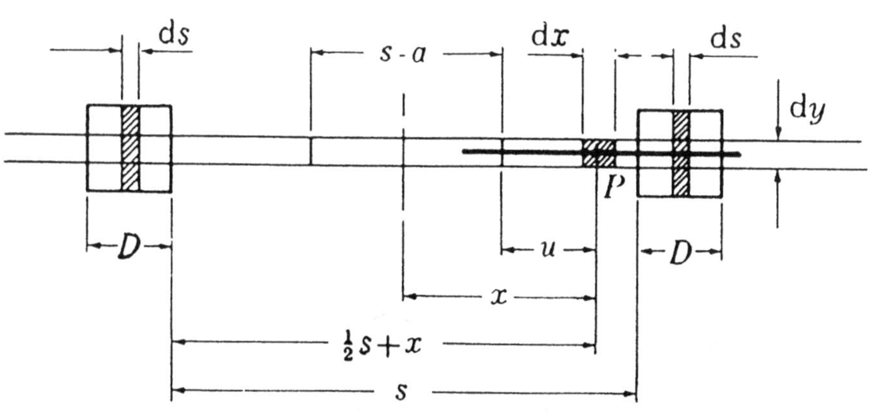

Figure 6.4 A flaw bridged by a particle with an adjacent arresting particle.

just two particles, of distance $s + D$ apart, that just touch the crack is given by

$$F(s, D) = (\rho_p D)^2 \exp[-\rho_p D(s + D)] \tag{6.33}$$

The shortest flaw centred at P (see Figure 6.4) that just intersects both particles to form a crack is $s + 2x$. Therefore, the probability that a crack is formed by a flaw centred within the element $dx\,dy$ at P which intersects both particles is

$$\int_{s+2x}^{\infty} q(a)\,da(dx\,dy) \tag{6.34}$$

and the probability $H(s)\,dy$ that a flaw centred anywhere along the thin strip of width dy intersects both particles is given by

$$H(s) = 2 \int_0^{\infty} \int_{s+2x}^{\infty} q(a)\,da\,dx \tag{6.35}$$

A fracture initiated at a flaw smaller than s, but larger than s/k^2, will be arrested by the two particles to form a crack. A flaw of length a will not touch either particle if its centre is within a median strip of length $(s - a)$ (see Figure 6.4). Hence, the probability of finding a flaw whose length is in the range a to $a + da$ and which does not touch either particle is

$$q(a)(s - a)\,da\,dy \tag{6.36}$$

In addition, a flaw centred within a small element $dx\,dy$ at P can be bridged, by the particle on the right, to form an effective flaw of length between a and $a + da$ if x is greater than $(s - a)/2$. The probability of finding such a flaw in the element $dx\,dy$ at P is

$$q(a + 2u)\,da(dx\,dy) \tag{6.37}$$

and the probability of finding one in both strips of width dy extending to infinity is

$$2 \int_0^{\infty} q(a + 2u)\,du(dx\,dy) \tag{6.38}$$

Therefore, the probability $Q(a, s)\,da\,dy$ of finding a flaw whose effective size is in the range a to $a + da$ in the total strip of width dy, is given by

$$Q(a, s) = \left[q(a)(s - a) + 2 \int_0^{\infty} q(a + 2u)\,du \right] \tag{6.39}$$

and the probability $E(s)\,dy$ that a micro-fracture initiated from such a flaw can be arrested to form a crack is given by

$$E(s) = \int_{s/k^2}^{s} Q(a, s)\,da \tag{6.40}$$

The probability $f(s)\,ds$ of finding a crack of length between s and $s + ds$ within a unit area of the plate is given by

$$f(s) = F(s, D)[E(s) + H(s)] \qquad (6.41)$$

The derivation of the probability of fracture from a crack at a given stress level follows directly from Hunt and McCartney (1979) as outlined in section 6.2.2. A fracture will be initiated from a crack if s is greater than a critical value given by

$$s(\sigma) = \frac{2}{\pi}\left(\frac{K_p}{\sigma}\right)^2 \qquad (6.42)$$

Hence, the probability $P_c(\sigma)$ that failure will occur at a stress σ or less is given by

$$P_c(\sigma) = 1 - \exp\left[-V\int_{s(\sigma)}^{\infty} f(s)\,ds\right] \qquad (6.43)$$

The last step is to consider the probability of failure by a fracture initiated at a flaw that is not arrested by adjacent particles. A fracture initiated at a flaw will continue to propagate through the second phase particle if the inequality (6.32) is not satisfied. Hence, the probability $b(a)\,da$ of finding a flaw (within a unit area of the plate) of size between a and $a + da$ that is not arrested is given by

$$b(a) = \int_{k^2 a}^{\infty} P(s, D)Q(a, s)\,ds \qquad (6.44)$$

The probability $P_f(\sigma)$ of failure by fracture from a flaw at a stress σ can be obtained in the same manner as the probability $P_c(\sigma)$ of failure by fracture from a crack and is given by

$$P_f(\sigma) = 1 - \exp\left[-V\int_{a(\sigma)}^{\infty} b(a)\,da\right] \qquad (6.45)$$

where $a(\sigma) = (2/\pi)(K_m/\sigma)^2$.

The probability $P(\sigma)$ of failure by either mechanism can be obtained from the joint probabilities of survival of the two modes of failure and is given by

$$P(\sigma) = 1 - [1 - P_c(\sigma)][1 - P_f(\sigma)]$$

$$= 1 - \exp\left\{-V\left[\int_{s(\sigma)}^{\infty} f(s)\,ds + \int_{a(\sigma)}^{\infty} b(a)\,da\right]\right\} \qquad (6.46)$$

The calculation of the probability of failure assuming flaws cut any intersecting particle is similar to the above analysis (Hu et al., 1985). The probability of failure has exactly the same form as eqn 6.46 except that in the cutting case

$$b(a) = q(a)\{[\rho_p D(k^2 - 1)a - 1]\exp(-\rho_p D^2) + 2\}$$
$$\times \exp\{-\rho_p D[(k^2 - 1)a + D]\} \quad (6.47)$$

The failure probability for two phase materials has been calculated (Hu et al., 1985) assuming that the flaw size is given by the Pareto distribution (eqn 6.27). If it is assumed that the flaws cut the second-phase particles, then the Weibull distribution is recovered as the limiting case when the fracture toughness of the particles approaches that of the matrix. The bridging case has a slightly different limit, because the bridging particles alter the effective flaw size distribution.

Results are presented for two different particle densities of 250 and 1500 per unit volume each with a sparse distribution of flaws of 10 per unit volume. The flaw size has a Pareto distribution with $m = 10$, and $a_0 = 0.064$; the size of the second phase particles is taken as $D = 0.02$. Typical flaw and particle distributions generated randomly by computer are shown in Figure 6.5. Considerable overlapping of the particles occurs at the higher density, and the volume fraction is considerably less than the indicated 0.6. The probability that no particles occupy the volume of a particle is $\exp(-\rho D^2)$. Hence, the probable volume fraction of the matrix in the two-phase material is

$$v_m = \exp(-\rho_p D^2) \quad (6.48)$$

and the volume of the second-phase particles is

$$v_p = 1 - v_m = 1 - \exp(-\rho_p D^2) \quad (6.49)$$

Therefore, the actual volume fractions of the second-phase particles in the two materials shown in Figure 6.5 are 0.095 and 0.45, compared with the indicated values of 0.1 and 0.6, respectively.

The failure probability for a material with a density of 250 second-phase particles per unit volume is shown in Figure 6.6 for a specimen 20 times the unit volume for a matrix with unit fracture toughness with the ratio $K_p/K_m = 4$. The strong second-phase particles significantly increase the strength of the material. Because few flaws are intersected at the comparatively low density of flaws, the results for the model that assumes the flaws cut through the intersecting particles are very similar to the bridging model. At higher flaw or second-phase particle densities there is more difference between the two models, with the bridging model giving the highest strengths. For example, the mean strength for the bridging model of a 20-volume specimen ($K_p/K_m = 4$), with a density of particles of 1500, is 19% greater than a similar specimen where it is assumed that the flaws cut through the intersecting particles.

There is no synergistic coupling between the flaws and the particles in the model and the increase in mean strength of the two-phase material is

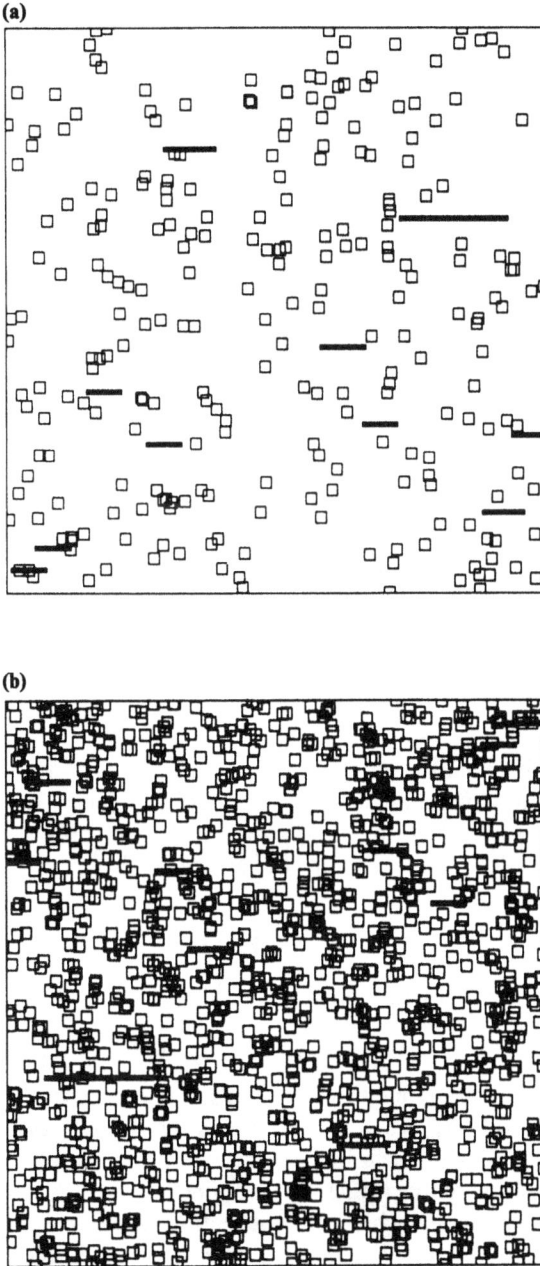

Figure 6.5 Computer simulation of a unit volume with two densities of particles: (a) $\rho_p = 250$, (b) $\rho_p = 1500$; $\rho = 10$, $m = 10$, $a_0 = 0.064$, and $D = 0.02$.

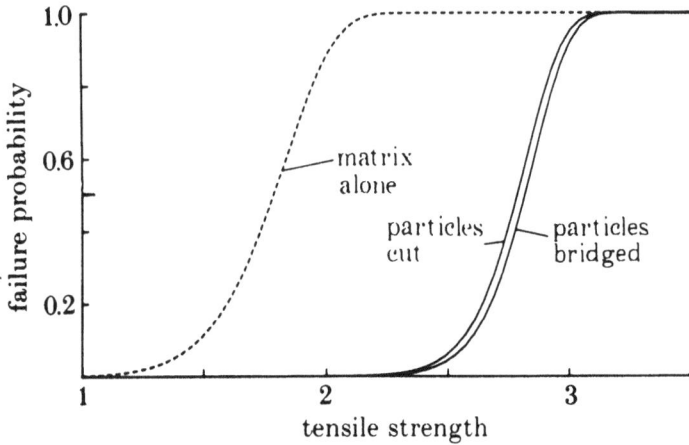

Figure 6.6 Failure probability of a two-phase material ($\rho_p = 250$, $\rho = 10$, $m = 10$, $a_0 = 0.064$, $D = 0.02$, $K_p/K_m = 4$, and $V = 20$).

very close to the value obtained from the rule of mixtures. Assuming that the specific work of fracture of the two-phase material is given by the rule of mixtures, the average fracture toughness K_c based on energy considerations is

$$K_c = [(1 - v_p)K_m^2 + v_p K_p^2]^{1/2} \qquad (6.50)$$

The flaw size distribution for the model, where the particles are cut by the intersecting flaws, is unaffected by the second phase. In the bridging model, the distortion of the flaw size distribution only occurs at high particle/flaw ratios. Thus it can be assumed that the mean fracture strength is

$$\bar{\sigma} = K_c \sqrt{2/(\pi \bar{a})} \qquad (6.51)$$

where $\bar{a}$ is the mean flaw size. Hence, according to the rule of mixtures,

$$\bar{\sigma}/\bar{\sigma}_m = [(1 - v_p) + v_p(K_p/K_m)^2]^{1/2} \qquad (6.52)$$

where $\bar{\sigma}_m$ is the mean strength of the matrix. For a second-phase volume fraction of 0.95 and $K_p/K_m = 4$, the ratio in the strength of the two-phase material to the matrix, according to the rule of mixtures, is 1.56, which compares closely with the value 1.57 obtained from the statistical model for a 20-volume specimen where the intersecting particles are cut. Naturally, the maximum discrepancy between the rule of mixtures and the statistical model occurs at a second-phase volume fraction of about 0.5 when it is 13% for a 20-volume specimen. The discrepancy increases with specimen size, because as the size gets bigger it is the extreme of the flaw size distribution that dominates.

The rule of mixtures applies equally well to three dimensions as to two. Hence, the agreement between the rule of mixtures and the statistical model suggests that the model, while a two-dimensional one, has relevance for real three-dimensional materials.

Since failure does not generally occur until a number of cracks are formed, it is possible to calculate the expected number of cracks in a specimen. In counting the number of cracks, those formed because a flaw is bridged by two particles will be excluded. Cracks can only form from flaws that are bigger than the critical size $a(\sigma)$ for that stress level. The expected number of flaws $N_f(\sigma)$ that are bigger than this size and not touching two particles is given by

$$N_f(\sigma) = V \int_{a(\sigma)}^{\infty} q^*(a) \, da \qquad (6.53)$$

where $q^*(a)$ differs from $q(a)$, because some flaws are bridged to form smaller flaws and is given by

$$q^*(\sigma) = \int_{a}^{\infty} F(s, D)Q(a, s) \, ds \qquad (6.54)$$

if the flaws are bridged by intersecting particles and by

$$q^*(a) = q(a)\{1 - [1 - \exp(-\rho_p D^2)]^2\} \qquad (6.55)$$

if the flaws cut the bridging particles.

The first micro-fracture from a flaw not arrested would cause failure, and so we must exclude from our count of the number of flaws that are initiated those that would not be arrested, because we know a priori that these could not have existed in the specimen. The number of flaws M_f that would not be arrested is given by

$$M_f = V \int_{a(\sigma)}^{\infty} b(a) \, da \qquad (6.56)$$

Hence, the expected number of cracks, N_c, to be formed when a specimen is stressed is given by

$$N_c = V \int_{a(\sigma)}^{\infty} [q^*(a) - b(a)] \, da \qquad (6.57)$$

The number of cracks expected to form before failure in a 20-volume specimen is shown in Figure 6.7 as a function of the stress level. As is to be expected, more cracks form in the material where the flaws cut intersecting particles. The total number of flaws in the sample volume is 200. The ability of the model to predict the development of microcracks is probably more important than its ability to predict the probability of fracture, which can always be empirically modelled by the Weibull distribution.

The analysis for uniaxially stressed specimens can be extended to bending and other inhomogeneously stressed specimens. As before, the specimen is divided into small elements each small enough to contain at most a single

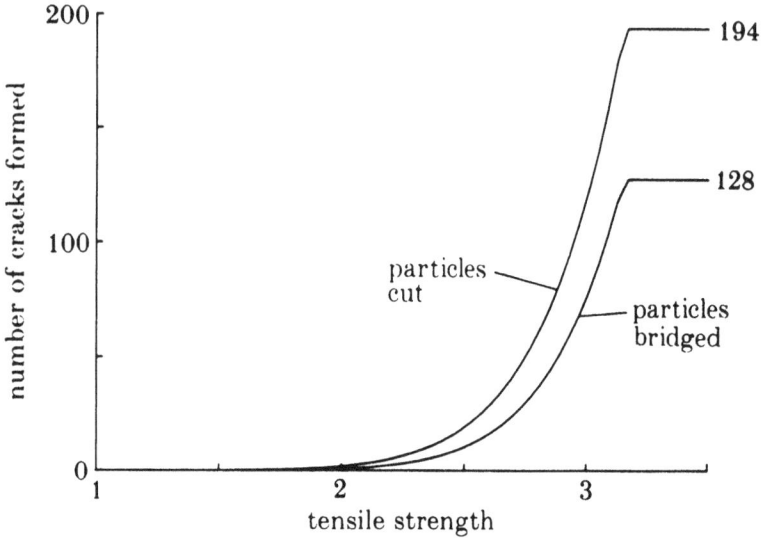

Figure 6.7 Number of cracks formed in a two-phase material ($\rho_p = 250$, $\rho = 10$, $m = 10$, $a_0 = 0.064$, $D = 0.02$, $K_p/K_m = 4$, and $V = 20$).

flaw. It is also assumed that the variation of stress across the average particle spacing is small.

The stress in a rectangular beam under pure bending at any element r, whose distance from the neutral axis line is given by

$$\sigma_r = \sigma_{max} u \tag{6.58}$$

where σ_{max} is the maximum stress at the surface of the beam and $u = 2y/W$ is the non-dimensional position of the element. It is assumed that fracture only occurs in the tension half of the beam. The probability $P_y(\sigma_y)$ that the element at a distance y from the neutral axis fails at a stress less than or equal to σ_y is given by

$$P_r(\sigma_r) = 1 - \exp - \left[\int_{s(\sigma_r)}^{\infty} f(s)\, ds + \int_{a(\sigma_r)}^{\infty} b(a)\, da \right] dV \tag{6.59}$$

Hence, the probability of failure of the beam at a maximum stress equal to or less than σ_{max}, $P(\sigma_{max}, V)$, is given by

$$P(\sigma_{max}, V) = 1 - \exp - \frac{V}{2} \int_0^1 \left[\int_{s(\sigma_r)}^{\infty} f(s)\, ds + \int_{a(\sigma_r)}^{\infty} b(a)\, da \right] du \tag{6.60}$$

The bending/tensile strength ratio for two-phase materials whose flaw size is given by the Pareto distribution is shown in Figure 6.8. The two-phase material shows a decrease in strength ratio as the specimen size increases, unlike a single-phase material. However, the decrease in strength ratio may not be entirely due to the presence of a second phase. The strength ratio

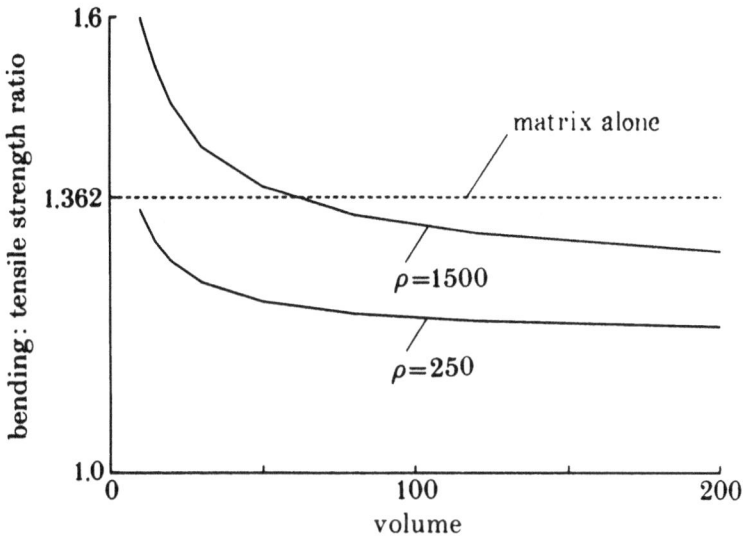

Figure 6.8 Bending-tensile strength ratio as a function of the size of a two-phase material ($\rho_p = 250$, $\rho = 10$, $m = 10$, $a_0 = 0.064$, $D = 0.02$, $K_p/K_m = 4$).

for a single-phase material whose flaw size distribution has a faster rate of decay with size than an inverse power law also shows a ratio that decreases with specimen size. A single-phase material with a flaw size following the Laplacian distribution is shown in Figure 6.9, where the decrease in strength

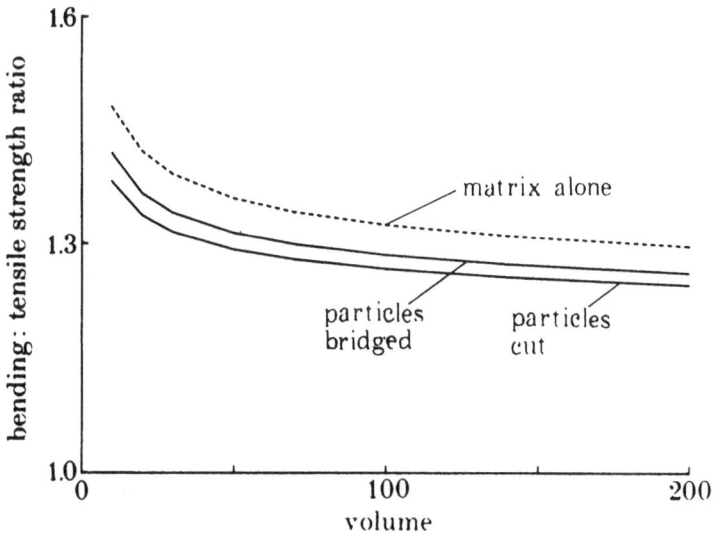

Figure 6.9 Bending-tensile strength ratio as a function of the size of a two-phase material where the flaws have a Laplacian size distribution ($\rho_p = 250$, $\rho = 10$, $m = 10$, $\bar{a} = 0.08$, $D = 0.02$, $K_p/K_m = 4$).

ratio is similar to that for a two-phase material whose flaw size follows the Pareto distribution and where the mean flaw size is the same.

6.3.2   Computer simulation of fracture in concrete modelled as a two-phase material

The above analysis shows that when a second phase is introduced in a brittle material the first crack does not necessarily lead to immediate fracture. Many cracks may form in the specimen as the stress increases before final failure. However, in a large specimen when an arrested crack is reinitiated at a second-phase particle, the fracture is then completely unstable. In practice, small cementitious specimens tested in a stiff testing machine can fracture stably. However, microcracks can form unstably. A fracture formed by a succession of microcracks that propagate unstably and are then arrested, absorbs more energy than one that propagates stably, and hence it has a higher fracture toughness. The second phase need not be actually tougher than the matrix to arrest a microcrack. Weak interfaces normal to the expected crack path can form efficient crack arrestors by the Cook–Gordon (1964) mechanism. Tensile fractures in concrete initiate primarily from flaws in aggregate/mortar interfaces and propagate through the mortar matrix until they meet another aggregate/mortar interface, when they usually propagate around the interface rather than across the aggregate (Mindess, 1983). Concrete has a higher fracture toughness than the constituent parts of the fracture path. The fracture toughness of mortar is about $0.6\,\text{MPa}\sqrt{\text{m}}$ and the toughness of the aggregate/mortar interface is only about $0.2\,\text{MPa}\sqrt{\text{m}}$, whereas concrete has a toughness of about $1\,\text{MPa}\sqrt{\text{m}}$ (Ziegeldorf, 1983). The fracture surface in concrete is very rough and the true surface area is greater than the projected area. Some of the toughness of concrete certainly comes from the additional fracture surface area, but a significant portion comes from unstable propagation of micro-fractures releasing energy in elastic waves.

In the analysis presented here (Hu *et al.*, 1986a) the growth of a fracture through the mortar and around the aggregate is followed by computer simulation similar to that used by Zaitsev (1983). The model is two-dimensional. The material parameters are based on the experimental results of Horvath and Petersson (1984). The concrete is composed of a matrix with a polygonal shaped aggregate of four to six edges whose Young's modulus is the same as that of the mortar. It is assumed that the corners of the aggregate lie on a circle that determines the size of the aggregate. The distribution in size between two limits d_{min} and d_{max} is assumed to be uniform, though any other distribution is possible. The number of sides is chosen randomly between four and six and the corners of the aggregate are positioned randomly. The centres of the aggregates are positioned randomly until the volume fraction reaches the required value; if two aggregates overlap a

new choice is made for its centre. It is assumed that mortar/aggregate interface flaws dominate and that there is at most one interface on each aggregate.
Every specimen is assumed to have the same number of flaws. When the
number of aggregates in a specimen have been determined, the probability,
p, that any aggregate has a flaw is calculated. To determine whether a particular aggregate has a flaw, a random number between 0 and 1 is generated
and if this number is less than p, a flaw is assigned to the aggregate. The
face on which the flaw appears is chosen randomly. A typical specimen of
the two-dimensional model is shown in Figure 6.10(a).

Cracks are assumed to propagate either along the mortar/aggregate
interface or in the mortar in a direction perpendicular to the applied stress.
In an isotropic homogeneous brittle material a crack will grow so that
there is local symmetry at its tip and the mode II stress intensity factor
zero (see section 1.7.1). After being deflected by an aggregate the crack
path will curve so that it becomes normal to the applied stress. Hence, the
assumption that it always propagates normal to the applied stress will not

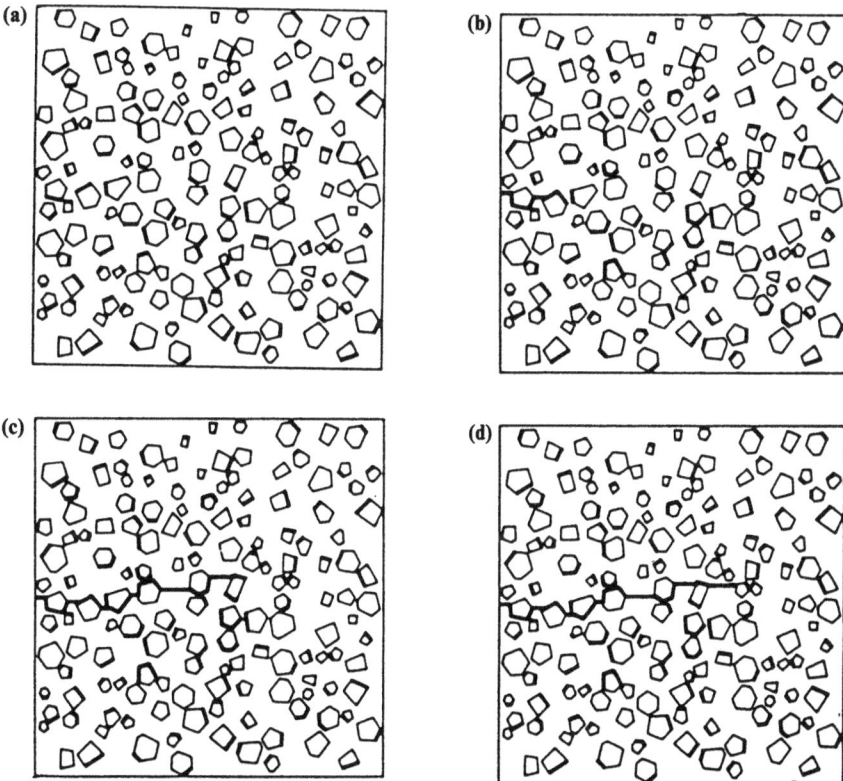

Figure 6.10 Crack development in a sample $100 \times 100\,\text{mm}$ two-dimensional concrete model:
(a) before loading; (b) at maximum load; (c) after large load drop; (d) final failure.

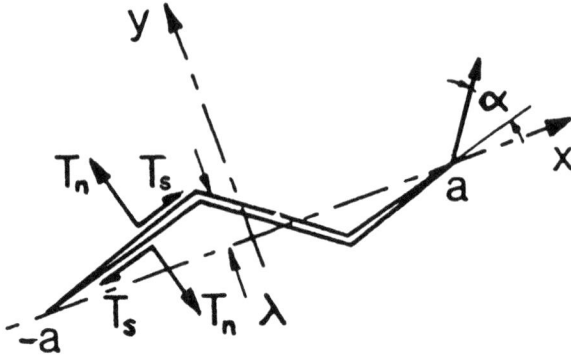

Figure 6.11 Crack configuration.

introduce a significant or systematic error. Since K_{II} is not necessarily zero, it is assumed that the crack grows when the modulus of the effective stress intensity factor, K_e, which is given by

$$K_e = (K_I^2 + K_{II}^2)^{1/2} \tag{6.61}$$

is greater than the fracture toughness of either the mortar or the mortar/ aggregate interface. It is assumed that there is no interaction between cracks and, since the mortar and aggregate have the same modulus, the aggregate does not disturb the stress distribution. The stress intensity factors are calculated from a first order solution for kinked or curved cracks (Cotterell and Rice, 1980) and for the tip at the right of the crack shown in Figure 6.11

$$K_I - iK_{II} = \frac{1}{\sqrt{\pi a}} \int_{-a}^{a} (q_I - iq_{II}) \left(\frac{a+t}{a-t} \right)^{1/2} dt \tag{6.62a}$$

where

$$q_I = T_n - \tfrac{3}{2}\lambda'(a)T_s + \lambda(x)T_s + 2\lambda'(x)T_s \tag{6.62b}$$

$$q_{II} = T_s + \lambda(x)T_n + \tfrac{1}{2}\lambda'(a)T_n \tag{6.62c}$$

and T_n, T_s are the normal and shear tractions at the crack surfaces and $\lambda(x)$ describes the crack profile; the prime indicates a derivative with respect to x. This first order solution is accurate to within 5% if the slope $\lambda'(x)$ is less than 15°. The expression for the stress intensity factor at the left-hand tip is similar. For a small kink at the mortar/aggregate interface, the stress intensity factor is given by

$$\begin{bmatrix} K_I \\ K_{II} \end{bmatrix} = \begin{bmatrix} C_{11} & C_{12} \\ C_{21} & C_{22} \end{bmatrix} \begin{bmatrix} k_I \\ k_{II} \end{bmatrix} \tag{6.63}$$

where k_I, k_{II} are the mode I and mode II stress intensity factors at the tip of the crack before it kinks, and

$$C_{11} = \tfrac{1}{4}[3\cos\alpha/2 + \cos 3\alpha/2]$$
$$C_{12} = -\tfrac{3}{4}[\sin\alpha/2 + \sin 3\alpha/2]$$
$$C_{21} = \tfrac{1}{4}[\sin\alpha/2 + \sin 3\alpha/2]$$
$$C_{22} = \tfrac{1}{4}[\cos\alpha/2 + 3\sin 3\alpha/2]$$

(6.64)

The expressions given above are for cracks in infinite plates and have been modified for cracks that are near to or intersect an edge, assuming that the geometrical correction factors applied to cracks normal to the applied stress in finite plates can be applied to the kinked cracks if the projected length of the crack a_p is used (Hu et al., 1986a).

If the stress is increased incrementally the history of the crack development can be traced. When the stress intensity factor at any crack tip exceeds the fracture toughness, the crack will grow unstably until it meets an aggregate. At this stage the crack may be arrested if the aggregate/mortar interface is at a low angle to the applied stress, or it may propagate along the mortar/aggregate interface and into the mortar once more. Under these conditions the ultimate strength of the specimen will be reached when one crack propagates unstably right across the specimen. If the specimen is loaded in a rigid machine under fixed grip conditions, there will be a decrease in stress as a crack propagates. This reduction in stress can cause a crack to arrest that would have otherwise caused failure under constant stress and it is possible for failure to occur after the maximum load. Before the maximum load is reached there will be many growing cracks, but afterwards localization occurs. In this model, the size of the specimen is necessarily small and the fracture process localizes to a single crack.

The extension of the specimen can be calculated from the increase in strain energy due to the growing cracks. The strain energy stored per unit thickness in a specimen of width W and length L is

$$\Lambda = \frac{\sigma^2}{2E}WL + \sum_{i=1}^{n}\int_0^{s_i} G_i\,ds$$

(6.65)

where

$$G_i = \frac{(1-\nu^2)}{E}K_e^2$$

(6.66)

is the plane strain crack extension force for the ith crack and the integral is taken along the crack. For the purpose of calculating the stress-displacement relationship, it is sufficiently accurate to replace the real crack by a straight

Table 6.1 Mechanical properties of model aggregate

Young's modulus (GPa)	Poisson's ratio	Fracture toughness	
		mortar	mortar/aggregate (MPa$\sqrt{m}$)
27	0.25	0.6	0.2

crack equal in length to the projected crack length, a_p. Hence

$$G_i = \frac{(1 - \nu^2)}{E} F_i^2 \sigma^2 \pi a_{ip} \tag{6.67}$$

and the strain energy stored becomes

$$\Lambda = \frac{\sigma^2}{2E} WL + \frac{(1 - \nu^2)}{4E} \sigma^2 \pi \sum_{i=1}^{n} \bar{F}_i a_{ip}^2 \tag{6.68}$$

where $\bar{F}_i$ is the average value of the finite width correction factor for the ith crack. The stress-displacement $(\sigma - \Delta L)$ relationship

$$\sigma = \frac{E\Delta L}{L} \left[1 + \frac{(1 - \nu^2)}{2WL} \pi \sum_{i=1}^{n} \bar{F}_i^2 a_{ip}^2 \right]^{-1} \tag{6.69}$$

is obtained by equating the strain energy stored to the work done.

The fracture energy, G_f, can be calculated from the area under the stress-elongation curve. Some of the work of fracture goes into forming micro-cracks away from the pupative fracture plane. The work done away from the fracture plane, which is quite small in the small samples that have been modelled here, has been excluded by taking a line parallel to the initial linear response of the specimen through the point of maximum stress and calculating the fracture energy, G_F, from the area of the stress-displacement curve to the right of this line.

The material properties used in the computer simulation of fracture are given in Table 6.1. To study the statistical distribution in strength and toughness, 60 computer experiments were performed. A typical stress-displacement curve (based on the specimen shown in Figure 6.10a) is shown in Figure 6.12. The corresponding crack development is shown in Figure 6.10b–d. The distribution in fracture strength for a 100×100 mm specimen is shown in Figure 6.13. Because the distribution in initial flaw size is small, the distribution in the stress at which the first crack occurs is narrow. The aggregate almost doubles the mean strength of the concrete. The reduction in stress intensity factor caused by the deflection of the crack path by the aggregate causes arrest, despite the toughness of the interfaces being considerably less than the matrix. In the comparatively small specimens modelled here, there is a wide range in the tensile strengths. The fracture toughness of the concrete has been calculated from the fracture

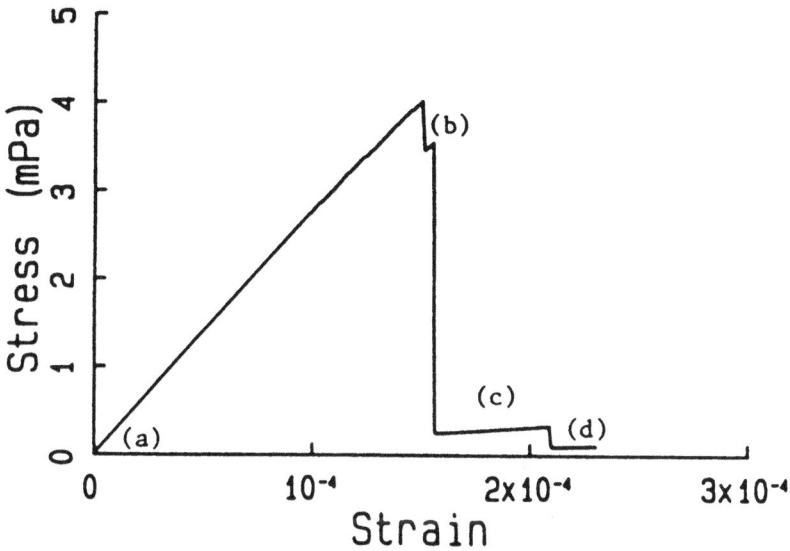

Figure 6.12 The stress-displacement curve for the specimen shown in Figure 6.10.

energy and is shown in Figure 6.14. The mean fracture toughness is 0.97 MPa√m which is very close to experimentally observed values. There is a large variation in fracture toughness in line with the large variation in strength. The excellent agreement between the calculated value and the experimentally determined value is to some extent fortuitous, but the

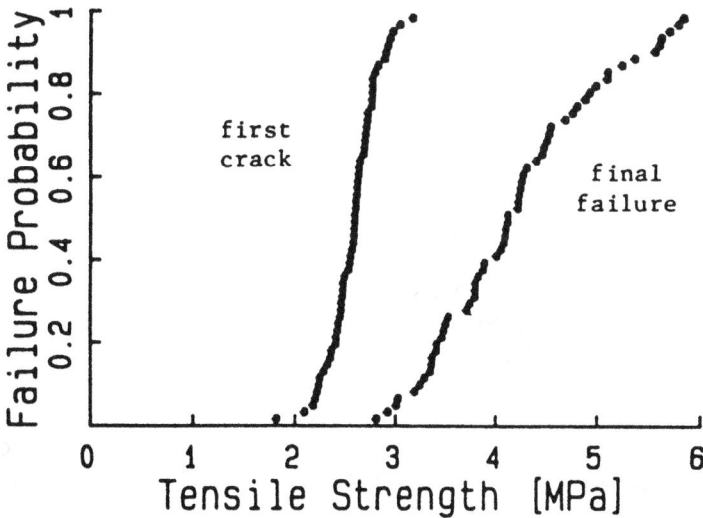

Figure 6.13 The strength distribution of the concrete specimens.

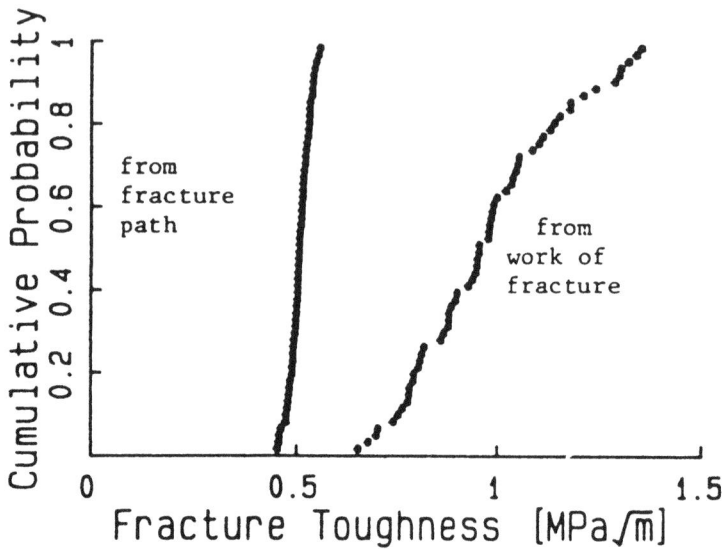

Figure 6.14 The fracture toughness distribution of the concrete specimens.

model does show that the arrest of unstable microcracks can be a significant source of toughness in concrete.

6.4   Statistics and size effect in cementitious materials

Fracture in cementitious materials is preceded by the formation of a localized strain-softened FPZ either at a notch, stress concentration, or defect. Since the size of the FPZ is large, a defect in an otherwise smooth component only locates the position for strain localization without significantly altering the strength. It is the condition of the material at the pupative FPZ that controls the fracture strength. Before the maximum load is reached, micro-cracking in a smooth component will occur outside the pupative FPZ, but such microcracking plays no significant part in the fracture process.

Mazars *et al.* (1991) have used the concept of damage to analyse the fracture of cementitious materials. In this concept, the stiffness of an element decreases from its virgin value to zero as the damage, D, increases from zero to unity. They assume that no damage occurs until the strain reaches a threshold value. After the threshold has been reached, it is assumed that damage increases with non-local strain. With a deterministic threshold for strain this approach gives similar results to other non-local damage models. However, Mazars *et al.* (1991) found that with a deterministic threshold to damage the strength of small beams was underestimated. By introducing a probabilistic threshold strain for damage that is based on

Weibull theory they obtain better agreement with experimental data. A probabilistic threshold for damage has little effect on the predicted strength variation for the larger beams.

Bažant and Xi (1991) have also considered the effect of statistical variation on the size effect law (Bažant et al., 1986). They argue that there are two asymptotic limits to the SEL for notched bend specimens: for large beams where the FPZ is small compared with the size of the beam, the variation in strength follows LEFM and $\sigma_N \propto W^{-1/2}$; for small beams where the FPZ dominates, the strength is according to the Weibull distribution and $\sigma_N \propto W^{-n/m}$, where n is the number of spatial dimensions. Bažant and Xi (1991) also give a simple empirical equation

$$\sigma_N = \frac{Bf_t}{\sqrt{\beta^{2n/m} + \beta}} \qquad (6.70)$$

that has these limits. A more exact SEL can be obtained from the fictitious crack line model (Cotterell et al., 1995).

The FPZ has a narrow flame-like shape with a maximum width that is not very size dependent (Bažant and Lin, 1988). Hence, for the purpose of determining the statistical strain-softening characteristics, the FPZ is assumed to be a narrow zone, whose width is a material constant independent of the size of the specimen. Outside of the FPZ, it is assumed that the material is elastic and suffers no damage even in any region of high compressive stress. Once the strain-softening relationship is determined, the FPZ is modelled by a fictitious extension to the true crack. A simple linear stress-displacement relationship is chosen for the strain-softening behaviour of the FPZ. As the FPZ grows so its chance of encountering a zone of weakness increases and the average stress carried will diminish. To avoid using a stochastic method, it is assumed that the stress threshold, f_t, necessary to initiate strain-softening depends on the volume, V_p, of the FPZ, and is proportional to the Weibull mean strength so that

$$\frac{f_t}{f_0} = \left(\frac{V_{p0}}{V_p} \right)^{1/m} \qquad (6.71)$$

where f_0 is the stress threshold for a standard FPZ of volume, V_{p0}. The fracture energy of cementitious materials, G_{If}, is usually assumed to be a material constant independent of size. Experiments indicate that in fact the fracture energy increases with specimen size (see section 3.5) and this increase in fracture energy with size must to some extent offset the decrease in f_t. However, here the fracture energy, G_{If}, is assumed to be a constant. Thus, combined with the assumption that the stress-displacement relationship is linear, the critical crack tip opening displacement δ_f is given by

$$\delta_f = \frac{2G_{If}}{f_t} \qquad (6.72)$$

The statistical effect of size on the strength of three-point bend specimens has been obtained using the fictitious crack model and the K-superposition theory (Cotterell et al., 1995). The faces of the fictitious crack are assumed to be straight, so that only the CTOD or the length of the fictitious crack, d_p, are unknown. Since the stress in the FPZ depends on its size, it is easiest to drive the program with the length of the fictitious crack, d_p. The values of the beam depth, W, initial notch depth, a_0, and, d_p, together with an initial guess of the crack tip opening displacement, δ_i, are used to calculate the load from the condition that the total stress intensity factor at the fictitious crack tip shall be zero, taking into account the variation in the mean ultimate strength of the cementitious material determined by eqn 6.71. An iterative routine similar to subroutine PROFILE shown in Figure 4.10 is used to find the actual value of the crack tip opening displacement, and then another subroutine calculates the corresponding load. A further subroutine finds the maximum load sustainable by the beam. In all cases maximum load occurs for $\delta_t < \delta_f$, though as the beam depth increases, so δ_t increases and approaches δ_f in the limit.

Since it has been argued that mortar and concrete do not follow Weibull's theory, supposed measurements of the Weibull modulus directly from a plot of eqn 6.13 cannot give the exact value for eqn 6.71. However, it is thought that Weibull moduli of 10 and 20 which bound the value measured by Zech and Wittmann (1977) are realistic bounds for cementitious materials. Three different classes of beams have been analysed:

(i) Beams with deterministic strain-softening characteristics ($m = \infty$).
(ii) Beams of constant width, $m = 20, 10$.
(iii) Beams whose width is proportional to their depth, $m = 20, 10$.

All the results have been normalized by the strength and size of the beams whose non-dimensional depth (W/l_{ch}) is unity.

Two sets of beams have been analysed. In the first beam the relative notch depth, a_0/W, is 0.3 and a logarithmic plot of the normalized strength against the normalized beam depth is shown in Figure 6.15. There is little statistical effect in large notched specimens because the FPZ is almost fully developed at final fracture and it is small compared with the notch size. Under these conditions it is the fracture energy, G_{If}, that controls the fracture not the stress-displacement relationship. In the limit for an infinitely large beam, the fracture occurs when

$$K_a = (EG_{If})^{1/2} \tag{6.73}$$

and the slope of the logarithmic plot tends to $-\frac{1}{2}$ as the beam size increases, as predicted by the size effect law (Bažant et al., 1986). For small specimens fracture occurs before the FPZ has developed very much and the FPZ is comparable or even larger than the length of the notch. Under these conditions the actual strain-softening relationship is unimportant. Since

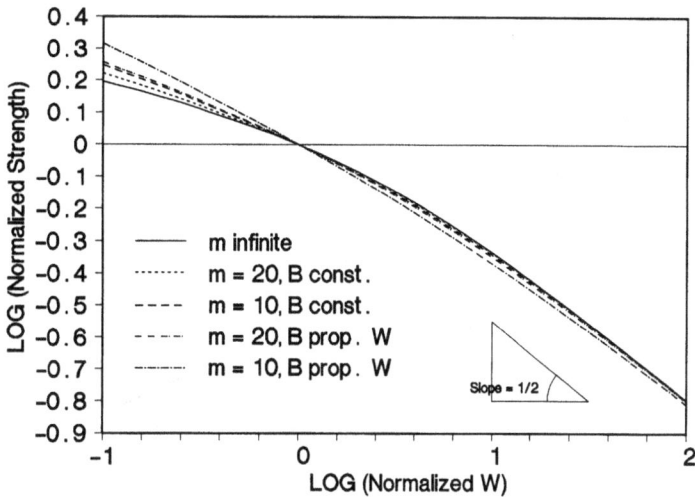

Figure 6.15 Normalized strength of notched beams $a_0/W = 0.3$.

physically the size of the FPZ must decrease as the specimens get smaller, the strength of small specimens is statistically greater than indicated from a deterministic model. Obviously, if the width of the specimens is in proportion to their depth then the increase in strength is greater than if the width of the specimens is kept constant, because the decrease in FPZ volume is greater.

These results are very similar to those given by the approximate relationship (see eqn 6.70) of Bažant and Xi (1991). However, here n is interpreted somewhat differently, and is taken as 2 for full geometric similarity and 1 for beams of constant width.

The second set of beams analysed is representative of plain, un-notched, beams. Because of a fundamental difference in the behaviour of notched and plain beams, a true un-notched beam cannot be analysed. In a notched beam, a FPZ starts to form immediately load is applied to the beam but, in a plain beam, a FPZ does not initiate until the maximum elastic stress attains the critical stress, f_t. If this stress is given by eqn 6.71, the critical stress is infinite at the initiation of a FPZ and the strength of the beam is infinite. In practice the inhomogeneity, inherent in cementitious materials, will lead to an early initiation of a FPZ. This early stage in the formation of a FPZ is not important and results have been obtained for beams that have a small defect or crack whose absolute size is the same for all beams. The non-dimensional defect size chosen is 10^{-2}, so that the smallest beam ($\bar{W} = 0.1$) has a defect whose relative size is $a_0/W = 10^{-1}$, and the largest beam ($\bar{W} = 100$) has a defect with $a/W = 10^{-4}$. A logarithmic plot of the normalized strength of such beams is shown in Figure 6.16. The beams whose strain-softening characteristics are deterministic tend to a non-dimensional strength of unity as they get larger. The slight dip in strength

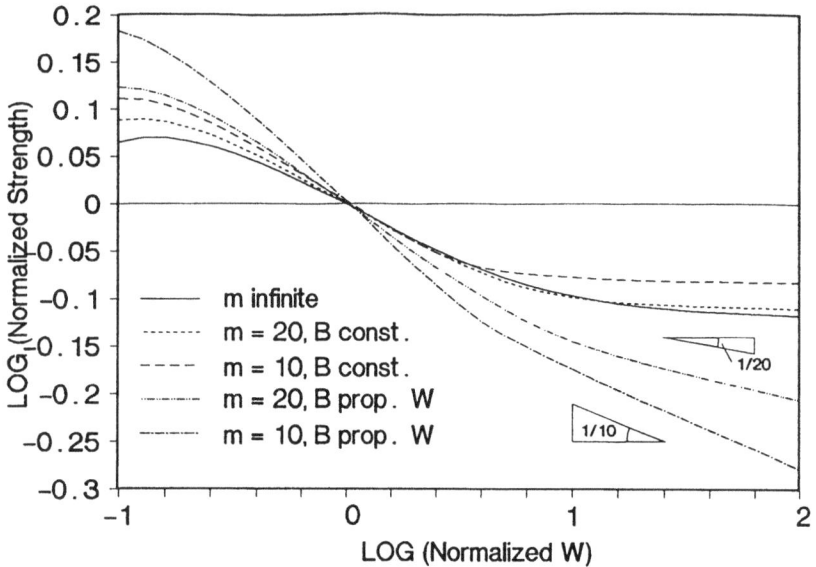

Figure 6.16 Normalized strength of 'plain' beams.

of these deterministic beams for $\bar{W} < 0.16$ is caused by the assumed defect, for a perfect beam the strength would continue to increase. The effect on the mean strength of the statistical strain-softening relationship is much more marked in these 'plain' specimens. The strength of the small specimens is increased by the statistical variation, as was the case for the notched beams shown in Figure 6.15. The size of the FPZ for 'plain' beams reaches its most developed state for near unit sized beams and then decreases to an asymptotic value as the beams increase in size. If the width of the beams is constant, the volume of the FPZ decreases causing the statistical mean strength of the large beams to be larger than the deterministic ones. However, the length of the FPZ decreases less slowly than the depth of the beam increases. Thus, if the width of the beam is kept in proportion to the depth of the beam, the FPZ increases in volume with size and the statistical strength is less than the deterministic value. For large beams the length of the FPZ is constant and the volume, therefore, in proportion to the depth of the beam. Thus, for beams whose width is in proportion to their depth, the slope of the logarithmic plot of strength against beam size tends to $-1/m$. This latter asymptotic trend was also noted by Mazars *et al.* (1991).

6.5 The statistics of fibre reinforced cementitious materials

Fibre reinforced cementitious materials form a diverse group from steel bar reinforced concrete, where in tension most of the load is taken by the reinforcement, through continuous fibre reinforced materials, where the fibres take a very significant portion of the load and cause multiple cracking, to short fibre reinforced materials, that do not greatly increase the strength of the cementitious material, but whose main purpose is to provide toughness as the fibres pull-out. Hence, the statistical behaviour is diverse. In the extreme case where there is a high density of strong continuous fibres, it is the strength of the fibres that determines the strength of the composite, so classic bundle theory is considered first.

6.5.1   The strength of bundles

In bundle theory it is usually assumed that the stress-elongation curve of each fibre is identical, but that the strength of each fibre is distributed according to some cumulative probability function $P(\sigma)$ (Daniels, 1945, 1989). The load F carried by the bundle of fibres, assuming that each fibre carries the same stress, σ, is given by

$$F = \sigma A N \qquad (6.74)$$

where A is the cross-sectional area of each fibre and N is the number of fibres remaining intact. The expected number of fibres intact at a stress σ is given by

$$N = N_0[1 - P(\sigma)] \qquad (6.75)$$

where N_0 is the number of fibres in the bundle. The expected strength of the bundle occurs when the differential of the force F sustained by the bundle is zero, or

$$\frac{\mathrm{d}F}{\mathrm{d}\sigma} = 0 = A N_0\{[1 - P(\sigma^*)] - \sigma^* P(\sigma^*)\} \qquad (6.76)$$

The distribution in the actual number of fibres intact, at the fibre stress, σ^*, that gives the maximum bundle stress, is a binomial one which tends to a normal distribution as the number of fibres becomes large. However, the mean maximum strength only converges very slowly to the asymptotic value given by

$$\mu(\sigma^*) = \sigma^*[1 - P(\sigma^*)] \qquad (6.77)$$

and Daniels (1989) has given a correction term for the asymptotic value that is correct within the $O(N^{-1/3})$. For large N the expected maximum strength, σ_b, is given by

$$\sigma_b = \mu(\sigma^*) + 0.99615 N^{-2/3}[\mu(\sigma^*)]^{2/3}[-\mu''(\sigma^*)]^{-1/3} \qquad (6.78)$$

where $\mu''(\sigma)$ is the second derivative with respect to σ. The standard deviation in the maximum strength has no distortion and is given by

$$s_b = \sigma^*\left\{\frac{P(\sigma^*)[1 - P(\sigma^*)]}{N_0}\right\}^{1/2} \qquad (6.79)$$

As is intuitively obvious, the variation in strength of a bundle decreases as the number of fibres is increased. If the strength of each fibre is assumed to be given by the Weibull distribution,

$$P(\sigma) = 1 - \exp -\frac{l}{l_0} \left(\frac{\sigma}{\sigma_0} \right)^m \tag{6.80}$$

the stress, σ^*, in each fibre at the expected strength of the bundle is given by

$$\sigma^* = \sigma_0 \left(\frac{l_0}{ml} \right)^{1/m} \tag{6.81}$$

Hence, the expected bundle strength for a large number of fibres is given by

$$\sigma_b = \sigma_0 \left(\frac{l_0}{ml} \right)^{1/m} [1 + 0.99615 N^{-2/3} \exp(2/3m)] \exp -(1/m) \tag{6.82}$$

The expected strength is always less than the average strength of an individual fibre and it is only when the Weibull modulus, m, becomes large that expected strength approaches the average strength of the fibres. Although the number of fibres in a bundle dominates the variation in the strength of the bundle, the standard deviation is also dependent on the Weibull modulus.

6.5.2   Type I continuous fibre reinforced cementitious materials

In Type I reinforced composites the fibres alone can more than withstand the load necessary to crack the matrix, and multiple matrix cracks form. The behaviour of these composites depends on the spacing, L_m (see eqn 3.61), of the multiple cracks. The spacing of matrix cracks is determined by two key parameters (Curtin, 1991; Phoenix, 1993)

$$\Delta_c = l_0^{1/(m+1)} \left[\frac{\sigma_0 d_f}{2\tau_b} \right]^{m/(m+1)}$$

$$\Sigma_c = \sigma_0^{m/(m+1)} \left[\frac{2\tau_b l_0}{d_f} \right]^{1/(m+1)} \tag{6.83}$$

These parameters define the expected fibre length, Δ_c, in which there will be exactly one flaw that will fail at a stress of Σ_c. The sub-Type I composites are: (1) Type IA, widely spaced multiple cracks: $L_m > \Delta_c$, $E_f \approx E_c$, $v_f \ll E_c/E_f$, and/or τ_b is small; (2) Type IB, closely spaced multiple cracks: $L_m \ll \Delta_c$, $v_f \gg v_{fc}$, $E_m \approx E_c$ and/or τ_b is large. The mechanical difference between Types IA and IB is in the variation in stress in the fibres (see Figure 6.17). In Type IA the stress transfer length is large and there is a relatively large variation in stress between the matrix cracks. After the matrix cracks in Type IB composites, the load is carried mainly by the fibres even away from the matrix cracks and there is little variation in fibre

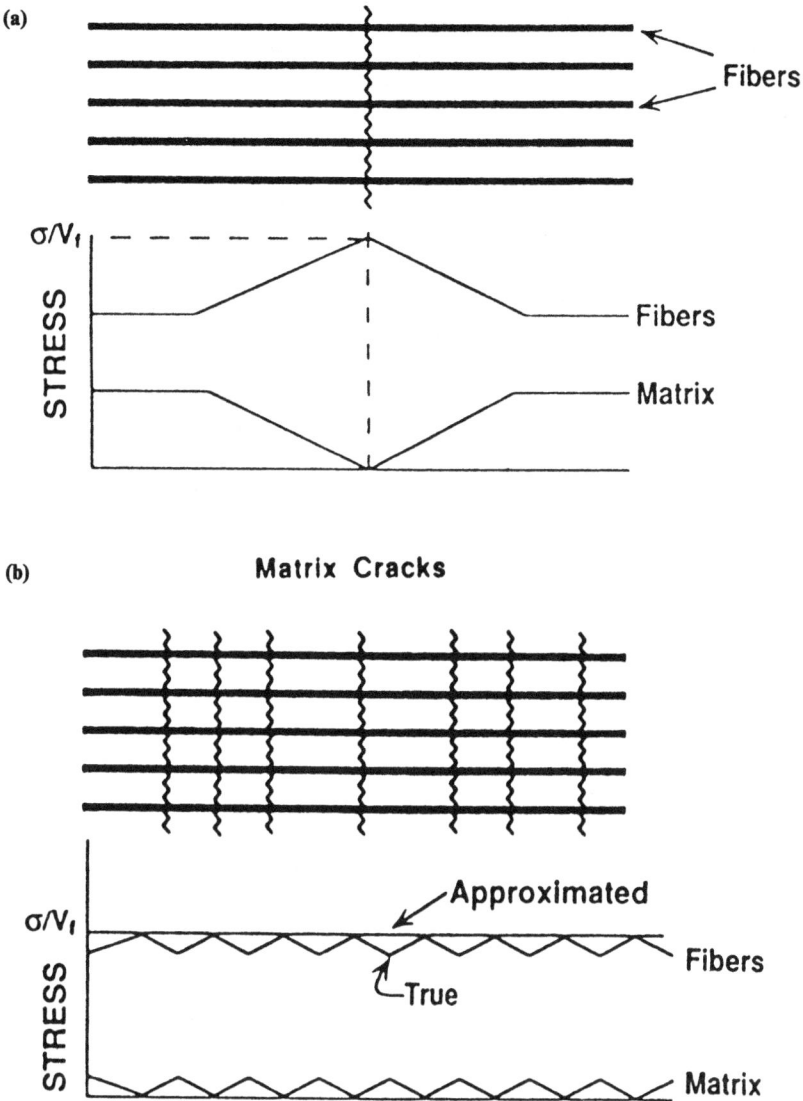

Figure 6.17 Fibre and matrix stress along composites reinforced with continuous fibres for (a) Type IA composites, $v_f \gg v_c$; (b) Type IB, $v_f \approx v_c$ (After Curtin, 1993).

stress. Since few cementitious composites would be of sub-type IB, which is more relevant to ceramic composites, this sub-type is not covered here and the reader is referred to the excellent review by Phoenix (1993).

The presentation of the statistics in this section assumes that there is equal load sharing across a composite in common with the classic bundle theory (Daniels, 1945, 1989). However, the 'characteristic bundle' does not occupy the whole length of the composite, but only a length over which there is load transfer between the fibres and matrix near a crack. In the most recent advances in the statistics of brittle matrix composites, the load is not considered to be shared equally across the composite and 'chain of bundles' both across and along the length of the composite are considered (Phoenix and Raj, 1992; Curtin, 1993b; Phoenix, 1993). However, in this discussion the characteristic bundle is assumed to encompass the width of the specimen.

The stress in the fibres is a maximum at the matrix crack and, if the strength of the fibres is deterministic, then failure occurs at this point. There is no fibre pull-out and the toughness of the composite comes from the work of debonding and the fracture energy of the matrix and fibres. For Type IA continuous fibre reinforced composites, fibre pull-out can only contribute to the toughness if there is a distribution in the strength of the fibres. Thus, it can be beneficial if the reinforcing fibres have a range of strengths. In contrast pull-out will occur in Type IB composites even if their strength is deterministic (Sutcu, 1989; Curtin, 1991).

Assuming a constant interface shear stress, τ_b, the stress in a fibre a distance z from a matrix crack decreases linearly and can be written approximately[2] as (Thouless and Evans, 1988; Sutcu, 1989)

$$\sigma_f(z) = \frac{4\tau_b}{d_f}[L_s - z] \qquad (6.84)$$

where d_f is the diameter of the fibres and L_s is a sampling length defined by

$$L_s = \frac{\sigma_{fmax}d_f}{4\tau_b} \qquad (6.85)$$

Broken fibres a distance greater than L_s from the matrix crack cannot be pulled out at the current stress level. The probability of a fibre failing under a maximum fibre stress, σ_{fm}, can be obtained from the Weibull distribution (eqn 6.80) and is given by (Thouless and Evans, 1988)

$$P(\sigma_{fm}) = 1 - \exp-\left[\frac{1}{m+1}\left(\frac{\sigma_{fm}}{\Sigma_c}\right)^{m+1}\right] \qquad (6.86)$$

The mean overall bundle stress, $\mu(\sigma_{fm}) = \sigma_{fb}/v_f$, at the matrix crack as $N \to \infty$ is given by (Phoenix, 1993)

$$\mu(\sigma_{fm}) = \sigma_{fm}[1 - P(\sigma_{fm})] + \frac{\langle L(\sigma_{fm})\rangle}{L_s}P(\sigma_{fm}) \qquad (6.87)$$

---

[2] Obviously the fibre stress does not decrease to zero, but to

$$\sigma_{fb} = \frac{\sigma_{fb}(0)}{\left[1 + \dfrac{E_m}{E_f}(1 - v_f)\right]}$$

However, the approximation leads to considerable simplification and does not significantly affect the results (Thouless and Evans, 1988).

where $L(\sigma_{fm})$ is the pull-out distance of a broken fibre and $\langle \rangle$ denotes the expected value. In eqn 6.87, the first term is the stress taken by the unbroken fibres and the second is the pull-out stress on the broken fibres. Phoenix (1993) shows that the mean bundle stress, $\mu(\sigma_{fm})$, is given approximately by

$$\mu(\sigma_{fm}) \approx \sigma_{fm}\left[1 - \frac{m+1}{m+2}P(\sigma_{fm})\right] \tag{6.88}$$

The fibre stress, σ_{fm}^*, at which the mean bundle stress reaches its maximum value is given by

$$\sigma_{fm}^* = \Sigma_c\left(\frac{m+1}{m}\right)^{1/(m+1)} \tag{6.89}$$

and an approximation to the maximum mean bundle strength, μ^*, is given by

$$\mu^* \approx \sigma_{fm}^*\left[\frac{1}{m+2} + \left(\frac{m+1}{m+2}\right)\exp-\frac{1}{m}\right] \tag{6.90}$$

Equation 6.90 gives the asymptotic maximum mean bundle strength at $N = \infty$, as discussed in section 6.5.1, this asymptotic value is only approached very slowly. Phoenix (1993) gives the correction to this value that reduces the error to $O(N^{-1/3})$.

The average pull-out length at final failure, L_{pav}, at a single matrix crack is (using the definition of Δ_c) given by (Sutcu, 1989)

$$L_{pav} = \frac{\alpha}{2}(2\pi)^{1/(m+1)}\Delta_c \tag{6.91}$$

where the coefficient α is given within $\pm 5\%$ for $2 < m < 60$ by the simple formula

$$\alpha = 1.1\frac{m}{(m+1)^2} \tag{6.92}$$

The average pull-out length obviously decreases to zero as m becomes large. The work of fibre pull-out can be obtained by integrating the fibre force, as a function of the pull-out, over the pull-out length. The work of pull-out per unit area of composite fracture surface, for a single matrix cracking, is given by (Sutcu, 1989)

$$w_p = \frac{\beta v_f}{4}(2\pi)^{2/(m+1)}\Sigma_c\Delta_c \tag{6.93}$$

A simpler non-dimensional version of eqn 6.93 can be obtained by non-dimensionalizing the work of pull-out of fibres, whose length L_{s0} is the sampling length of a fibre whose strength is σ_0 (see eqn 6.85) and is given by

$$\bar{w}_p = w_p\Big/\left(\frac{\sigma_0 L_{s0}}{2}\right) = \beta\left(\pi\frac{l_0}{L_{s0}}\right)^{2/(m+1)} \tag{6.94}$$

The coefficient β is given within $\pm 2.5\%$ for $m > 3$ by the simple formula (Sutcu, 1989)

$$\beta = \frac{2.14(m-1)}{m^2(m+2)}[1 - \exp -0.387m] \qquad (6.95)$$

As already discussed, the pull-out contribution to the toughness decreases to zero as m tends to infinity. It should be noted that Sutcu's (1989) analysis of the work of pull-out, when closely spaced multiple cracking occurs in Type IB composites, does not appear to be as accurate as that given by Curtin (1991, 1993a).

6.5.3   Type II discontinuous fibre reinforced cementitious materials

In most practical cementitious composites reinforced with discontinuous fibres there is little fibre fracture and fibres simply pull out. The average post matrix cracking stress-displacement relationship for plain fibres during pull-out has already been discussed for both Coulomb and constant interfacial friction (see section 3.9.5).

Cementitious materials are often reinforced by fibres whose Young's modulus is less than that of the matrix, for example, cement mortar reinforced with cellulose or polypropylene fibres. The rule of mixtures predicts no increase in strength for a Type II composite reinforced by such fibres. However, some reinforcement does occur in materials that contain pores and other defects, because the fibres can bridge flaws in the mortar and reduce the effective stress intensity factors at the tips of the flaws (Andonian *et al.*, 1979). The fibres bridging flaws will have a higher strain than the nominal strain on the composite and so can carry higher stress even if they are of lower elastic modulus than the matrix. The fibres will also make the fracture more stable.

The possible statistical effects on fracture of fibre reinforced cementitious composites can be examined from a simple extension to Weibull's theory, as interpreted by Hunt and McCartney (1979) and presented in section 6.2.1 (Hu *et al.*, 1991). The statistics of the fracture of a cementitious composite are presented for a two-dimensional body loaded in simple tension. Flaws are assumed to be distributed randomly with an areal density, ρ, that can be modelled as equivalent through-the-thickness cracks whose length, a, has the Pareto distribution given by eqn 6.27. The matrix is assumed to be brittle and have a fracture toughness, K_{Ic}. Thus, the unreinforced strength of the matrix is given by the Weibull distribution. The fibres are assumed to be aligned, to have a fixed length, l, and to be randomly distributed with an areal density, ρ_f. It is assumed that the fibres pull-out rather than fracture. If the flaws are small, the fracture will be unstable after they have opened only slightly. Thus, the average force exerted by a bridging fibre is approximately its maximum value, F_f, given by

$$F_f = \frac{\pi l d_f \tau_b}{4} \tag{6.96}$$

If the fibre density, ρ_f, is high, the number of fibres bridging a crack will be close to the expected number, $\rho_f \pi a l$. The average maximum stress exerted by the fibre stress to close a crack is given by

$$\sigma_{fb} = \frac{\rho_f \pi l^2 d_f \tau_b}{4} = v_f \tau_b \left(\frac{l}{d_f} \right) \tag{6.97}$$

Thus, an upper bound to the average strength of the composite is given by

$$\sigma = \sigma_m + \sigma_{fb} \tag{6.98}$$

where σ_m is the strength of the matrix alone. If the density of fibres is moderate, the number bridging any crack will vary. The largest crack may not be the most critical and there is the possibility of stable crack growth due to a statistical R-curve.

The number of fibres, n, bridging a crack will vary so that the bridging stress (assuming the fibres are smeared over the surface) can be written as

$$\sigma_{fb}(n, a) = \frac{n F_f}{at} \tag{6.99}$$

where t is the thickness of the plate. The maximum number of bridging fibres that will enable a crack to propagate under a stress σ is given by

$$n_{max} = \frac{at}{F_f} \left[\sigma - K_{Ic} \left(\frac{2}{a\pi} \right)^{1/2} \right] \tag{6.100}$$

The probability of finding n fibres bridging a crack is given by Poisson's distribution. If $q(a)\,da$ is the probability of finding a crack of length between a and da in the volume dV, the probability of finding such a crack that has less than n_{max} bridging fibres and can therefore propagate is given by

$$\sum_{n=0}^{n_{max}} q(a) \frac{(\rho_f la)^n}{n!} \exp(-\rho_f la) \, da \, dV \tag{6.101}$$

Hence, the probability for first cracking is given by

$$P_{fc}(\sigma) = 1 - \exp\left[-V \int_{a(\sigma)}^{\infty} q(a) \sum_{n=0}^{n_{max}} \frac{(\rho_f la)^n}{n!} \exp(-\rho_f la) \, da \right] \tag{6.102}$$

If there are less than n_{max} fibres bridging a crack of size a, the crack can grow. However, it may be arrested if the number of fibres, $n + \Delta n$, bridging the crack after an extension Δa, is greater than N_{max}, given by

$$N_{max} = \frac{(a + \Delta a)t}{F_f} \left[\sigma - K_{Ic} \left(\frac{2}{a\pi} \right)^{1/2} \right] \tag{6.103}$$

that will cause the crack to arrest. Thus, for a given stress σ, the probability, $Q(a, \Delta a) \, da \, dV$, of finding a crack of size a in dV that can initiate and propagate further after it has grown to $a + \Delta a$, is given by

$$Q(a, \Delta a) = q(a) \exp[-\rho_f l(a + \Delta a)] \sum_{n=0}^{n_{\max}} \frac{(\rho_f la)^n}{n!} \sum_{\Delta n=0}^{N_{\max} - n} \frac{[\rho_f l \Delta a]^{\Delta n}}{\Delta n!} \quad (6.104)$$

It can be shown that there must be a finite non-zero value of Δa for which the integral

$$P(\sigma, \Delta a) = 1 - \exp\left[-V \int_{a(\sigma)}^{\infty} Q(a, \Delta a) \, da\right] \quad (6.105)$$

has a minimum. Hence, if Δa_c is that value, the probability of failure of the composite is given by

$$P(\sigma, \Delta a_c) = 1 - \exp\left[-V \int_{a(\sigma)}^{\infty} Q(a, \Delta a_c) \, da\right]$$

$$= P(\sigma, \Delta a)_{\min} \quad (6.106)$$

Since $P(\sigma, \Delta a)_{\min}$ gives the maximum stress level for a fixed failure probability, eqn 6.106 indicates the probability of the unstable fracture of a multiple crack system. Therefore, Δa_c is the statistically averaged maximum stable crack growth increment.

The most dubious assumption, in the above derivation of the strength of short fibre reinforced brittle matrices, is the smearing of the fibre bridging stresses over the crack surfaces. To test whether the assumption is justified, the theoretical strengths were compared with a computer simulation (Hu et al., 1986b, 1991). The composite chosen is representative of cellulose fibre reinforced cement mortar whose properties are given in Table 6.2. For comparison with computer simulation, a two-dimensional version of the above was used (Hu et al., 1991). A specimen $100 \times 100 \times 0.1 \, \text{mm}$ was analysed. The median first crack strength was calculated from eqn 6.102. The median strength of the composite, for a particular value of Δa, is

Table 6.2 Properties of a short fibre reinforced mortar used in statistical analysis

	Fracture toughness of matrix	$K_{Ic} = 0.6 \, \text{MPa}\sqrt{\text{m}}$
Matrix properties:	Reference crack size	$a_0 = 2 \, \text{mm}$
	Weibull modulus	$m = 8$
	Flaw density	$\rho = 0.003 \, \text{mm}^{-2}$
Fibre properties:	Fibre diameter	$d_f = 0.1 \, \text{mm}$
	Fibre length	$l = 5 \, \text{mm}$
	Fibre bond strength	$\tau_b = 4 \, \text{MPa}$
	Fibre densities	$\rho_f = 0.1, 0.2 \, \text{mm}^{-2}$
	Fibre volume fraction	$v_f = 0.393, 0.785$

Table 6.3 Median strengths of short fibre reinforced mortar

Fibre density (mm^{-2})	Expected strength from eqn 6.98 (MPa)	Median first cracking strength (MPa)	Median ultimate strength (MPa)	Average stable crack growth (mm)
0	6.71	6.71	6.71	0
0.1	14.56	8.51	9.60	10
0.2	22.42	9.67	13.2	40

found by an iterative routine and Δa increased until the ultimate strength is obtained at Δa_c. The crack growth before the ultimate strength is reached is large, 10 mm for $\rho_f = 0.1\,mm^{-2}$ and 40 mm for $\rho_f = 0.2\,mm^{-2}$, but the increase in strength over the first cracking strength is only moderate. Table 6.3 gives the median first cracking stress and ultimate strength of the simulated composite. Note that the ultimate strength of a composite reinforced with a moderate volume fraction of fibres is considerably less than that given by eqn 6.98 for high volume fractions.

Sixty computer experiments were also run for the composites given in Table 6.2 (Hu et al., 1986b). A typical flaw and fibre distribution (the actual number of fibres has been reduced by a factor of 10 for clarity) before loading is shown in Figure 6.18a. Finite width corrections for the cracks were used to calculate the stress intensity factors due to the applied stress. Finite length corrections were not used. The bridging fibres were modelled as point forces located symmetrically halfway along the shortest half of the fibre to avoid the problem of an infinite stress intensity factor when a fibre is at the crack tip. The stress intensity factors due to the bridging fibres have been calculated from the expression for point forces given by Tada et al. (1973) for cracks in infinite plates with a finite width correction (Hu et al., 1991). The cracks present at final failure are shown in Figure 6.18b. The distribution in first cracking and final failure for the two fibre densities are shown in Figure 6.19. The Weibull moduli for these distributions is shown in Table 6.4. The stable crack growth is caused by a statistical R-curve and, as discussed in section 6.2.2, this increases the effective Weibull modulus of the material. Since for high volume fractions of fibres there is little stable crack growth prior to fracture, the statistical R-curve is less pronounced at higher volume fractions. Thus, for $\rho_f = 0.2\,mm^{-2}$ the effective Weibull modulus for final fracture, while much larger than that of the matrix, is less than that for $\rho_f = 0.1\,mm^{-2}$. The probabilities of failure for the two different fibre densities are compared with the theoretical predictions in Figure 6.20. The failure probability of the unreinforced mortar is indicated by the lines W_1 and the high density limiting probabilities are indicated by the lines W_2. The median strength of the composites is underestimated by the theory by about 15%. Nevertheless, the smeared fibre theoretical model does give a simple method that predicts the strength distribution reasonably well.

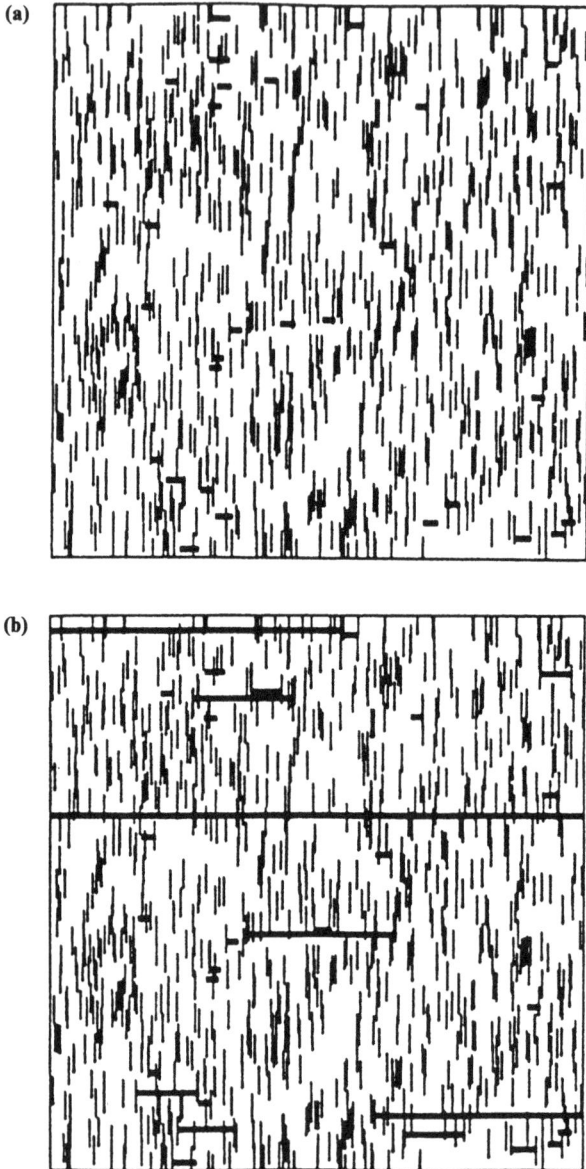

Figure 6.18 Crack development in a sample 100×100 mm two-dimensional fibre reinforced mortar: (a) before loading; (b) at final failure.

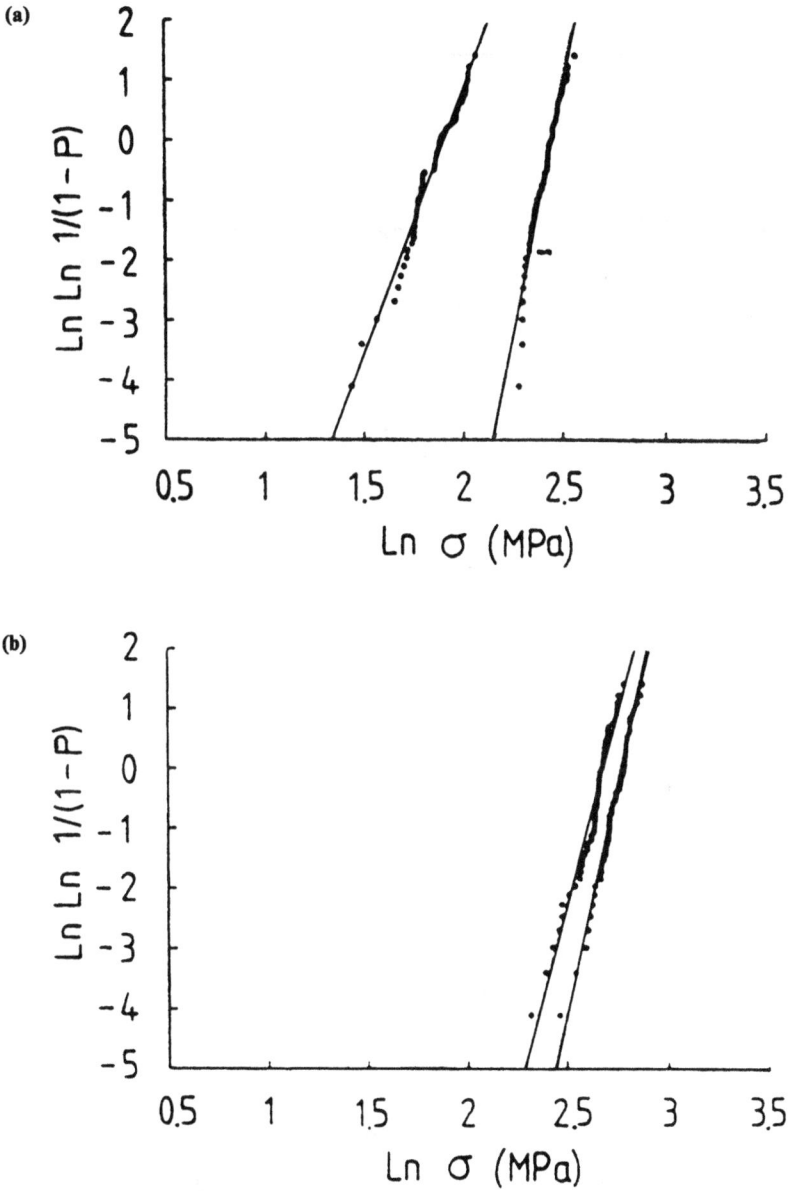

Figure 6.19 First crack and final failure strength distributions: (a) $\rho_f = 0.1\,\text{mm}^{-2}$; (b) $\rho_f = 0.2\,\text{mm}^{-2}$.

Table 6.4 Weibull moduli for short fibre reinforced mortar

Fibre density (mm^{-2})	Weibull modulus	
	First cracking	Final fracture
0.1	8.74	16.7
0.2	12.6	14.8

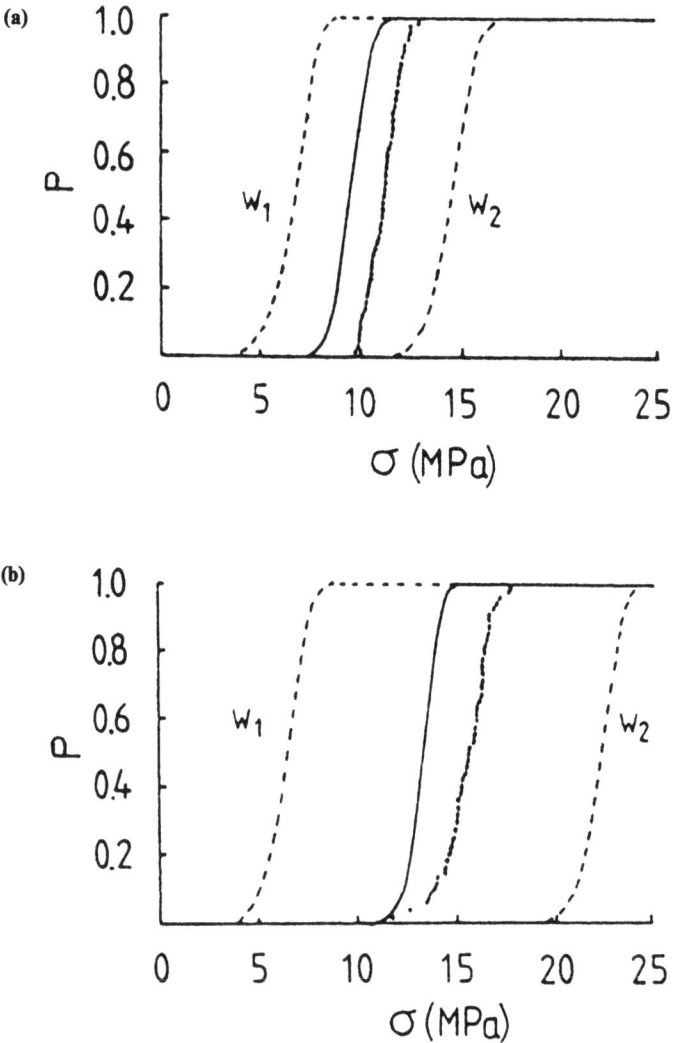

Figure 6.20 Comparison of strength distributions from theory — (eqn 6.108), and computer simulation ······: (a) $\rho_f = 0.1 \, mm^{-2}$; (b) $\rho_f = 0.2 \, mm^{-2}$.

6.6 Summary

Cementitious materials are only quasi-brittle and weakest-link statistics do not have a direct application. However, though the Weibull distribution was derived from a weakest-link concept, it can still be used in an empirical fashion for the reliability of supposedly identical components. What the Weibull distribution cannot do is predict the strength dependence on size in cementitious materials. The natural inhomogeneity of cementitious materials makes them somewhat crack tolerant. Random microcracking in cementitious materials is relatively unimportant, though eventually it does lead to localization of damage to the pupative fracture process zone. In practice the region of localization is more likely to be determined by design stress concentrations rather than material weaknesses. Therefore, the dominant volume is the region of localization, rather than the volume as a whole as is assumed in classic Weibull theory.

Reinforcement of cementitious materials with fibres makes them not only stronger but also more reliable which is often more important. Load sharing can occur between fibres as is assumed in classic bundle theory. For practical reasons useful fibre volume fraction that can be used with cementitious materials is relatively small. Thus, if the fibre bond strength or the fibre length is large and the fibres fracture before pulling out, it is paradoxically an advantage to have fibre strengths with a wide distribution in strength. Although reinforcing fibres can make the work of fracture large, such work can only be fully utilized in situations where there are large stress gradients, due to large fibre free cracks, stress concentrations, or, in small beams, bending stress. The development of an R-curve due to fibre pull-out from natural defects in uniformly stressed components is due only to the random distribution of the fibres. Thus, in fibre reinforced cementitious materials, a randomness in fibre distribution or strength can be an advantage.

7 Time-dependent fracture behaviour of cementitious materials

7.1 Introduction

It is well recognized that the strength degradation of cementitious materials and fibre cements is dependent on the loading rate as well as the time under which a sustained or cyclic load is applied (Mindess and Nadeau, 1977; Mindess, 1985; Hu *et al.*, 1989). There are two main factors which control these time-dependent strength characteristics. One factor is the flaw statistics in terms of flaw density and flaw size distribution; the other factor is the slow crack growth process determined by the chemical reactive species at the flaw tip (Beaudon, 1986; Tait and Garrett, 1986). For slow loading rates both factors determine the strength characteristics of these materials. To avoid dealing with the statistics of pre-existing flaws the conventional approach is to consider the growth of a single crack only. This is acceptable for a uniform stress field since the largest flaw is the one that determines the strength according to the Weibull theory (1951). However, many cementitious structural applications have a non-uniform stress field and the single crack approach is not appropriate. A time-dependent statistical fracture mechanics theory is necessary and is given in the next section.

7.1.1 Slow crack growth in cementitious materials

Many glasses and brittle ceramics, such as aluminas and silicon nitrides, suffer slow crack growth as a result of attack by moisture or humidity in the air under the application of an external stress. Such environmental species effectively lower the surface energy of the material by chemically enhanced bond rupture (Evans, 1972; Atkins and Mai, 1985). In cementitious materials, such as cement paste, mortar and concrete, cracks grow from the flaws under stress in the presence of a wet environment (Mindess and Nadeau, 1977; Mindess, 1985; Wittmann, 1985). It is the rate or time-dependent fracture process at the flaw tip that determines the slow crack growth and hence the residual strength of the material.

The best way to describe the environmentally assisted slow crack growth is to relate the crack velocity (da/dt) to the applied stress intensity factor (K_a). A schematic log–log plot of da/dt *versus* K_a is shown in Figure 7.1 for ceramic materials. There are three regions of crack growth. In Region I the

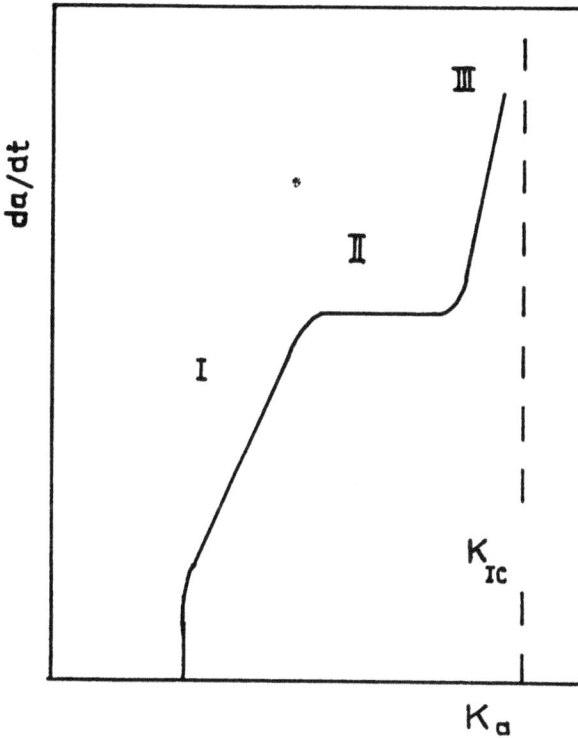

Figure 7.1 Schematic dependence of time-dependent crack growth on stress intensity factor.

crack velocity is a power law function of the crack tip stress intensity factor, that is,

$$\frac{\mathrm{d}a}{\mathrm{d}t} = AK_a^n \qquad (7.1)$$

where A and n, often called the corrosion exponent, are constants that depend on the environment-material system. In Region I crack growth is controlled by the stress enhanced chemical reaction rate. The crack velocity is independent of the applied stress intensity factor and dependent only on the rate of diffusion or mass transport controlled processes of the chemical species in Region II. The last region, Region III, has a similar power law function as in Region I, but crack growth is mechanically controlled with very little environmental effect (Atkins and Mai, 1985; Wiederhorn et al., 1982).

In cementitious materials, only Region I has been observed with few reports of Regions II and III. Hence only slow crack growth in Region I is considered in this chapter. Since cementitious materials, and in particular concrete, are very heterogeneous in structure, as distinct from ceramics and glasses, the slow crack growth parameters, A and n, are not really constant. This phenomenon is seldom found in glasses and the more brittle ceramics. For example, the microscopic fracture path in hardened cement paste is tortuous both within the hydration products and around the unhydrated cement grains which effectively bridge the crack faces (Baldie and Pratt, 1986). The effects of crack tortuosity and grain bridging become more obvious in mortars and concretes and they exert a closure stress intensity K_r on the crack tip (van Mier, 1991). Hence, the applied crack tip stress intensity K_a is reduced by K_r to an effective crack tip stress intensity factor K_t. Since it is K_t and not K_a that drives the crack, it is not surprising that A and n are not true constants. However, if K_t is used instead of K_a the log–log plot is a straight line and A and n, referred to K_t, are constants.

7.2 Modelling time-dependent crack growth in brittle materials

The strength of brittle materials in glasses, ceramics and cementitious matrices varies with the stress rate ($\dot{\sigma}_a$), cyclic fatigue and the time under which a constant stress is applied. In the first case, the material is subjected to different constant stress rates often termed 'dynamic fatigue' by the ceramics community. Cyclic fatigue refers to experiments in which the material is under the application of a repeated stress of given magnitude and test frequency. Fracture that occurs with time under a constant stress is often called 'static fatigue' or 'delayed fracture'. All these tests involve an element of time and the strength results obtained simply reflect the time-dependent crack growth behaviour described by eqn 7.1.

7.2.1 Conventional single crack theory

The conventional approach to the prediction of the time-dependent strength of brittle materials neglects the flaw statistics and considers only the growth of a crack from a single flaw size, a_i, to a critical flaw size, $a_f = 1/\pi(K_{Ic}/\sigma\phi)^2$, whose rate is governed by the fundamental environment assisted crack growth law for the particular material concerned (Atkins and Mai, 1985). Assuming that there is no crack tip shielding effect eqn 7.1 becomes

$$\frac{\mathrm{d}a}{\mathrm{d}t} = A(\sigma\phi\sqrt{\pi a})^n \tag{7.2}$$

where σ is the nominal stress at the flaw and ϕ is a geometrical form factor. A flaw of size a_i will grow under uniform stress to a size a in a time, t, given by

$$\int_{a_i}^{a} \frac{\mathrm{d}a}{a^{n/2}} = A(\sigma\phi\sqrt{\pi})^n t \tag{7.3}$$

so that

$$t = \frac{2}{(n-2)A(\sigma\phi\sqrt{\pi})^n} \left[\frac{1}{a_i^{(n-2)/2}} - \frac{1}{a^{(n-2)/2}} \right] \tag{7.4}$$

If $(a_i/a_f)^{n/(2-1)} \ll 1$ the lifetime, t_f, for a specimen containing a flaw of size a_i under static fatigue is given by

$$t_f \sigma^n = \frac{2a_i^{(1-n/2)}}{(n-2)(\phi\sqrt{\pi})^n A} \qquad (7.5)$$

Evans and Wiederhorn (1974) have combined the lifetimes given by eqn 7.5 with the Weibull inert strength distribution given by eqn 6.5, rewritten here as

$$P = 1 - \exp{-\left(\frac{\sigma}{\sigma_*}\right)^m} \qquad (7.6)$$

where $\sigma_* = \sigma_0/\rho V$ is the normalizing parameter, to obtain the probability of finding a flaw bigger than a_i, on the assumption that the flaw that causes fracture in an inert environment also causes time-dependent fracture. Both m and σ_* can be determined from inert strength experiments. Hence, Evans and Wiederhorn (1974) obtained expressions for (a) static fatigue or constant applied stress, (b) dynamic fatigue or constant stress rate and (c) cyclic fatigue as follows.

Static fatigue or constant applied stress. The relationship between applied stress, σ_a, and the time to failure, t_f

$$t_f \sigma_a^n = \lambda_s^s \qquad (7.7)$$

where

$$\lambda_s^s = \frac{2\sigma_*^{n-2} \left[\ln\left(\frac{1}{1-P}\right) \right]^{(n-2)/m}}{A(n-2)\phi^2 \pi K_{Ic}^{n-2}} \qquad (7.8)$$

The time to failure, t_f, is distributed according to the Weibull distribution with a modulus m_s^* given by

$$m_s^* = \frac{m}{n-2} \qquad (7.9)$$

Dynamic fatigue or constant stress rate. Similarly the fracture strength, σ_f, under constant stress rate, $\dot{\sigma}_a$, experiments is given by

$$\sigma_f^{n+1} = \dot{\sigma}_a \lambda_d^s \qquad (7.10)$$

where

$$\lambda_d^s = (n+1)\lambda_s^s \qquad (7.11)$$

The strength probability for dynamic fatigue has the same form as a Weibull distribution and the failure probability, $P(\dot{\sigma}_a)$, for any particular stress rate can be written as

$$P(\sigma_f) = 1 - \exp{-\left(\frac{\sigma_f}{\sigma_*^d}\right)^{m_d^*}} \tag{7.12}$$

where

$$m_d^* = m\frac{(n+1)}{(n-2)} \tag{7.13a}$$

$$\sigma_*^d = (\dot{\sigma}_a C_T)^{1/(n+1)} \tag{7.13b}$$

and

$$C_T = \frac{2(n+1)}{(n-2)A\phi^2\pi}\left(\frac{\sigma_*}{K_{Ic}}\right)^{n-2} \tag{7.13c}$$

Clearly the effective Weibull modulus m_d^* is always bigger than m in the presence of slow crack growth. This dependence of the Weibull modulus on the stress rate has been reported widely. Physically, this implies that all the flaws will grow to a more uniform size distribution so that there is less scatter in the fracture strength and the Weibull modulus becomes larger.

Cyclic fatigue. The single crack approach to cyclic fatigue, both rotational bending and reversed bending, has also been given by Evans and Fuller (1974). It is assumed that the material does not suffer mechanical fatigue induced damage, and that crack growth is simply a time-dependent effect. In its original form the Weibull inert strength distribution eqn 7.6 was not included. However, this can be easily done as given in the following expressions. Thus, for rotational bending, where $\sigma_a(t) = \sigma_c \sin(\omega t)$,

$$t_f \sigma_{max}^n = \lambda_c^s \tag{7.14}$$

where σ_{max} is the maximum stress,

$$\lambda_c^s = \frac{\lambda_s^s}{\psi(n)} \tag{7.15}$$

and $\psi(n) = (1/\pi)\int_0^{\pi/2} \sin^n\theta\,d\theta$ is related to the gamma function by

$$\psi(n) = \frac{1}{\sqrt{4\pi}}\frac{\Gamma[(n+1)/2]}{\Gamma[(n+2)/2]}$$

For more general mechanical fatigue where $\sigma_a(t) = \sigma_0 + \sigma_c \sin(\omega t)$ and $\zeta = \sigma_c/\sigma_0$,

$$\lambda_c^s = \frac{\lambda_s^s}{G(n,\zeta)} \tag{7.16}$$

where

$$G(n,\zeta) = \frac{(1+\zeta)^n}{2\pi}\int_0^{2\pi} f(n,\zeta)\,d\theta$$

and

$$f(n, \zeta) = [1 + \zeta \sin \theta]^n \quad \text{if} \quad [1 + \zeta \sin \theta] > 0$$
$$f(n, \zeta) = 0 \quad \text{if} \quad [1 + \zeta \sin \theta] < 0$$

The number of cycles to failure $N = f t_f$, where f is the frequency of loading, is distributed according to the Weibull distribution

$$P(N) = 1 - \exp -\left(\frac{N}{N_*}\right)^{m_c^*} \tag{7.17}$$

where

$$m_c^* = \frac{m}{n - 2} \tag{7.18a}$$

and

$$N_* = \frac{f C_T}{(n + 1) G(n, \zeta) \sigma_{max}^n} \tag{7.18b}$$

All these single-crack equations are applicable to all specimens with small flaws since the geometric factor ϕ is independent of the specimen geometry. The slow crack growth exponent, n, can be evaluated from simple static or dynamic fatigue. It is then possible to predict the lifetimes under static or dynamic and cyclic fatigue from the relationship for the parameter λ.

7.2.2 Statistical theory of time-dependent fracture

Since it is much easier to carry out flexural than uniaxial strength experiments, the statistical theory developed in this section will be for specimens subjected to pure bending only. However, the theory can be simply extended to other loading configurations. For simplicity, uniform specimens with rectangular cross-sections or circular cross-sections are considered. It is also assumed that the inherent flaws are distributed within the volume V of the material, whose size variation follows the Pareto distribution given by eqn 6.27. The density of the flaws is assumed to be small so that there is no interaction between flaws. For cementitious materials it is assumed that the environmental species will slowly diffuse through to interact with the flaws.

The single-crack theory predicts that fracture is independent of the specimen or loading conditions, but a statistical fracture mechanics approach shows that the flaw which causes instantaneous fracture is not necessarily the same as the flaw that causes time-dependent fracture except if the stress is constant. Thus, in this fuller treatment of the statistics of time-dependent fracture the growth of all flaws is studied rather than a single flaw, since crack growth alters the existing crack size distribution (Hu et al., 1988). The failure probability is obtained from the current crack size distribution

$q(a, t)$ which from eqn 6.22 is given by

$$P(\sigma_a, t) = 1 - \exp\left[-\int_V \int_{a(\sigma)}^{\infty} q(a, t) \, \mathrm{d}a \, \mathrm{d}V\right] \quad (7.19)$$

where $a(\sigma) = 1/\pi(K_{\mathrm{Ic}}/\sigma\phi)^2$. Equation 7.19 can be solved by noting that if a_i is the flaw which grew to a size a in time t then

$$q(a, t) \, \mathrm{d}a = q(a_i) \, \mathrm{d}a_i \quad (7.20)$$

Rearranging eqn 7.4, the crack of size a can be expressed in terms of the original flaw size a_i from which it grew, and the failure probability becomes

$$P(\sigma_a, t) = 1 - \exp\left[-\int_V \int_{a_i(a)}^{\infty} q(a_i) \, \mathrm{d}a_i \, \mathrm{d}V\right] \quad (7.21)$$

Hu et al. (1988) have used eqn 7.21 to obtain a more exact statistical strength of beams. Their results for static fatigue, dynamic fatigue and cyclic fatigue are summarized below. Details of the calculation together with other results are given by Hu et al. (1988).

Static fatigue or constant applied stress (rectangular cross-section). Under a constant stress, σ_a, the crack size, a, at any time, t, can be expressed in terms of the original flaw size a_i by

$$a = a_i\left[1 - \frac{n - 2}{2} A(\sigma\phi\sqrt{\pi})^n t a_i^{(n-2)/2}\right]^{2/(2-n)} \quad (7.22)$$

Thus, with the assumption $AK_{\mathrm{Ic}}^{n-2}\sigma_a^2 t_f \gg 1$, the parameter λ_s for a true statistical theory is given in terms of the single crack parameter λ_s^s by

$$\lambda_s = t_f \sigma_a^n = \lambda_s^s\left[\frac{mn + n - 2}{(m+1)(n-2)}\right]^{(n-2)/m} \quad (7.23)$$

Dynamic fatigue or constant stress rate (rectangular cross-section). Under a constant stress rate, $\dot{\sigma}_a$, the crack size, a, at any time, t, can be expressed in terms of the original flaw size a_i by

$$a = a_i\left[1 - \frac{n - 2}{2(n+1)} A(\dot{\sigma}t\phi\sqrt{\pi})^n t a_i^{(n-2)/2}\right]^{2/(2-n)} \quad (7.24)$$

Thus, the parameter λ_d for a true statistical theory, assuming that $AK_{\mathrm{Ic}}^{n-2}\dot{\sigma}_a^2 t_f^3 \gg 1$, is given in terms of the single crack parameter λ_s^s by

$$\lambda_d = \frac{\sigma_f^{n+1}}{\dot{\sigma}_a} = (n+1)\lambda_s = (n+1)\lambda_s^s\left[\frac{mn + n - 2}{(m+1)(n-2)}\right]^{(n-2)/m} \quad (7.25)$$

Cyclic fatigue (circular cross-section). Under a rotational bending, $\sigma_a = \sigma_c \sin \omega t$, assuming cracks only grow under the tensile half of the stress cycle, the crack size, a, at any time, t, can be expressed in terms of the original flaw size a_i by

$$a = a_i \left[1 - \frac{n-2}{2} A(\sigma\phi\sqrt{\pi})^n t\psi(n) a_i^{(n-2)/2} \right]^{2/(2-n)} \tag{7.26}$$

Assuming that $AK_{Ic}^{n-2}\sigma_c^2\psi(n)t_f \gg 1$, the parameter λ_c for a true statistical theory is given in terms of the single crack parameter λ_s^s by

$$\lambda_c = t_f\sigma_c^n = \lambda_c^s \left[\frac{mn+2n-4}{2\pi(n-2)} \beta\left(\frac{m+1}{2}, \frac{3}{2}\right) \right]^{(n-2)/m}$$

$$= \frac{\lambda_s^s}{\psi(n)} \left[\frac{mn+2n-4}{2\pi(n-2)} \beta\left(\frac{m+1}{2}, \frac{3}{2}\right) \right]^{(n-2)/m} \tag{7.27}$$

where β is the beta function. It is commonly believed that glasses, ceramics and cementitious matrices do not suffer true cyclic-fatigue-induced damage because there is no associated plastic flow at the crack tip (Gurney and Pearson, 1948; Mai and Gurney, 1975; Tait and Garrett, 1986). However, it is not clear that there are no mechanical fatigue effects in these brittle materials. It has been found recently that ceramics and cementitious materials that possess crack-resistance curve characteristics do suffer mechanical fatigue in addition to the environment effect (Mai et al., 1992). At present it is difficult to calculate the crack velocity component due to mechanical fatigue. Clearly, lifetimes predicted from the environment effect alone as given above will be longer than the experimental data if there is a genuine mechanical fatigue. If predictions and experiments agree then the mechanical fatigue is negligible.

The time-dependent strength predictions derived in this section differ from those obtained by earlier authors (Evans and Fuller, 1974; Jakus et al., 1978; Helfinstine, 1980). These authors assumed that the flaw which causes failure in an inert atmosphere would also cause time-dependent failure in a hostile environment. This assumption is only true for uniformly stressed solids and not for non-uniformly stressed materials under bending. In the latter case the flaw which would cause failure in an inert atmosphere is not necessarily the same as the flaw that leads to time-dependent fracture. Therefore, although probability has been linked to the lifetime predictions by previous authors (Evans and Wiederhorn, 1974), the theory is essentially that for a single flaw and is not for multiple flaws.

7.2.3 Comparison of single crack and statistical fracture theories

The method of incorporating the failure probability into the single-crack based equations is empirical. It is useful therefore to compare the lifetime predictions for single crack theory, t_f^s, with the more rigorous statistical-fracture-based predictions, t_f, given in section 7.2.2. For constant applied stresses the predicted lifetimes from the single-crack approach are always smaller than the statistical theory—for large n and small m values the difference can be as much as 60%. In cyclic bending every flaw goes through tension and compression in each cycle and the predicted lifetimes from the statistical theory are orders of magnitude less than those obtained using the single-crack approach. The physical reason for this large difference is that in the statistical theory there are always flaws subjected to the cyclic stress at any time, and if n is large the time required to failure is small. In contrast, in the single-crack theory, the flaw is subjected to the cyclic stress and crack growth only occurs in the tension half of each cycle. Therefore, if rotation bending experiments are carried out on circular cross-sectional samples the data will provide a 'litmus test' for the relative accuracy of the two theories.

The prediction of lifetimes due to cyclic fatigue, t_{fc}, can be obtained from the lifetime data, t_{fs}, due to sustained stresses provided that there is no fatigue-induced damage at the flaw tip. In rotation bending of beams whose cross-section is circular the single crack theory (Evans and Fuller, 1974) gives

$$\alpha^s = \frac{t_{fc}^s}{t_{fs}^s} = \frac{1}{\psi(n)}\left(\frac{\sigma_a}{\sigma_c}\right)^n \tag{7.28}$$

From the more exact statistical theory the ratio is (Hu et al., 1988)

$$\alpha = \frac{t_{fc}}{t_{fs}} = \alpha^s \left[\beta\left(\frac{mn+n-2}{2(n-2)},\frac{3}{2}\right)\left(\frac{mn+2n-4}{2\pi(n-2)}\right)\right]^{(n-2)/m} \tag{7.29}$$

The lifetime ratios given by eqn 7.28 or 7.29 are the same whether the flaws are distributed over the surface or within the volume. Table 7.1 compares the ratio of lifetimes as given by eqns 7.28 and 7.29 when $\sigma_a = \sigma_c$. According to the

Table 7.1 Ratio of predicted lifetimes for rotation bending to constant sustained stress

m	n	Single-crack theory α^s	Statistical fracture theory α
5	10	8.13	4.15×10^{-1}
5	20	11.40	1.61×10^{-2}
5	40	16.00	1.87×10^{-5}
5	80	22.50	1.81×10^{-11}
10	10	8.13	1.40
10	20	11.40	2.39×10^{-1}
10	40	16.00	5.07×10^{-3}
10	80	22.50	1.63×10^{-6}
15	10	8.13	2.26
15	20	11.40	6.84×10^{-1}
15	40	16.00	4.53×10^{-2}
15	60	22.50	1.42×10^{-4}
20	10	8.13	2.94
20	20	11.40	1.22
20	40	16.00	1.50×10^{-1}
20	60	22.50	1.64×10^{-3}

single-crack theory the lifetime under cyclic loading is always larger than that under constant stress. The ratio of the lifetimes, α^s, only depends on the corrosion exponent n and is independent of the Weibull modulus m. On the other hand, the true statistical fracture theory shows that the lifetime under cyclic loading can be smaller than that under constant stress if the corrosion coefficient, n, is large or the Weibull modulus, m, is small. For many brittle materials such as glasses ($m \approx 5$, $n \approx 20$) and ceramics ($m \approx 20, n \approx 50$) the cyclic fatigue lifetimes are always less than the static fatigue lifetimes. The experimental results of Gurney and Pearson (1948) and Williams (1956) support the theoretical prediction of the true statistical theory for these materials.

7.2.4   Time-dependent creep strain

There are two components to the creep strain. One is caused by the homogeneous creep of the bulk material and the other is caused by the cumulative crack opening displacement of all the pre-existing flaws which extend during creep. The total creep is the sum of these two components. For practical purposes, in glasses, ceramics and cementitious materials at temperatures less than half the melting point, it is not necessary to consider the bulk material creep component. Only the component due to the growth of the pre-existing cracks has to be calculated. This concept of analysis is not new and has been applied to creep in polymers immersed in organic solvents (Wiedmann and Williams, 1975). In this case creep elongation is caused by the opening of the surface crazes which grow according to some time-dependent law similar to eqn 7.1.

Creep strain can be obtained from simple energy-balance consideration (Hu *et al.*, 1986, 1988). Consider a rectangular beam of length L and cross-section BW, containing volume-distributed flaws and subjected to a constant applied bending moment M_a. Only those flaws in the tensile half of the beam will grow and contribute to creep. Assuming that the cracks are penny-shaped with radius a, at time t the strain energy, $d\Lambda_c$, stored in an elemental volume $LB\,dy$ at a distance y from the neutral axis due to cracks whose size lies between a and $(a + da)$, is given by

$$d\Lambda_c = BL\,dy\,q(a, t)\,da \int_0^a \frac{G\pi a}{2}\,da$$

$$= BL\,dy\,q(a, t)\,da \frac{(1 - \nu^2)}{3E}\sigma^2(y)a^3 \tag{7.30}$$

where $\sigma(y) = 2\sigma_a y/W$ is the normal stress acting at a depth y below the neutral axis, and $G = (2\sigma^2(1 - \nu^2)a)/\pi E$ is the potential energy release rate. The total strain energy stored for all flaws from a to infinity is

$$\Lambda_c = \frac{BL(1 - \nu^2)}{3E} \int_0^{W/2} \left(\frac{2\sigma_a y}{W}\right)^2 \int_a^{a_c(y)} q(a, t)a^3\,da\,dy \tag{7.31}$$

where

$$a_c(y) = \frac{1}{16\pi}\left(\frac{K_{Ic}W}{\sigma_a y}\right)^2$$

is the size of the crack that will cause failure at time t. Cracks larger than a_c must be excluded from the calculation of the strain energy, because they would cause failure. Equation 7.31 can be solved numerically making use of the relationship given in eqn 7.20. The strain energy, Λ_{nc}, stored in the uncracked beam is

$$\Lambda_{nc} = \left(\frac{WBL}{6}\right)\left(\frac{\sigma_a^2}{E}\right) \tag{7.32}$$

and the total strain energy stored is

$$\Lambda = \Lambda_{nc} + \Lambda_c \tag{7.33}$$

The external work W_e performed by the bending moment M_a is given by

$$W_e = \tfrac{1}{2}M_a\theta = \frac{M_a L}{2R} = \left(\frac{WBL\sigma_a}{6}\right)\epsilon_a \tag{7.34}$$

where R is the radius of curvature of the beam and ϵ_a is the strain in the surface of the beam. Hence, the strain in the surface of the beam is given by

$$\epsilon_a(\sigma_a, t) = \Lambda\left(\frac{6}{WBL\sigma_a}\right) = \frac{\sigma_a}{E} + \Lambda_c\left(\frac{6}{WBL\sigma_a}\right) \tag{7.35}$$

Creep strain cannot be predicted from the single-crack theory.

7.2.5 Application to cementitious materials

The statistical fracture theory presented in section 7.2.3 to predict time-dependent strength characteristics of brittle materials is strictly valid only for single phase homogeneous materials. However, just as the Weibull strength theory, though strictly not applicable to heterogeneous materials, can be applied empirically to the fracture of cementitious materials, so too can the time-dependent theory for a brittle material. The extension of the time-dependent theory to two-phase materials is covered in section 7.2.6. To use the time-dependent statistical fracture theory it is necessary to determine the Weibull inert strength distribution and the environment-assisted slow crack growth law given by eqns 7.1 and 7.3 to determine the parameters σ_*, m, A and n. The fracture toughness K_{Ic} can be estimated from separate experiments.

A hardened cement paste is the most homogeneous cementitious material. The slow crack growth law

$$\frac{da}{dt} = 10^{16}K_a^{36} \quad (m/s, MPa\sqrt{m}) \tag{7.36}$$

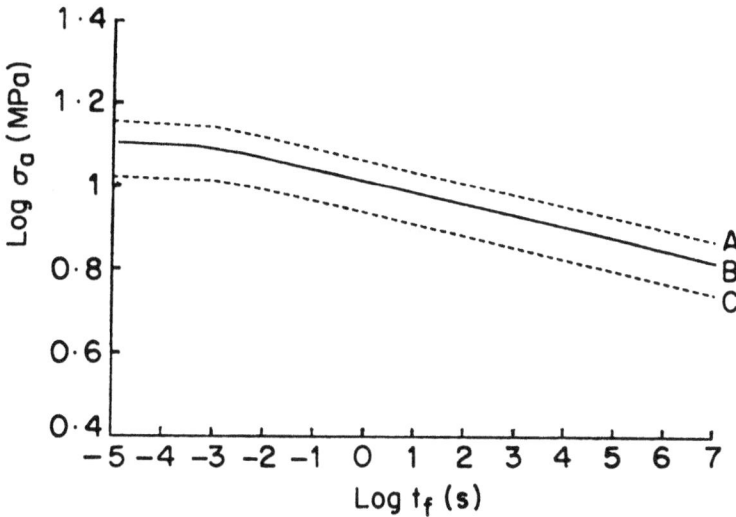

Figure 7.2 Time-dependent static fatigue strength predictions for a cement paste. Curve A: $P = 0.9$; curve B: $P = 0.5$; curve C: $P = 0.1$.

and fracture toughness, $K_{Ic} = 0.34\,\text{MPa}\sqrt{m}$, were obtained for a wet cement paste (water/cement ratio $= 0.5$, aged for 3–4 months) by Nadeau *et al.* (1974). No inert strength experiments were performed but the flexural strength was 12.7 MPa at 50% failure probability. Assuming $m = 10$ for cement paste (Hu *et al.*, 1985), σ_* is estimated from eqn 7.6 to be 13.17 MPa. Using these five parameters and assuming that $a_f/a_i \gg 1$, the time-dependent strength predictions for static fatigue and dynamic fatigue are shown in Figures 7.2 and 7.3 for rectangular cross-section specimens containing volume-distributed flaws subjected to pure bending. The curves shown are for three failure probabilities of $P = 0.1, 0.5$ and 0.9. For $t_f > 10^{-2}\,\text{s}$ and $\dot{\sigma}_a < 10^2\,\text{MPa/s}$, straight lines are obtained in the log–log plots because $a_f/a_i \gg 1$ and the predictions agree with those obtained from eqns 7.23 and 7.25. The plateaux shown in Figures 7.2 and 7.3 is because the size of the crack at failure is not large in comparison with the initial flaw size.

Creep curves can also be estimated by the method outlined in section 7.2.4, but, because of the large corrosion exponent, there is negligible creep strain due to flaw growth for this particular cement paste. The corrosion exponent for polymer modified cement paste is much smaller ($n \approx 8$) and for such pastes the creep strain is significant. In this case, the creep strain at failure at low applied stress can exceed that at high stress because cracks can grow to longer lengths at low stress final fracture.

7.2.6 Statistical time-dependent fracture in two-phase materials

Concrete and mortar can be modelled approximately as two-phase materials. There are two approximate ways to estimate the time-dependent strength

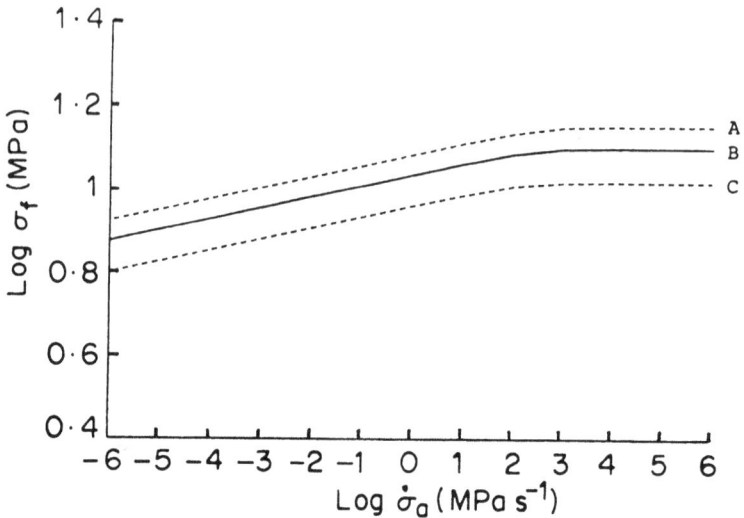

Figure 7.3 Time-dependent dynamic fatigue strength predictions for a cement paste. Curve A: $P = 0.9$; curve B: $P = 0.5$; curve C: $P = 0.1$.

behaviour of two-phase materials: one method is to develop an appropriate model to consider the effect of the second phase particles on the environment-assisted growth of the statistically distributed flaws in the matrix; the other method applies the theory developed in section 7.2.2 for a single-phase material to the two-phase material assuming equivalent slow crack growth parameters A and n for the latter material.

The analysis of the time-dependent strength of a two-phase material (Hu *et al.*, 1992) derives from the time independent two-phase theory (Hu *et al.*, 1985) presented in section 6.3.1. Only the behaviour of uniformly stressed specimens under static or dynamic fatigue are considered, but the analysis could be extended to cover non-uniformly stressed specimens. As in time independent statistics there are two models; the second phase particles are assumed to either bridge any intersecting matrix flaws or to be cut by any intersecting flaw. Only the bridging model is presented here; details of the cutting model can be found in Hu *et al.* (1992).

A number of simplified assumptions have been made in the analysis. The matrix and the particles are assumed to have the same elastic properties and the bond between them is perfect. Hence any stress variation due to the particles is neglected. The second phase particles are also assumed to be insensitive to slow crack growth. Thus when a crack meets a second phase particle it is arrested at that tip. If both crack tips meet a particle the crack is immobilized. The pre-existing flaws are assumed to have a low density so that there are no interactions between neighbouring flaws or cracks. The flaw sizes are distributed according to the Pareto distribution function (see eqn 6.27) in which the equivalent cracks are normal to the applied stress.

If the distance between two particles of size D is s, and k is the ratio of the fracture toughness of the particle K_p to that of the matrix K_m, then a crack is arrested according to eqn 6.32 where now the crack size is time-dependent. Thus, any crack that becomes unstable whose length is between s and s/k^2 will be arrested and is called a 'stabilized' crack. If one tip of a slowly growing crack encounters a second phase particle, that tip is arrested, but the other can still continue to grow. In the two-phase material, both the particles and the pre-existing cracks are assumed to be randomly distributed independently of each other. Let $f(s, t)$ denote the distribution function of the stabilized cracks formed by arresting matrix cracks, and $b(a(t), t)$ denote the distribution function of the cracks at time t. The failure probability of the two-phase material of volume V under a uniform tension at that moment is given by a time-dependent version of eqn 6.46

$$P(\sigma, t) = 1 - \exp\left[-V\left(\int_{s(\sigma)}^{\infty} f(s, t)\, ds + \int_{a(\sigma)}^{\infty} b(a, t)\, da\right)\right] \quad (7.37)$$

The first integral is obtained for the unstable fracture of the stabilized cracks, and the second integral is obtained for the unstable fracture of the non-arrested cracks. For a homogeneous material $f(s, t)$ is zero. If both $f(s, t)$ and $b(a, t)$ can be solved in terms of the initial pre-existing flaw distribution function $q(a_i)$, the failure probability $P(\sigma, t)$ and the lifetime under a given stress level or time-dependent strength for a given stress rate can be obtained.

Time-dependent fracture under a constant stress. The growth of the cracks touching less than two particles is given by eqn 7.4, where the initial flaw size a_i at $t = 0$ will extend to a at time t for a given applied stress σ_a. The expression for the crack size, a, at time t is given by eqn 7.22 and eqn 7.20 gives the relationship between the distribution of the current crack size as compared with the distribution of the initial flaw size. These two equations are required to evaluate the distribution functions $f(s, t)$ and $b(a, t)$ in eqn 7.37.

Because both particles and pre-existing cracks are randomly distributed, there are cases where they overlap. A crack is said to be bridged if it is intersected by a particle. As a result, the true crack distribution will be different to $q(a)$ as discussed in section 6.3.1. Consider a stabilized crack of size s at time t, which touches two particles. This crack can be formed by a pre-existing flaw intersecting the particles at $t = 0$; or by a flaw a_i, growing to s by time t, or by a crack growing to the range of s and s/k^2 and then propagating unstably under the applied stress until it meets two particles. The probability of finding such a crack of size s is given by the joint probability of finding a separation s for the particles and of finding a crack which satisfies eqn 6.32.

The probability that there are just two particles, of distance $(s + D)$ apart, that just touch the crack, is given by $F(s, D)$ in eqn 6.33. Similarly, the probability $H(s)\,dy$ that a pre-existing crack centred anywhere along a thin strip of width dy is obtained from eqn 6.35. Therefore, $F(s, D)H(s)$ is the probability of finding a pre-existing crack touching two particles with a distance s apart at $t = 0$. Also, the probability $Q(a_i, s)\,da_i\,dy$ of finding a crack whose size is in the range a_i to $a_i + da_i$ in the strip of width dy at $t = 0$ is given by eqn 6.39 with $a = a_i$. Further, the probability $E(s)\,dy$ that microfracture initiated from such a crack which can be arrested to form a stabilized crack of size s is given by eqn 6.40 with a replaced by a_i. Hence, $F(s, D)E(s)$ is the probability of finding a stabilized crack formed by arresting the fracture of a crack at $t = 0$.

A crack of length between s/k^2 and $a_i(s/k^2, \sigma, t)$ at $t = 0$ will extend to the range s and s/k^2 by the time t according to eqn 7.4. At time t, such a crack will be stabilized by two particles of distance s apart given by eqn 6.40. The crack will grow following eqn 7.2 and will not be affected by the encountered particles, unless both tips of the crack are touching particles, in which case, there is no slow crack growth. The probability $E^*(s, t)\,dy$ at time t that a microfracture initiated from a crack, which will be arrested by particles to form a stabilized crack of size s, is given by

$$E^*(s, t) = \int_{a_i(s/k^2, \sigma, t)}^{s/k^2} Q(a_i, s)\,da_i \tag{7.38}$$

$F(s, D)E^*(s, t)$ is therefore the probability of finding a stabilized crack formed by arresting fracture of a crack at time t. Note that at $t = 0$, $a_i = s/k^2 = a$ and $E^* = 0$.

From these three joint probabilities, the probability $f(s, t)\,ds$ of finding a stabilized crack of length between s and $s + ds$ within a unit area at time t is given by

$$f(s, t) = F(s, D)[H(s) + E(s) + E^*(s, t)] \tag{7.39}$$

The physical meaning of $H(s)$, $E(s)$ and $E^*(s, t)$ is straightforward. $H(s)$ is the contribution to stabilized cracks formed by flaws by touching the particles at $t = 0$. $E(s)$ is the contribution to stabilized cracks formed by arresting unstable microfractures at $t = 0$, and $E^*(s, t)$ is the contribution through arresting unstable microfracture by time t. Therefore, the failure probability $P_c(\sigma, t)$ of the two-phase material at time t under a tensile stress σ_a due to the final fracture of the stabilized crack is given by the time-dependent version of eqn 6.43, but with $f(s)$ replaced by $f(s, t)$ of eqn 7.39.

Now consider the failure probability of the two-phase material due to fracture initiated at cracks which are not arrested by the tougher particles. The probability $b(a_i, t)\,da$ of finding a crack of size between a_i and $a_i + da_i$

within a unit volume which is not arrested at time t is given by the time-dependent version of eqn 6.44

$$b(a_i, t) = \int_{k^2 a(a_i, \sigma, t)}^{\infty} F(s, D) Q(a_i, s) \, ds \qquad (7.40)$$

where $a(a_i, \sigma, t)$ is given by eqn 7.22. The failure probability $P_f(\sigma, t)$ of the two-phase material under tension caused by the propagation of the cracks is given by a time-dependent version of eqn 6.45. Finally, the failure probability $P(\sigma, t)$ of the two-phase material of volume V under tension by either mechanism at time t is given by

$$P(\sigma, t) = 1 - (1 - P_f)(1 - P_c)$$

$$= 1 - \exp - V \left[\int_{s(\sigma)}^{\infty} f(s, t) \, ds + \int_{a_i(\sigma, t)}^{\infty} b(a_i, t) \, da_i \right] \qquad (7.41)$$

which allows the failure probability to be calculated for a given stress, σ_a, and time t if the slow crack growth parameters, A and n, the Pareto distribution function $q(a)$, and the fracture toughness K_p and K_m are known.

Time-dependent fracture under a constant stress rate. The slow crack growth (eqn 7.1) can be evaluated for the condition that the stress rate, $\dot{\sigma}_a$, is a constant. The crack size, a, at time t can be represented in terms of the initial flaw size, a_i, at $t = 0$ by eqn 7.25. The analysis follows that given for the static fatigue case and the probability of failure is given by

$$P(\dot{\sigma}, t) = 1 - \exp \left[-V \left(\int_{s(\dot{\sigma}, t)}^{\infty} f(s, t) \, ds + \int_{a(\dot{\sigma}, t)}^{\infty} b(a_i, t) \, da_i \right) \right] \qquad (7.42)$$

7.2.7 Comparison of the time-dependent statistics of homogeneous and heterogeneous materials

In time-dependent fracture in a two-phase material the second phase particles can be crack arrestors for both unstable and environmentally assisted slow crack growth, whereas in a single phase material all the flaws grow until one becomes a critical size. To examine the effectiveness of tough second phase particles on the time-dependent behaviour, Hu et al. (1992) examined the statistics of fracture under simple tension in a hypothetical mortar where the matrix properties were based on the properties of hardened cement paste taken from Nadeau et al. (1974). The statistics were examined for the two-phase model where second phase particles are cut rather than bridged, but the results would not be very significantly different for a bridging model. There is a slight difference in the statistical theory for the matrix alone used in this section as compared with that presented in section 7.2.2. In this section the exact time for the growth of a flaw to a size a is used

Table 7.2 Material properties of hypothetical mortar (based on the properties of hardened cement paste) (Nadeau *et al.*, 1974)

Weibull modulus of matrix	$m = 8$
Flaw density	$\rho = 0.003\,\mathrm{mm}^{-2}$
Reference flaw size in $q(a)$	$a_0 = 2\,\mathrm{mm}$
Fracture toughness of matrix	$K_m = 0.34\,\mathrm{MPam}^{1/2}$
Volume of specimen	$V = \mathrm{LBW} = 100 \times 100 \times 1\,\mathrm{mm}$
Crack growth law parameters	see eqn 7.36
Particle size	$D = 0.6\,\mathrm{mm}$
Fracture toughness of particle	$K_p = 0.68\,\mathrm{MPam}^{1/2}$
Density of particles $(v_p/V = 0.1)$	$\rho_p = 0.2778\,\mathrm{mm}^{-2}$
$(v_p/V = 0.2)$	$\rho_p = 0.5556\,\mathrm{mm}^{-2}$

instead of the approximation obtained if it is assumed that $a_f/a_i \gg 1$ to be consistent with the two-phase model.

The failure probability curves $P(\sigma_a, t)$ for a constant applied stress, $\sigma_a = 3.6\,\mathrm{MPa}$, are shown in Figure 7.4. Curve A gives the failure probability of the hardened cement paste matrix, and curves B and C are for the hypothetical mortars with particulate volume fractions of 10% and 20%, respectively. It can be seen that the failure probability of the two-phase material is much smaller than the matrix material. Also, with a two-phase material the probability can reach a plateau value, if the applied stress is below an inherent threshold value, at which there is no further increase in

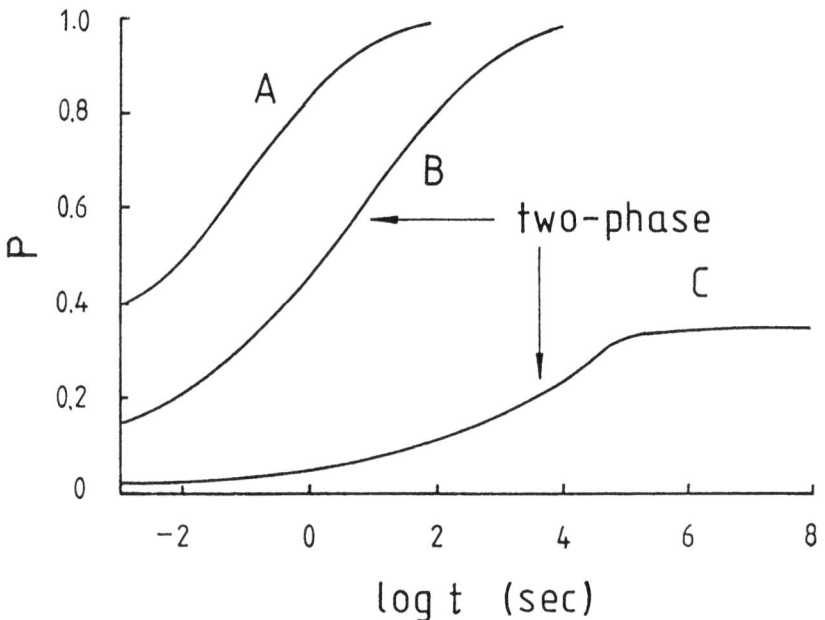

Figure 7.4 Failure probability for hypothetical mortars under static fatigue (applied stress 3.6 MPa: curve A, $v_p = 0\%$; curve B, $v_p = 10\%$; curve C, $v_p = 20\%$).

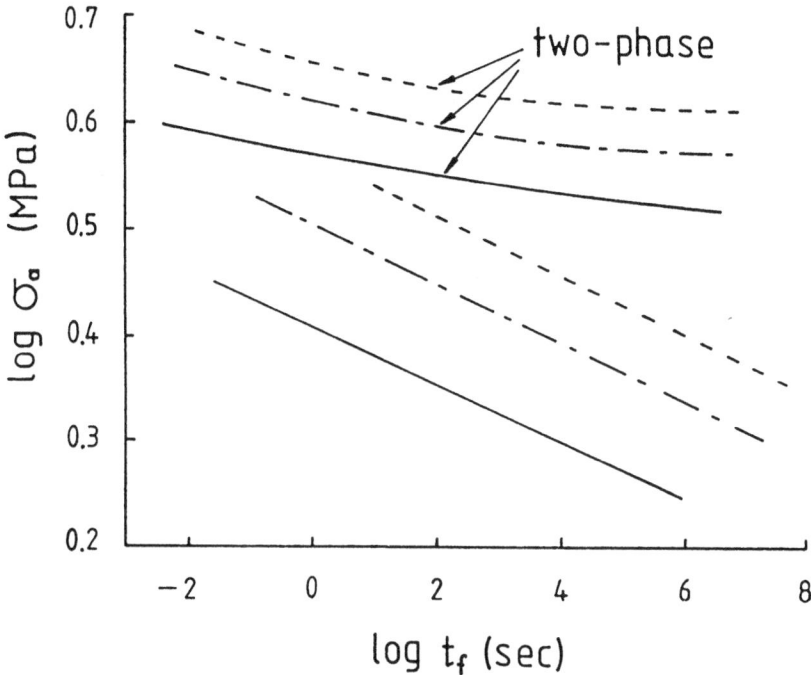

Figure 7.5 Time-dependent strength of hypothetical hardened paste, $v_p = 0$, and mortar, $v_p = 20\%$, under static fatigue, for failure probabilities of 10% —————, 50% — – — and 90% - - - - -.

failure probability with time. The plateau occurs when all the matrix cracks have been arrested by particles. Naturally, in a real material the second phase particles could not completely arrest a crack, and in time, once a crack had met a particle it would either propagate around or through it. However, second phase particles can greatly reduce the rate of further crack growth.

The statistics of single phase materials given in section 7.2.2 show that in a $\log(\sigma_a)$–$\log(t_f)$ plot, the results for static fatigue should fall on a straight line. The results for the hardened cement paste and the hypothetical mortar with $v_p = 20\%$ are shown in Figure 7.5 for three levels of failure probability 0.1, 0.5 and 0.9. For the applied stress range considered, the matrix material follows the straight line relationship predicted, but the same cannot be said of the two-phase material. However, if a straight line relationship is forced to apply, then an effective slow crack growth exponent parameter n^* can be obtained which is larger than n. This result is in accordance with the larger n values reported for a range of two-phase materials.

The $\log(\sigma_f)$–$\log(\dot{\sigma}_a)$ plot for dynamic fatigue is a straight line according to the approximate theory for single phase materials. In Figure 7.6 the log–log plots for hardened cement paste and the hypothetical mortar with $v_p = 20\%$ are shown; the hardened cement paste has a straight line relationship for

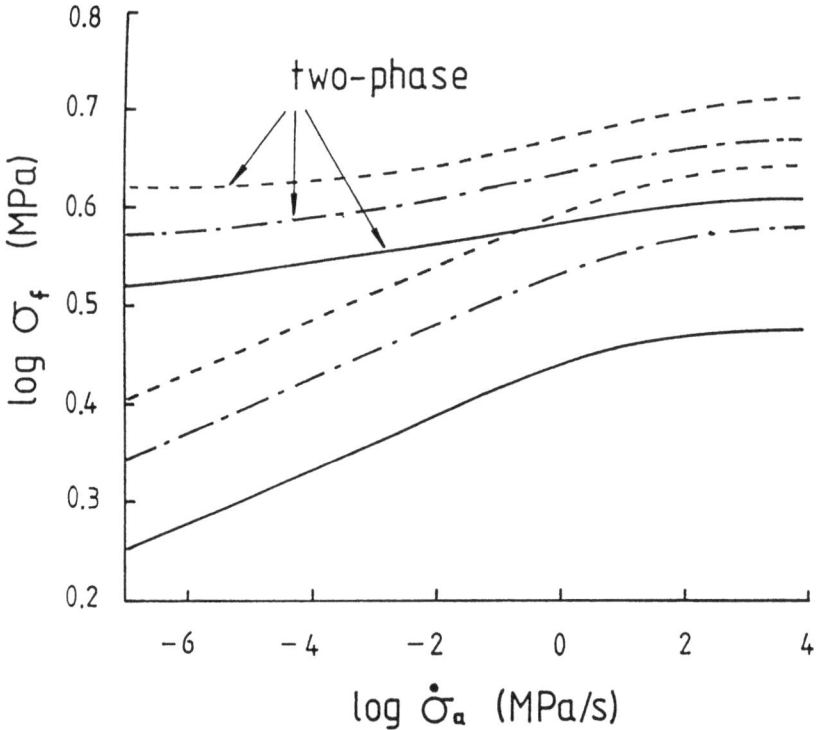

Figure 7.6 Time-dependent strength of hypothetical hardened paste, $v_p = 0$, and mortar, $v_p = 20\%$, under dynamic fatigue, for failure probabilities of 10% ——, 50% — – — and 90% - - - - -.

stress rate less than about 10 MPa/s when the assumption $a_f/a_i \gg 1$ is valid, but for larger stress rates the plot is curved. Even the plot for the hypothetical mortar is approximately straight for $\dot{\sigma}_a < 10$ MPa/s.

It is easier to perform a test at constant stress rate than one at constant stress and it has been shown that the static fatigue performance can be predicted from dynamic data using the single crack theory relationship of eqn 7.11 (which is identical to the relationship predicted by the fuller statistical treatment given in section 7.2.2). Hence, it is interesting to examine whether the same theoretical relationship applies to two-phase materials as is suggested by experimental evidence. If the best straight line is drawn through the theoretical dynamic fatigue predictions for 50% fracture probability (for stress rates less than 10 MPa/s), the single phase statistic predicts that

$$\sigma_a^{(n^* + 1)} = \dot{\sigma}_a \lambda_d^*$$

or

$$\log \sigma = \frac{1}{n^* + 1} \log \dot{\sigma} + \frac{1}{n^* + 1} \log \lambda_d^* \tag{7.43}$$

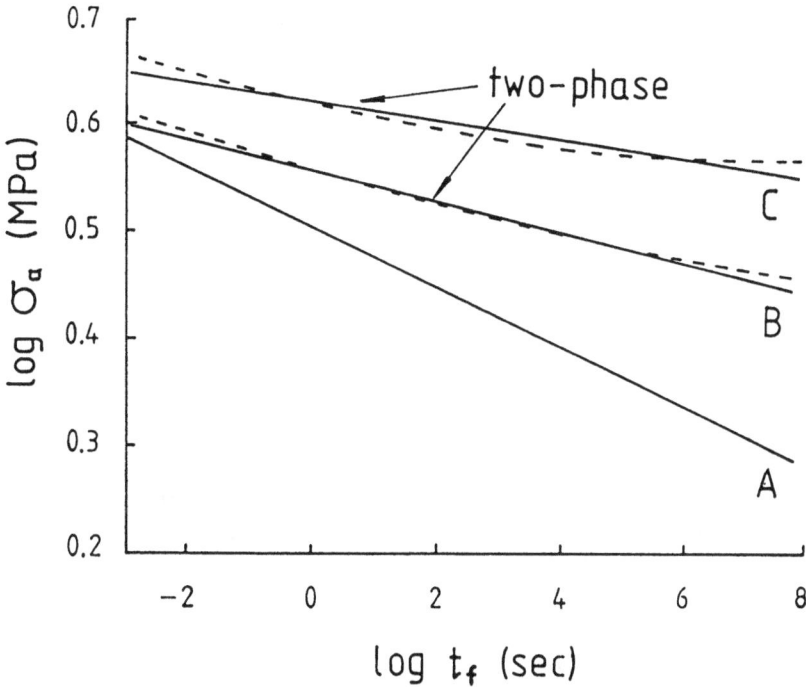

Figure 7.7 Comparison of time-dependent strength for a 50% failure probability for hypothetical two-phase materials: A, $v_p = 0\%$; B, $v_p = 10\%$; C, $v_p = 20\%$. (- - - -) exact solution; (——) single phase predictions from dynamic fatigue data.

Hence the effective corrosion coefficient n^* can be obtained from the slope and the effective dynamic fatigue parameter λ_d^* obtained from the intercept of the straight line. The effective static fatigue parameter, λ_s^*, can then be estimated from eqn 7.11. Thus, using the λ_s^* value obtained from the dynamic fatigue data, the single-phase lifetime predictions are compared with the theoretical two-phase results in Figure 7.7. The agreement is excellent. Thus, despite the fact that the tougher second phase particles interfere with the time-dependent slow crack growth and stabilize some unstable fractures of cracks, eqn 7.10 still holds over a wide range. The significance of the above results is that the lifetimes of a heterogeneous material under static fatigue can be predicted from the dynamic fatigue results which are easier to obtain experimentally.

The ability to apply the simple equations for single phase materials to predict static fatigue behaviour from dynamic fatigue data despite inhomogeneities leads to the assumption that the method may also be applied to cyclic fatigue, provided any mechanical fatigue effects are negligible. Evidence of the validity of this method comes from Tait (1984). He showed that in dry mortar the equivalent crack velocity under cyclic loading can be accurately predicted from crack velocity data obtained from static

Table 7.3 Cyclic fatigue stresses used in the tests by Saito and Imai (1983)

Series	Maximum stress (MPa)	$\zeta = \sigma_c/\sigma_0$
1	2.5	0.8072
2	2.582	0.8129
3	2.667	0.8182
4	2.748	0.8232
5	2.832	0.8280
6	2.916	0.8325

fatigue tests. Since the cyclic crack velocity is independent of cyclic frequency, provided the equivalent parameters A and n are obtained for the heterogeneous mortar, it can be treated as if it were a single-phase homogeneous material. With more complex structures, such as in concrete, it is interesting to determine if and under what conditions the single-phase statistical fracture theory can apply to predict the lifetimes.

According to the single crack statistical theory the failure probability, $P(\sigma_f)$, for dynamic fatigue is given by eqn 7.12 and the corresponding life probability $P(N)$ under general cyclic fatigue by eqn 7.17. If m_d^*, σ_*^d determined from constant stress rate tests using eqn 7.12 and m estimated from high rate inert strength experiments, the fatigue life N for a given cyclic stress can be predicted from eqn 7.17.

Direct tensile fatigue experiments were conducted by Saito and Imai (1983) on plain concrete (Portland cement, river sand, and crushed stone aggregate with a maximum size of 20 mm). A series of six fatigue tests were performed on specimens $160 \times 100 \times 70$ mm at a cyclic frequency of 40 Hz using the stresses given in Table 7.3. The plot of the probability of failure strength in fast direct tensile tests (inert strength), shown in Figure 7.8, can be fitted to the Weibull equation with a modulus $m = 20.3$. Because constant stress rate tests were not performed, a different technique has to be used to that outlined above. Equation 7.17 is first applied to the data obtained in the third test series (this series was chosen because it had the greatest number of tests). The effective Weibull modulus m^* obtained from this test series was 0.515. Knowing that $m = 20.3$, the corrosion exponent n can be calculated from eqn 7.18a to be 41.4. The reference number of cycles N^* can also be calculated. From the values of m^* and N^* obtained from series 3, the predicted lifetimes for the other series have been calculated. The experimental results, shown in Figure 7.9, are in good agreement with the predicted results.

As a further example of the prediction of cyclic lifetimes, the tensile fatigue experiments on lightweight concrete of Saito (1984), shown in Figure 7.10, are considered. The test specimens were similar to those used by Saito and Imai (1983) and the same cyclic frequency of 40 Hz was employed. The stresses used in these tests are shown in Table 7.4. Since no inert strength

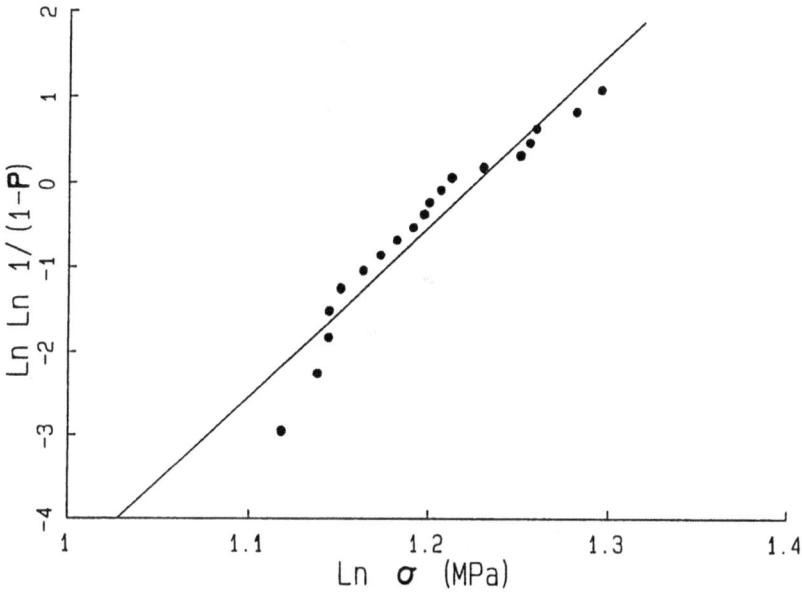

Figure 7.8 Inert strength probability for concrete. (After Saito and Imai, 1983.)

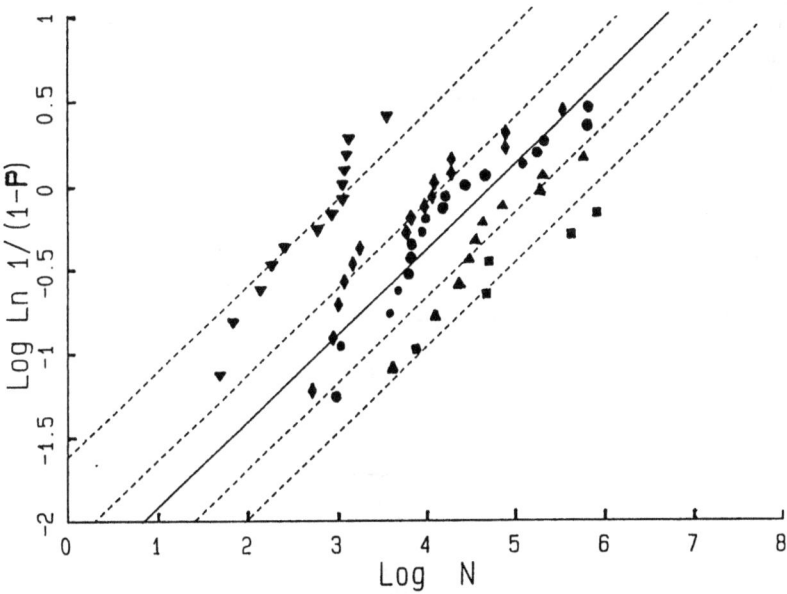

Figure 7.9 Comparison of single phase predictions of the cyclic fatigue of concrete with experimental results. (——) best fit; (- - - -) predictions. $S =$ (▼) 0.875, (◆) 0.825, (●) 0.8, (▲) 0.775, (■) 0.75 (after Saito and Imai, 1983.)

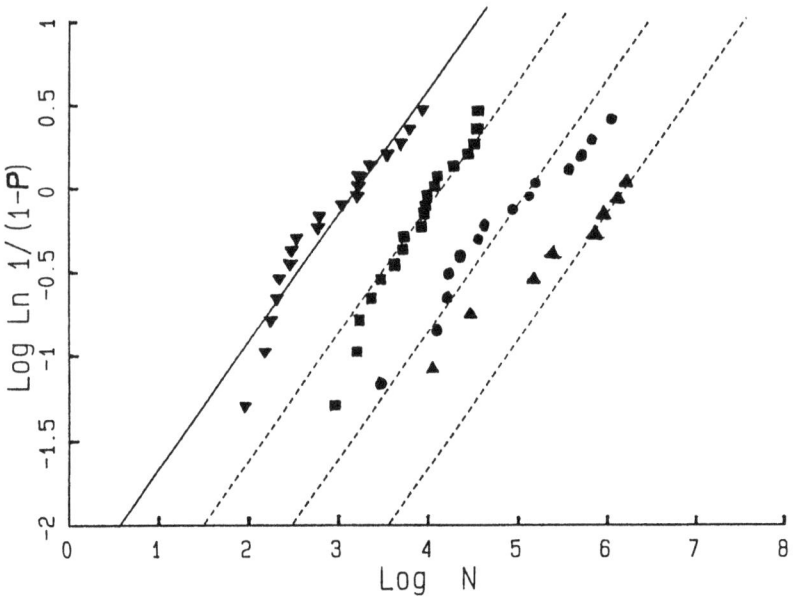

Figure 7.10 Comparison of single phase predictions of the cyclic fatigue of lightweight concrete with experimental results. (——) best fit; (- - - -) predictions. $S = (\blacktriangledown)$ 0.919, ($\blacksquare$) 0.871, ($\bullet$) 0.823, ($\blacktriangle$) 0.774 (after Saito, 1984).

data are given for the lightweight concrete either the same corrosion coefficient as that for the previous concrete experiments has to be assumed, or two series analysed simultaneously to give both m and n. In either case m and n are approximately 30 and 40, respectively. The predictions of the lifetimes shown in Figure 7.10 in each case agree very well with experimental data.

Although the statistical fracture theory originally developed for the single-phase material predicts the lifetimes of the more heterogeneous plain and lightweight concretes subjected to cyclic stress reasonably well in the two examples given above, the approach is largely empirical because cracks in these latter materials are most likely stabilized by sands and aggregates. Hence, the Weibull parameters (m and σ_*) obtained are only statistical averages and so are the parameters (A and n) in the slow crack growth

Table 7.4 Cyclic fatigue stresses used in the tests by Saito (1984)

Series	Maximum stress (MPa)	$\zeta = \sigma_c/\sigma_0$
1	2.579	0.8126
2	2.744	0.8228
3	2.904	0.8318
4	3.062	0.8398

equation. This conclusion is consistent with the discussion of the relationship between dynamic and static fatigue.

It is not possible to infer from the above experimental results that the time-dependent fracture of plain and lightweight concrete under cyclic stress is purely caused by the environment-assisted crack growth. To do so it is necessary to be able to predict the cyclic fatigue results from either constant stress rate or constant sustained stress data. While Tait (1984) did exactly this for mortar, thus showing there was no true mechanical fatigue, the picture is not so clear for concrete. Murdock and Kesler (1958) have given some evidence that true mechanical fatigue did occur in bending tests on plain concrete beams (see Figure 7.11). For a given maximum stress it can be shown (Hu et al., 1989) that, if there is no mechanical fatigue, the smaller the stress ratio, R, the longer the fatigue life. But the opposite trend is observed in Figure 7.11, indicating that the slow crack growth eqn 7.1 needs to be modified to account for the cyclic crack velocity component due to the true mechanical fatigue effect. Thus

$$\frac{da}{dt} = AK_a^n + A'\Delta K_a^{n'} \tag{7.44}$$

If the minimum stress in a cycle σ_{min} is fixed, then da/dt increases with the maximum stress σ_{max}. However, if σ_{max} is fixed and σ_{min} is reduced, the effect of the environment-assisted slow crack growth decreases and the effect of the

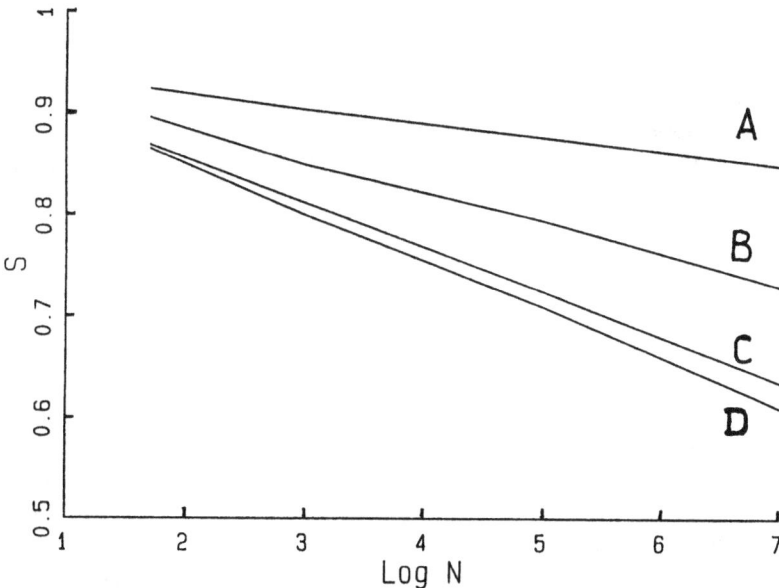

Figure 7.11 Effect of stress range ($S = \sigma_{max}/\sigma_{0.5}$) on the behaviour of plain concrete under bending fatigue at different stress ratio (R). A: $R = 0.75$, B: $R = 0.50$, C: $R = 0.25$, D: $R = 0.13$–0.18 (after Murdock and Kesler, 1958).

mechanical fatigue is enhanced. Whether the resultant da/dt increases or decreases depends on the relative magnitude of these two individual effects.

The parameter $G(n, \zeta)$ in the equation for cyclic fatigue life (see eqn 7.16) is essentially constant for $\zeta > 0.087$ $(R < 0.107)$. It is shown by eqn 7.16 that for these low R values the fatigue lifetime only depends on the maximum stress σ_{max}. Hence the fatigue data in Figure 7.11 indicate that, if $R < 0.2$, N is independent of R because $G(n, \zeta)$ is practically constant. For $R > 0.25$ the lifetime N increases with R for any given maximum stress, showing that the mechanical fatigue effect is reduced. Even so, it is not accurate to use eqn 7.16 to predict the lifetime under cyclic stress. If the fatigue lifetime N only depends on σ_{max} such as at low R ratios, no matter whether mechanical fatigue occurs, the equivalent A^* and n^* values can be used to recast eqn 7.44 to an equivalent of eqn 7.1. However, for large R ratios the second term in eqn 7.44 depends on R and not σ_{max} or K_{max} alone. The tensile fatigue data in Figures 7.9 and 7.10 are for $R = 0.1$ and both slow crack growth and mechanical fatigue effects are operative. Thus, $n = 40$ determined from these data is actually the equivalent n^*. Lifetime predictions can hence be made for different maximum stresses using this value of n^* as already shown.

7.3 Summary

The strength of cementitious matrices is time-dependent being controlled by the slow crack growth induced by environmental effect and the statistics of flaw size and flaw distribution. The single crack approach which ignores the flaw statistics is found to be deficient, particularly in non-uniform stress situations, in predicting the time-dependent strength characteristics. Because of the large inhomogeneities in mortars and concretes that interfere with crack growth a more rigorous statistical fracture mechanics treatment than that applied to the 'pseudo'-homogeneous cement pastes is required. However, for practical purposes, such a distinction is not always needed provided the equivalent slow crack growth law of the statistically distributed flaws is determined. But this approximation approach cannot accurately predict the lifetimes of concrete under cyclic stress with high stress ratio because the mechanical fatigue effect has not been included.

Before leaving this chapter it is important to point out that at very fast loading rates not discussed here, slow crack growth is minimal as there is not enough time to permit chemically activated processes to take place. The time-dependent strength characteristics are then mainly determined by inertial effects and the original flaw size and density distributions. Because of the recent Kobe earthquake disaster (Sakamoto and Indrawan, 1995) studies on the dynamic effects or very high loading rates on the strength and fracture toughness of concrete structures have multiplied.

8 Application of fracture mechanics to the design of structures

8.1 Introduction

Until comparatively recently concrete was either assumed to have negligible tensile strength or to fail in a brittle manner at a low stress. As long as the stresses in a structure are small this assumption is reasonable. However, the trend is to use concrete more efficiently and to subject it to higher stresses. Also in many cases, such as un-reinforced beams, mass structures like dams, punching shear in concrete decks, reinforcement bonds and anchorages, and pipes, tensile failure can govern the strength. It has been shown in the preceding chapters that there is a size effect in strength that cannot be explained in terms of either strength of materials or classic LEFM. The application of fracture mechanics to concrete structures has therefore received considerable attention in the last few years, particularly by RILEM (Elfgren, 1989).

The realization of the importance of fracture mechanics to concrete design has seen the beginning of code-type formulation of fracture mechanics concepts (Hilsdorf and Brameshuber, 1991). The new draft CEB-FIP Model code (1990) includes a section 2.1, *Concrete—Classification and Constitutive Relations*, which gives some fracture mechanics properties. The key parameter for fracture mechanics, the mode I fracture energy, G_{If}, is given empirically in terms of the mean compressive strength, f_c, and a coefficient, a_d, that depends on the aggregate size, d_a (see Table 8.1) by

$$G_{If} = a_d f_c^{0.7} \quad (\text{J/m}^2, \text{MPa}) \tag{8.1}$$

This empirical relationship was largely based on a round robin test series (Hillerborg, 1985). Table 8.1 also gives the CTOD at fracture. The fracture energy is size dependent to a certain extent, but it has been shown that it reaches a plateau value at a remaining ligament of about 300 mm (Wittmann et al., 1987). The mean tensile strength given empirically by

$$f_t = 0.3 f_c^{2/3} \quad (\text{MPa}, \text{MPa}) \tag{8.2}$$

The stress-strain relation is assumed to be linearly elastic up to a stress, $\sigma = 0.9 f_t$, with the Young's modulus given by

$$E = 10^4 f_c^{1/3} \quad (\text{MPa}, \text{MPa}) \tag{8.3}$$

At higher tensile stresses, the non-elastic deformation is assumed to decrease the stiffness and a strain of 0.00015 is assumed to be attained at the ultimate

Table 8.1 Coefficients for fracture parameters CEB-FIP (1990)

Aggregate size d_a (mm)	Coefficient a_d for G_{If}	Crack opening δ_f (mm)
8	4	0.12
16	6	0.15
32	10	0.25

strength of the concrete (see Figure 8.1). An empirical expression for the characteristic length, l_{ch}, can be found by combining eqns 8.1–8.3 and is given by

$$l_{ch} = \frac{G_{If} E}{f_t^2} = 110 a_d f_c^{-0.3} \qquad \text{(mm, MPa)} \qquad (8.4)$$

The stress-displacement strain-softening relationship is assumed to be bilinear (see Figure 3.8) with the break points (v, s) defined by

$$s = 0.15$$

$$v = \frac{G_{If}/\delta_f - 22(G_{If}/a_d)^{0.95}}{150(G_{If}/a_d)^{0.95}} \qquad (8.5)$$

The CEB-FIP Model Code 1990 is the first international concrete code that includes fracture mechanics data. However, the code-formulations

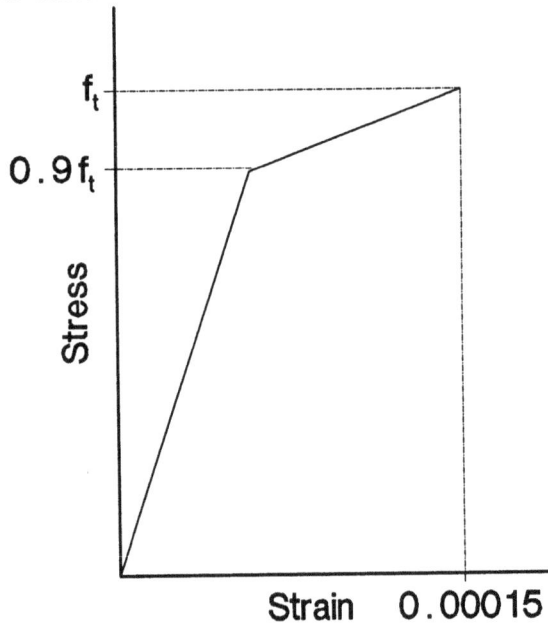

Figure 8.1 Stress-strain for uniaxial tension, $\epsilon < 0.00015$. (After Hilsdorf and Brameshuber, 1991.)

can be improved by experimental studies orientated towards practical engineering applications. In this chapter the application of fracture mechanics to practical engineering is discussed.

8.2 Application to monolithic structures

Although monolithic structures are mainly loaded in compression, it is usually cracks in local tension fields that cause problems. Dams are the largest monolithic structures and though the principal stresses are generally compressive, tensile stresses occur locally around openings and galleries. Reinforcement is usually provided in all areas where the tensile stresses exceed the allowable tensile stress and tensile cracks are more likely to be due to secondary rather than primary tensile stresses. Secondary tensile stresses can be caused by: temperature differences between the surface of the dam and the interior which can cause stresses as high as 4.5 MPa (Fanelli and Giuseppeti, 1990), abutment deformations exceeding that allowed for in the analysis, swelling of the concrete due to ageing or chemical processes, which can amount to 0.03% (Cervera *et al.*, 1990), and earthquakes. Although tensile cracks have been observed in all types of dams, buttress dams seem more prone to cracks than other types (Fanelli and Giuseppeti, 1990).

Unfortunately, fracture mechanics is normally only used either to assess the effect of cracks that have been discovered in existing structures, or for studies of catastrophic failures, rather than as a design tool. Often fracture mechanics is seen as only relevant to older structures because it is assumed that new structures, designed using state-of-the-art finite element packages, will have only low tensile stresses that are unlikely to be a problem (Boggs *et al.*, 1988). Classic LEFM can be applied in large monolithic structures whose dimensions and crack length are large compared with the characteristic length, l_{ch}. Here it must be noted that the characteristic length of the concrete used in dams is much larger than that of common concrete because of the large size of the aggregate (often up to 150 mm across) and l_{ch} can be as large as 1.7 m (Linsbauer, 1989a; Brühwiller and Wittmann, 1989b). Hence, the physical size of cracks in dams must be large before LEFM is applicable. Also, though dams are massive, the initial development of a crack cannot be modelled accurately by LEFM and the FPZ needs to be modelled by either the discrete or smeared cracks. In the case study of the fracture of a plinth in the Schoharie bridge, where there was no initial crack, modelling of the FPZ was essential (Swenson and Ingraffea, 1991). On the other hand, cracks discovered in dams are frequently very large and can be analysed using classic LEFM. The fracture mechanics applied to dams is described first.

8.2.1   The analysis of cracks in dams

Levy's rule is often used to determine the design profile of a gravity dam. According to this rule, the compressive stress on the upstream face of the dam must not be less than a threshold value, σ_u, to prevent water penetrating the dam face. The threshold value is given by (Linsbauer, 1989b)

$$\sigma_u = -\lambda \rho_w g h \qquad (8.6)$$

where ρ_w is the density of water and λ is a reduction factor to allow for pore-water pressure. If it is assumed that the normal compressive stress has a linear variation across the base of the dam, it can be shown, for a dam with a triangular profile, that the ratio of the base (B) to height (h) must be given by (Linsbauer, 1989b)

$$\frac{B}{h} \geq \frac{1}{\sqrt{\rho_c/\rho_w - \lambda}} \qquad (8.7)$$

where ρ_c is the density of concrete. Assuming that $\lambda = 0.85$, and $\rho_c = 2400\,\text{kg/m}^3$, then according to Levy's rule, $B/h \geq 0.8$. Although fracture mechanics is not usually applied as such in dam design, the effects of cracks are considered in gravity and arch dams using a strength of materials approach similar to that described in section 4.2 (Rescher, 1990). It is assumed that cracks may form at the upstream face of a gravity dam and be penetrated by water to the point where the compressive stress is equal to the hydrostatic pressure. The variation of the compressive stress over the remaining ligament of the dam is assumed to be linear (see Figure 8.2). The crack is opened by the water pressure on the upstream face of the dam and closed by the weight of the dam. There will be a FPZ at the end of the crack which will be strained so that at its tip there will be a closing stress equal to the tensile strength of the concrete f_t. Water under hydrostatic pressure $\rho_w g h$ will penetrate right to the tip of the FPZ, so that here the tensile stress is $f_t - \rho_w g h$. Since the size of the FPZ would be small compared with a long crack the variation over the FPZ can be modelled by a discontinuous change in the stress from the hydrostatic pressure to a tensile stress $f_t - \rho_w g h$. Outside of the FPZ the stress must be continuous and could be modelled as a linear variation starting from a crack tip stress of $f_t - \rho_w g h$, as suggested by Saouma et al. (1987). However, Saouma et al. (1987) go on to assume that f_t is zero perhaps because of difficulties that are mentioned below. With the compressive stress assumed to be distributed as shown in Figure 8.2, it can be shown from equilibrium that the remaining uncracked ligament of the dam, b, is given by

$$\frac{b}{3} = \frac{M_0 - \rho_w g h B^2/2}{mg - \rho_w g h B} \qquad (8.8)$$

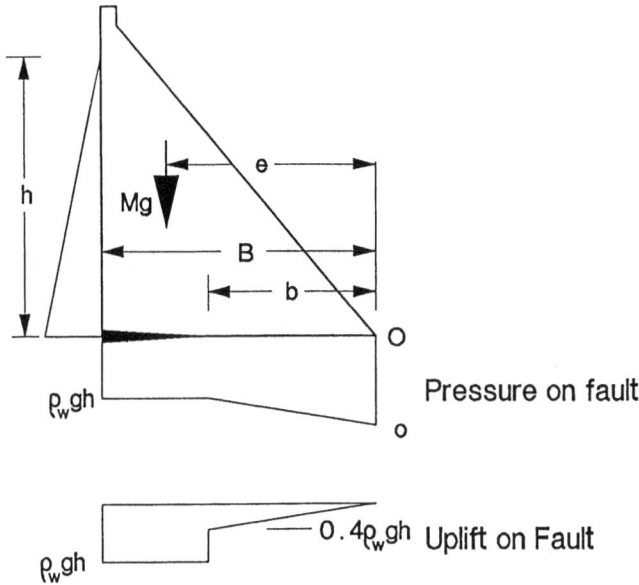

Figure 8.2 Forces acting on a section of a gravity dam.

where m is the mass (per unit width) of the dam above the particular horizontal section and M_0 is the moment (per unit width) of the weight of the dam about the toe (Corns *et al.*, 1988). For a dam of triangular profile, full with water, the remaining ligament is given by

$$\frac{b}{B} = \frac{2\left(\frac{\rho_c}{\rho_w}\right) - \left(\frac{h}{B}\right)^2 - 3}{\left(\frac{\rho_c}{\rho_w}\right) - 2} \tag{8.9}$$

The safety of the dam is then determined from the shear strength of the remaining ligament which must resist the hydrostatic thrust on the dam face. The dam is assumed to be subjected to a total uplift force, U, per unit width, due to the hydrostatic pressure in the cracked section and a reduced pressure varying from $0.4\rho_w gh$ at the crack tip to zero at the downstream face of the dam (see Figure 8.2). The shear strength, S, per unit width of the dam is then assumed to be

$$S = \tau_u b + \mu(Mg - U) \tag{8.10}$$

where τ_u is the shear strength of the concrete and μ is the coefficient of friction. Ingraffea (1990) gives a sample calculation for a generic gravity dam ($B/h = 0.8$) at a depth of 100 m. The equilibrium crack length, 32.5 m, can be obtained from eqn 8.9. Assuming that $\mu = 0.7$ and $\tau_u = 2.75\,\text{MPa}$, there is a factor of safety of 3.5 which would be less than the recommended value (Anonymous, 1976). A possible problem with the application of this method is the extreme sensitivity to the profile of the dam. Equation 8.6 predicts that crack penetration is impossible if $B/h > 0.845$ and that cracks will completely penetrate the dam when $B/h < 0.745$. If one attempts to improve on the approximate method by taking a representative tensile strength for the concrete instead of assuming that it is zero, as Saouma et al. (1987) did explicitly and every other author has done implicitly, then one finds that no equilibrium crack exists.[1] Similar strength of materials approaches have been used with arch dams (Rescher, 1990).

The effect of horizontal cracks initiated in the upstream face of gravity dams has also been considered using LEFM. A horizontal crack path is not a natural crack path for propagation under combined thrust on the upstream face of the dam by hydrostatic pressure in the crack, and gravity forces, because the stress intensity factor is a mixed mode. Linsbauer (1989b) has calculated the stress intensity factors at the tip of such a crack in a gravity dam with a triangular profile $(B/h = 0.8)$, using back face correction factors (Linsbauer and Rossmanith, 1984). Assuming that the appropriate fracture criterion is the maximum circumferential tensile stress criterion (Erdogan and Sih, 1963), Linsbauer (1989b) constructed curves of constant K_{Ic} as a function of the depth of the crack (see Figure 8.3). The 'strength of materials' approach could be considered equivalent to a LEFM criterion of $K_I = 0$, curves for which are given in Figure 8.3. At a depth of 100 m, the equilibrium crack length for $K_I = 0$ can be read from Figure 8.3 to be about 26 m. Ingraffea (1990) has used finite elements to analyse the same problem and obtained a crack length of 31 m. Both these crack lengths are similar to that obtained from the 'strength of materials' value of 32.5 m. However, though the agreement between the 'strength of materials' approach and LEFM is good, the sensitivity of the 'strength of materials' approach to the assumptions used creates the suspicion that the agreement might be fortuitous. There is also some doubt whether the limit criterion of $K_I = 0$ for the LEFM solutions is reasonable. Such a criterion does predict an increase in the critical crack length with depth in common with the 'strength of materials' approach. However, if either a constant K_I or constant K_{Ic} is used as a criterion, the solution of Linsbauer (1989b) predicts a decrease in critical crack length with depth, and Ingraffea's (1990) FEM solution predicts a maximum stable crack length at a particular depth (21 m at a depth of 40 m, using a criterion of $K_I = 2\,\text{MPa}\sqrt{m}$). Ingraffea (1990) also gives two other important qualifications to the comparison. If the elastic properties of the foundation and the concrete differ, then superposition cannot be used. Secondly the crack will not grow horizontally, but will propagate in a curvilinear manner downwards.

---

[1] The roots of the quadratic equation to determine the equilibrium crack length are imaginary.

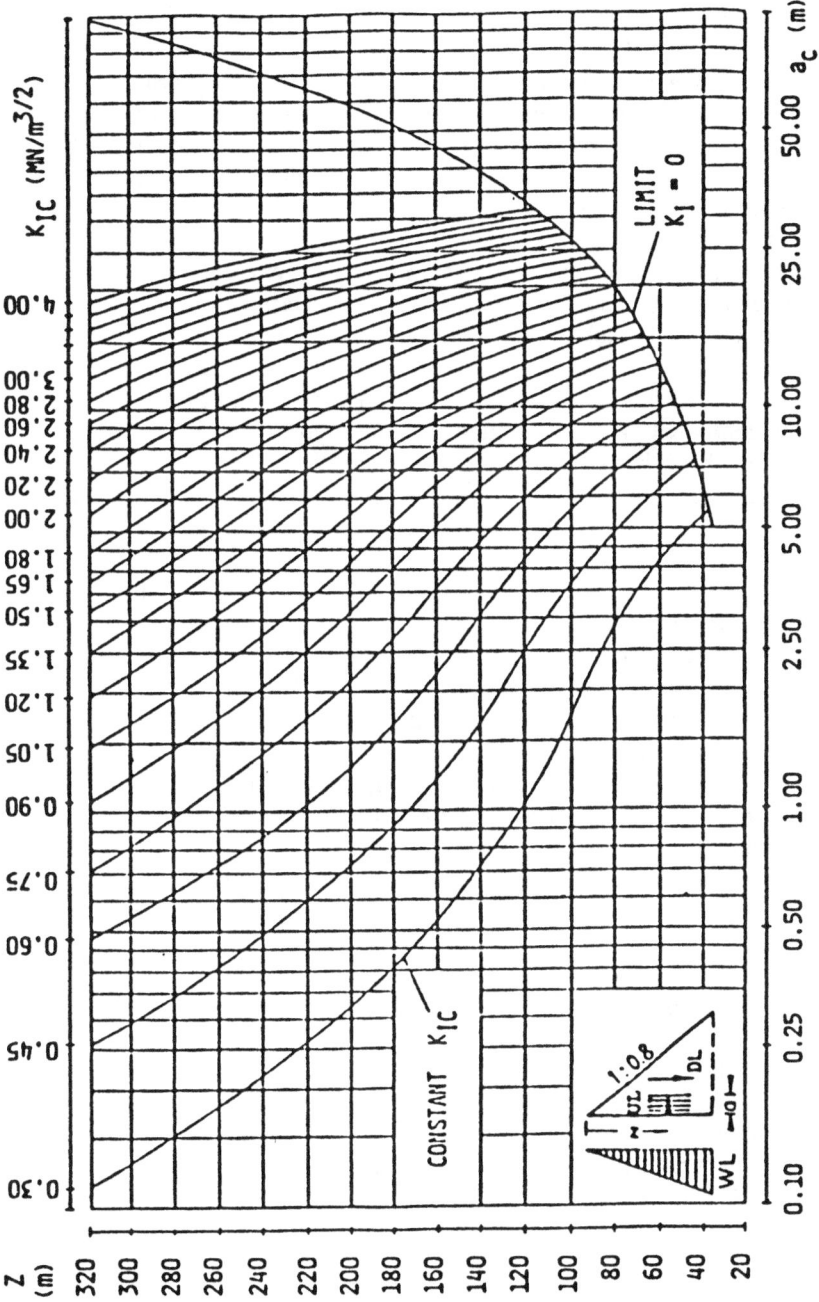

Figure 8.3 Stress intensity factor as a function of crack depth for a horizontal crack in a gravity dam of triangular profile ($B/h = 0.8$). (After Linsbauer, 1989b.)

The first application of LEFM to dams was made in 1981 in the USA to the Fontana gravity dam by Chappell and Ingraffea (1981), and in China to the investigation of thermally induced cracks in the Zhexi diamond head buttress dam by Yu (1981, 1989). Since then there have been a number of other LEFM analyses of other gravity and arch dams (Saouma et al., 1987; Linsbauer et al., 1989). The study of the Austrian Kölnbrein arch dam (Linsbauer et al., 1989) made use of a two-dimensional fracture element analysis code, FRANC, developed by Wawrzynek and Ingraffea (1987a,b). This code models both the initiation[2] and propagation of cracks by the discrete method. The maximum circumferential stress criterion of Erdogan and Sih (1963) is used to determine the crack path at each increment in crack extension and the code has automatic remeshing after each crack increment. The fracture in the Fontana dam was also re-analysed using FRANC (Ingraffea, 1990) and the predictions of the crack path are compared with the actual path in Figure 8.4 (the crack path predicted in the original analysis used a code that needed to be driven manually and a slightly different definition of the crack path direction was also used). Both of the theoretical predictions are in excellent agreement with the observed crack. Gravity dams and the buttresses of buttress dams are essentially two-dimensional. Arch dams are more inherently three-dimensional structures, but even here a two-dimensional analysis can be sufficient for practical purposes (Linsbauer et al., 1989). Full three-dimensional codes will no doubt be produced in the future, but the question of mixed mode propagation in three dimensions has not yet been completely answered.

In order to predict the safety of a dam, in which a crack is discovered or to calculate the safety factor of a dam design, it is necessary to know the material properties of the concrete. The concrete used in dams is characterized by a very large aggregate, which is often low strength gneiss. As a consequence fractures usually pass through the aggregate rather than around it and the tensile strength is smaller than that observed for normal concrete (Brühwiller and Wittmann, 1989b). In contrast the fracture energy, G_{If}, is two to three times that of common concrete (Brühwiller and Wittmann, 1989b), maybe because a large aggregate implies a wide fracture process zone and hence a large CTOD. The net result is that the characteristic length (up to about 1.7 m) is up to ten times that of common concrete. The large size poses problems for the measurement of G_{If} since it has a size dependence when the RILEM (1985) TC-50 method is used to determine it. However,

---

[2] On a smooth surface a crack is initiated when the modulus of rupture is exceeded with an immediate transition to LEFM by the insertion of a small crack. Thus, the initial stages of crack formation will not be highly accurate. However, since in dams cracks are usually long, the initiation phase is not critical.

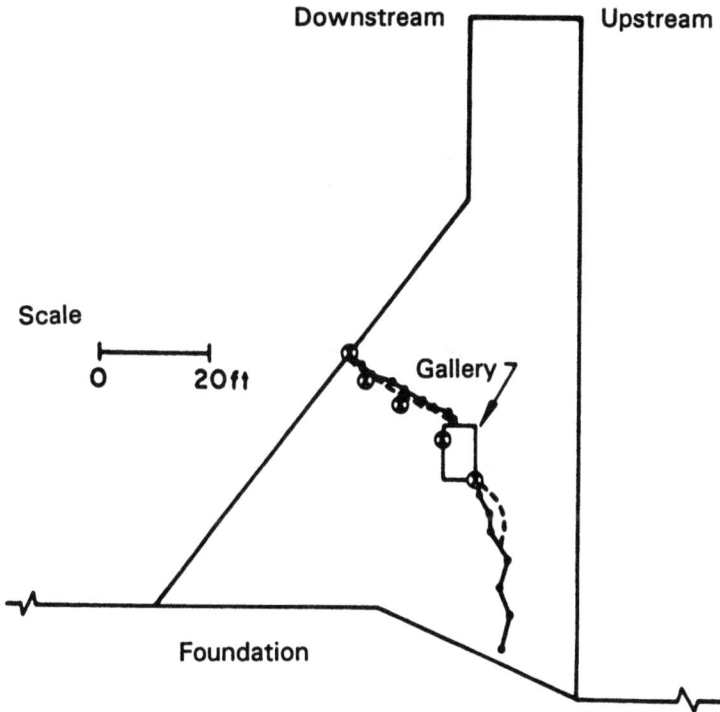

Figure 8.4 Comparison of predicted crack paths with that observed in the Fontana dam (- - - - - FEM, Chappel and Ingraffea (1981); - - - FEM, Ingraffea (1990); ⊗ Location of crack from drilling). (After Ingraffea, 1990.)

the wedge splitting test suggested by Brühwiller and Wittmann (1989a) is appropriate for concrete with large aggregate.

8.2.2   Case study of a fracture in a concrete bridge plinth

Swenson and Ingraffea (1991) have made a detailed fracture mechanics analysis of the fracture of the plinth of one of the piers in the Schoharie Creek Bridge (see Figure 8.5) which collapsed in 1987 killing ten people. The collapse of the bridge was due to the fracture of the plinth of Pier Three that had been undermined by scouring of its footing. The bridge was constructed in 1954. The plinth contained no structural reinforcement, but had some reinforcement against shrinkage and temperature. About a year after the bridge opened, vertical cracks were found in the upper surfaces of the plinths of all the piers. These cracks were attributed to excessive bending stresses in the plinths and they were reinforced with reinforced concrete beams dowelled to the plinth. The plinth of Pier Three fractured because of extensive scouring of its footing. The fracture did not initiate at

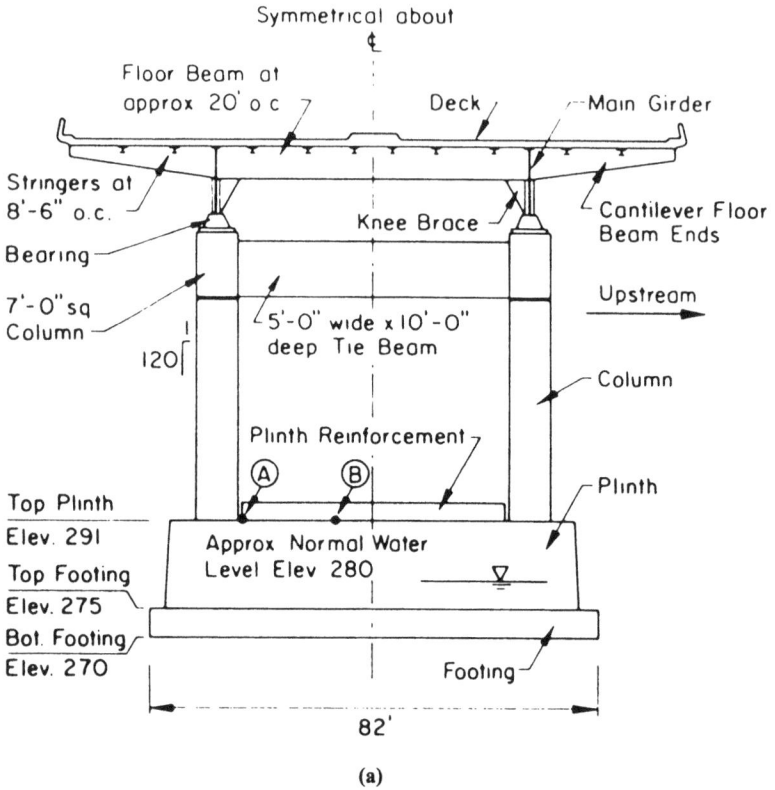

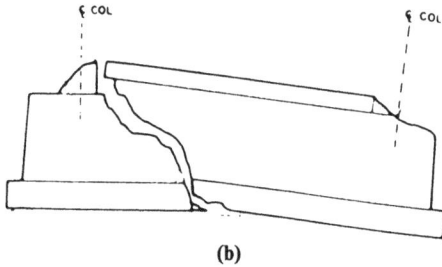

Figure 8.5 (a) Pier Three of the Schoharie Creek Bridge; (b) Fracture plinth of Pier Three. (After Wiss *et al.*, 1987.)

the pre-existing crack, location B in Figure 8.5, but at location A at the first set of dowels 0.5 m from the end of the reinforcing beam. The original failure report (Wiss *et al.*, 1987) used FEM to calculate the elastic stresses and found that at location A the tensile stress was 1.5 MPa (210 psi). Since the splitting tensile strength of the concrete was found to be 4.3 MPa (620 psi) the investigators looked around for a possible explanation of how the plinth failed when the calculated stress was so much less than the measured strength.[3] What was done illustrates the danger of non-fracture mechanists using fracture mechanics.[4] Bažant's size effect law (Bažant et al., 1986) was correctly used to argue that the modulus of rupture in the plinth would be smaller than the modulus of rupture measured in a small laboratory specimen, but the tensile splitting strength was confused with the modulus of rupture (Swenson and Ingraffea, 1991). It was not understood that the modulus of rupture does not represent the actual bending strength, but is the equivalent elastic bending stress at failure. Another problem with the original failure analysis was the failure to recognize that, near a singularity, the stresses calculated by FEM are mesh size dependent (Swenson and Ingraffea, 1991). A correct analysis of the problem has been given by Swenson and Ingraffea (1991), who modelled the development of the FPZ and crack at location A using the discrete model of Hillerborg et al. (1976). Not having the actual value of the fracture energy, G_{If}, Swenson and Ingraffea (1991) estimated its value from the size of the aggregate (25 mm) to be 241 J/m^2.[5] Since the fracture energy was not known accurately a linear stress-displacement relationship was assumed for the FPZ. Theoretically a FPZ must initiate, at the singularity at the re-entrant corner, for the smallest applied load, and to model accurately the early stages in the development of the FPZ the mesh size must be able to cope only with a high, but not infinite, stress gradient at the tip of the FPZ. Thus the analysis is objective and converges. Swenson and Ingraffea (1991) used quadratic triangular elements. With a mesh size of 3 inches, a FPZ of 4 inches was shown to form after a scour of 32 feet. The size of the zone does not increase rapidly with the length of scour, and was not much different after a scour of 36 feet. A scour of only 28 feet was assumed to have caused unstable crack propagation by Wiss et al. (1987). Even with a scour of 41 feet and equal column loads, unstable propagation was not predicted and the pier became kinematically unstable. Swenson and Ingraffea (1991) postulated that settlement in the piers led to a redistribution in the column loads so that scouring greater than 41 feet could take place without kinematic

---

[3] The authors of the original report on the failure should not be blamed too much for seeking to alter their material properties to give a satisfactory answer, since this is exactly what the father of fracture mechanics, Griffith, did himself. In his first paper, Griffith (1920) presented results from tests on glass to support his theory by annealing his specimens, arguing that it was necessary to anneal to remove the residual stresses. Unfortunately, there was an error in the theory of his first paper that caused a 44% error in fracture strength. In his second paper, Griffith (1925) argued that annealing had blunted his cracks in the glass; by reducing the annealing time he got agreement between his experiments and his theory. This very human failing in no way detracts from Griffith's genius.
[4] Alternatively, one could say it illustrates the failure of fracture mechanists to communicate their work to practising engineers.
[5] This value seems somewhat high. If the formula given by the CEB-FIB Model Code 1990 (see section 8.1) is used $G_{If} \approx 110$ J/m^2. Naturally the code is giving a lower bound to G_{If}, but it is unlikely that the most probable value will be more than twice this value.

instability. There was in fact direct evidence, from the disposition of the fractured plinth, that at least 44 feet of scouring had taken place before failure. At 44 feet of scour the analysis of Swenson and Ingraffea (1991) indicated that a true crack of greater than 4 inches would have formed with a FPZ of about 10 inches.[6] Calculation of the applied stress intensity factor at the tip of a true crack showed that instability would occur with a true crack less than 1 foot with a scour of 44 feet. Swenson and Ingraffea (1991) concluded that the plinth fractured after a scour of about 44 feet. They predicted the fracture path using LEFM and found good agreement with the actual fracture path. It should be noted that in inherently unstable geometries, that is those where the fracture of an ideal brittle material would be unstable, fracture always occurs before the FPZ is fully developed. The larger the component, the nearer to complete formation will be the FPZ at instability.

8.2.3 In situ *measurement of fracture properties*

The properties of concrete used in a structure are usually determined from field-cast specimens. However, there is a difference in properties between concrete cast and cured in test cylinders and the same concrete in a structure. Also, in the event of a structural failure many years after construction, the properties may not be available or suspect. For these reasons there is increasing interest in the *in situ* properties. The Break-Off (BO) test, reviewed by Naik (1991), was established as a method of measuring the *in situ* strength of concrete. In the BO test a cylindrical test specimen is separated from the concrete structure during pouring by a plastic cylinder. The test is performed using a special loading attachment that produces a transverse load (see Figure 8.6). Although the BO test is fundamentally a flexural test, the break-off load is usually correlated with the compressive strength of the concrete.

The BO test has been modified by Hashida *et al.* (1990) to enable the fracture toughness to be measured *in situ*. A pair of cores are drilled into the face of the concrete. A conventional BO test is performed on one of the cores, and a notch is then machined into the base of the other using a diamond saw (see Figure 8.7). The second core is broken off from the structure using the standard BO loading apparatus. A LVDT is used to measure the displacement during fracture and the acoustic emission is measured. The J-integral is calculated from the area (A) under the load-displacement curve; at the moment there is a sudden rapid increase in acoustic emission, from the

[6] For this degree of scouring, the condition of zero stress intensity factor was not met. In fact the stress intensity factor was about 80% of the fracture toughness. No reasons are stated for not determining the equilibrium condition but, since the finite element program was driven by the degree of scouring rather than the CTOD or true crack length, the program would encounter difficulties near instability.

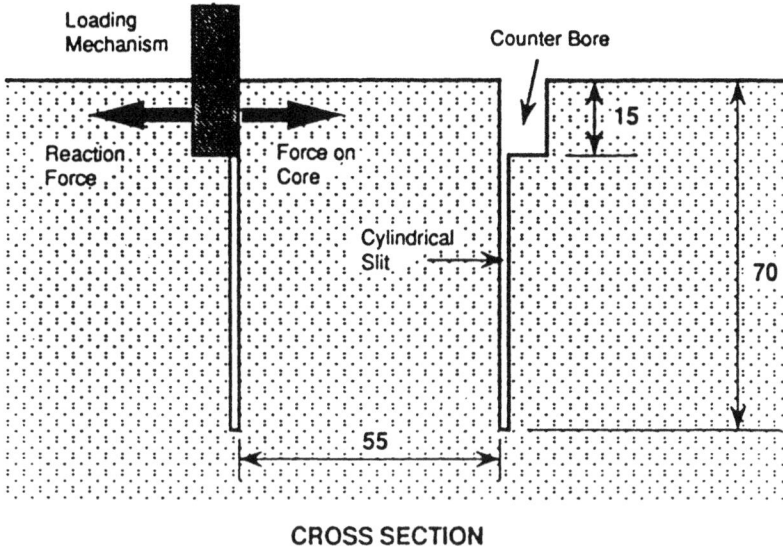

CROSS SECTION

Dimensions in millimeters

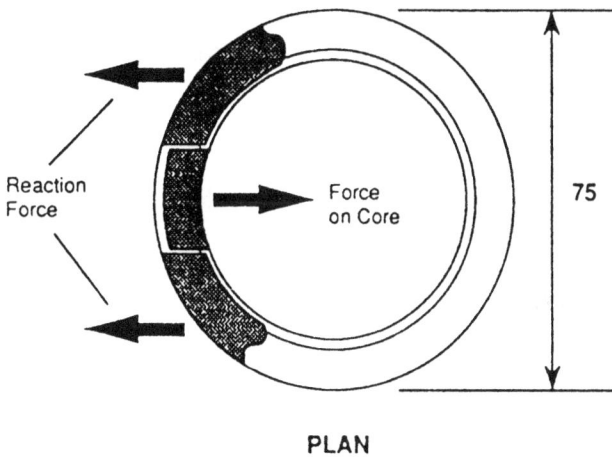

PLAN

Figure 8.6 The Break-Off Test. (After Naik, 1991.)

expression

$$J = \frac{\eta A}{Bb} \tag{8.11}$$

where b is the remaining ligament, B is the width of the notch, and η is estimated to be 2.5 for $a/D > 0.4$ from simple beam theory. The results from more standard fracture toughness specimens using the same technique

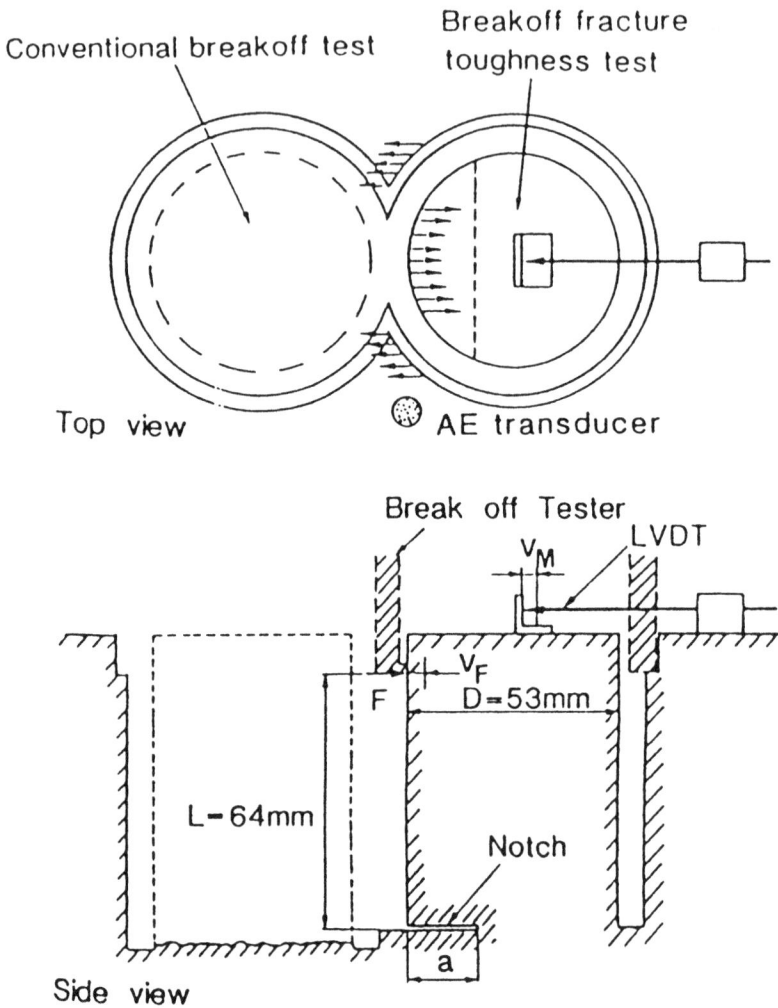

Figure 8.7 Break-off fracture toughness testing method. (After Hashida *et al.*, 1990.)

are used to suggest that the value measured in this way is independent of the specimen size. The break-off tests performed by Hashida *et al.* (1990) used core diameter of 53 mm and the concrete had a maximum aggregate size of 20 mm. The results from these tests gave similar fracture toughness values to those obtained from compact tension specimens with a ligament b of 110 mm. However, the use of fracture toughness for small size specimens is questionable and it would perhaps be preferable to measure the fracture energy by this method. Whether the fracture toughness or fracture energy is measured, there seems little point in performing *in situ* tests when the broken off core could be more easily tested in the laboratory using either the short rod or notch bend geometry.

A more interesting *in situ* test, suitable for massive structures such as dams, has been suggested by Saouma *et al.* (1991). In this test a borehole is drilled into the concrete and the borehole pressurized with a dilatometer probe which contains LVDTs to measure the hole dilatation. Acoustic emission sensors are mounted near the borehole on the surface to detect cracking. Initially, the borehole is subjected to cycles of pressure, too small to cause cracking, so that effective Young's modulus can be obtained from compliance measurements. The probe is then slowly pressurized until a large burst in acoustic emissions indicates the borehole has cracked. This pressure then gives the apparent tensile strength, f_{tapp}, of the concrete. The true strength cannot be determined directly because the state of stress at the position of the probe is not known exactly. The apparent strength of the concrete is given approximately by

$$f_{tapp} = f_t - 3\sigma_2 + \sigma_1 \qquad \text{if} \quad 3\sigma_2 > \sigma_1 > \sigma_2 \qquad (8.12a)$$

and

$$f_{tapp} = f_t - 3\sigma_1 + \sigma_2 \qquad \text{if} \quad \sigma_1 > 3\sigma_2 \qquad (8.12b)$$

where σ_1 is the maximum normal stress acting perpendicularly to the bore-hole and σ_2 is the minimum. In a dam, one or both of these normal stresses are usually compressive and this case f_{tapp} is greater than the true tensile strength.[7] It is possible, in theory, to estimate the maximum tensile stress that exists at the borehole from a measurement of the tensile strength of the core. However, the core is necessarily going to be quite small in diameter[8] and in practice the tensile strength obtained may not be accurate enough. The borehole is then subjected to larger and larger pressure cycles and compliance measurements used to determine the effective crack length at each cycle. The apparent fracture toughness, K_{Ic}^{app}, is then calculated from the effective crack length and the pressure. There are two reasons why Saouma *et al.* (1991) quote the fracture toughness as an apparent value. Firstly, the state of stress at the borehole is not known exactly and K_{Ic}^{app} is calculated from only the pressure applied. Secondly, since the effective crack method of measuring fracture toughness is very approximate, K_{Ic}^{app} depends upon the effective crack length. The results of the *in situ* measurements are given in Figure 8.8.

---

[7] In laboratory tests on blocks of concrete $910 \times 910 \times 1140$ mm, which were compressed across one pair of faces with a compressive stress of 0.69 MPa, the differences between f_{tapp} and f_t, obtained from a standard splitting test, were 0.71, 0.63 and 0.13 for the three different aggregate sizes of 19 mm, 38 mm and 76 mm, respectively (Saouma *et al.*, 1991). Obviously the results from the concrete with the two smaller aggregates agree with eqn 8.7. The large discrepancy for the largest aggregate size may have been due to an inadequate size for the splitting test.
[8] In field tests by Saouma *et al.* (1991), the borehole was cored with a standard HWD4 core drill which has a 61 mm diameter barrel.

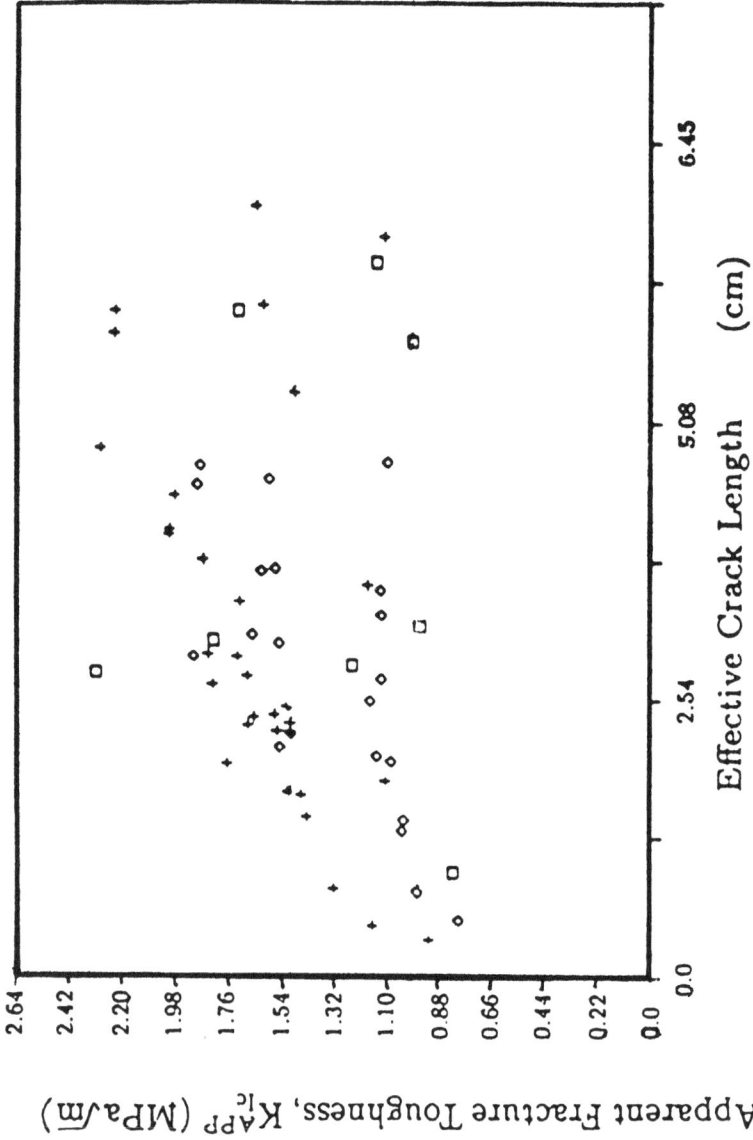

Figure 8.8 Apparent fracture toughness, measured *in situ* in a US dam, as a function of the effective crack length. □, 19-mm MSA; +, 39-mm MSA; ◇, 76-mm MSA. (After Saouma *et al.*, 1991.)

As might be expected, there is considerable scatter in the results, but there is a trend for the apparent fracture toughness to increase with the effective crack length. Such an increase in apparent fracture toughness is not surprising because the size of the FPZ must increase with the crack size and hence the apparent fracture toughness must also increase. However, the method pioneered by Sauoma *et al.* (1991) is worth pursuing, because it could very easily be modelled by the fictitious crack model to obtain a reasonable estimate of the fracture energy, G_{If}, if a suitable generic form of the strain-softening stress-displacement relationship is assumed.

8.3 Punching shear failure of slabs

One of the possible modes of failure for slabs is punching. This failure mode occurs when a slab is subjected to a concentrated load which eventually punches a slug from the slab. The current design approaches to punching are covered by a wide range of empirical formulae much derived from plastic limit analysis. These formulae have been reviewed by Stefanou (1993). The codes, though primarily strength based, defining nominal allowable stresses based on the concentrated load divided by a representative area of the concrete, do have allowances for size effect. For example, CP110 (1972) gives the effective shear strength, τ_u, as

$$\tau_u = \frac{0.27}{\gamma_m} \left(\frac{500}{d} \right)^{1/4} (100 v f_c)^{1/3} \tag{8.13}$$

where d is the effective depth of the slab (measured from the compression surface to the centre of any reinforcement), γ_m is a constant ($= 1.25$, for $f_c < 40\,\text{MPa}$), and v is the volume fraction of reinforcement. Bažant and Cao (1987) have discussed the size effect in punching from the concept of Bažant's (1984) SEL. The SEL applies to geometrically similar notched specimens. Since there is no notch in the punching test, it has to be assumed that a geometrically similar crack has developed stably at maximum load. Bažant and Cao's (1987) size effect plot (see Figure 8.9) shows that there is only a weak dependence of the shear strength on size and there is no sign of a trend to an asymptotic slope of $-1/2$ as the size gets larger which throws some doubt on the application of the SEL in punching. Gardner (1990) has compared the shear strength predicted by the codes and the prediction given by Bažant and Cao (1987) with the experimental results of others—the codes give a better prediction. In fact the size effect predicted by eqn 8.13 is stronger than that predicted by Bažant and Cao (1987).

There have been a few fracture mechanics analyses of punching using the smeared crack model (de Borst and Nauta, 1985; González-Vidosa *et al.*, 1988; Malvar, 1992). The analyses using the smeared crack model are aimed more at reproducing experimental results numerically rather than

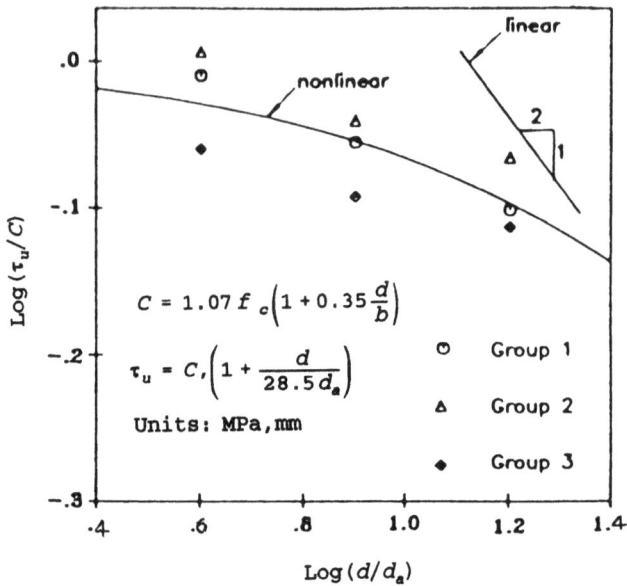

Figure 8.9 Shear strength of punching tests relative to the strength of the smallest specimen (d is the full slab thickness and d_a is the size of the aggregate). (After Bažant and Cao, 1987.)

establishing a more rational treatment of punching. The details of some of the numerical analyses are at best sketchy. Statements like 'a fracture energy of 100 N/m was assumed' (Malvar, 1992) give rise to the suspicions that the parameters in this paper and the others may have been adjusted in order to give good agreement with the experimental results.

The smeared crack model would appear to be too complex to establish general equations for the predictions of shear strength in punching and the time may be right for simpler treatments. It would appear that the early analysis of Kinnunen and Nylander (1960) could be re-examined in light of the fictitious crack model as a possible method of obtaining a simple rational approach to the important phenomenon of punching.

8.4   Reinforcement bonding and anchorage to concrete

The difference between the reinforcement and anchor bolts is that reinforcement is continuous, so that the tensile load is transmitted mainly by the reinforcement and the prime function of the concrete is to provide shear strength, whereas in anchor bolts there is a direct transfer of tensile load to the concrete through shear over the shank and compression under the head. However, there are similarities between the behaviour of reinforcement and anchors. Even with continuous reinforcement some tensile load is taken

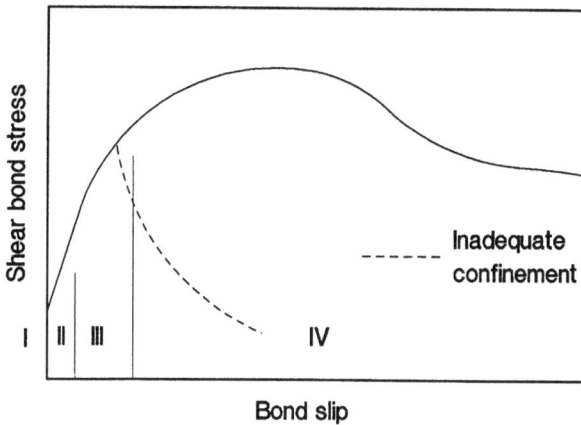

Figure 8.10 Schematic reinforcement bond stress-slip relationship. (After Gambarova *et al.*, 1989a.)

by the concrete between the cracks, resulting in what is known as tensile stiffening (Gilbert and Warner, 1978; Foegl and Mang, 1982; Gerstle *et al.*, 1982; Wu *et al.*, 1991). If anchor bolts are long, a considerable proportion of the load can be transferred to the concrete along the shank of the anchor. The reinforcement bond is considered first.

8.4.1 Reinforcement bond slip and tensile stiffening

Four stages (see Figure 8.10) have been characterized in the pull-out behaviour of reinforcement bar (Gambarova *et al.*, 1989a). At low shear stresses ($<0.5f_t$) the shear in stage I can be resisted by the chemical bond and no bond slip occurs (Gambarova and Karakoç, 1982). At higher shear stresses the chemical bond is broken by the wedging action of the ribs on the reinforcement and the bond becomes purely mechanical. In stage II, for bond shear stresses $0.7–1.5f_t$, secondary transverse cracks develop from the ribs. The stress at which the transverse cracks develop in this stage is dependent on the confining pressure. At higher shear stresses in stage III ($1–3f_t$), the concrete ahead of the ribs starts to crush and, under the hoop stress caused by the wedging action, longitudinal splits initiate (Gambarova and Karakoç, 1982). If the cover on the reinforcement is insufficient, primary transverse cracks can propagate from the outer surface of the concrete to the reinforcement bar (Gerstle *et al.*, 1982) and the longitudinal cracks extend to the surface. In the absence of transverse reinforcement the shear resistance drops to practically zero at this stage. However, if there is transverse reinforcement that provides a significant confining pressure, the shear resistance can increase to give stage IV. In this stage the bond shear stress can reach as high as 1/3 to $1/2f_c$. There are two aspects of the problem: the

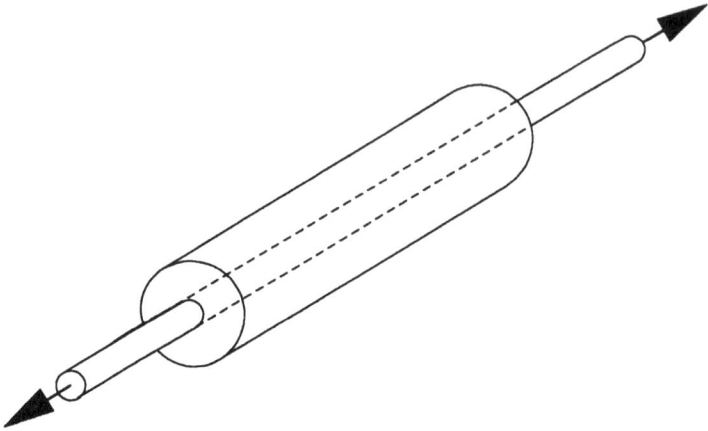

Figure 8.11 Pull–pull tension reinforcement specimen.

modelling of the bond-slip mechanisms and the incorporation of bond-slip interface elements in finite element programs.

The two main contributions to bond-slip up to stage IV are transverse cracking and crushing at the ribs. The mechanics of bond-slip have been modelled using either smeared cracks (Rots, 1985, 1988) or discrete cracks (Ingraffea *et al.*, 1984; Yao and Murray, 1993). The favoured geometry analysed is that of the pull–pull tension specimen (see Figure 8.11) which was the experimental geometry used by Goto and his co-workers (1971, 1979). The choice of this specimen geometry is perhaps unfortunate because it does not really correspond to the practical reinforcement situation. The cracking is confined to the loaded ends of the specimen and the secondary cracks here soon extend to the ends of the concrete confinement. Rots (1985, 1988) considers both the transverse and longitudinal cracks. The transverse cracks were modelled using the coaxial rotating crack concept and the longitudinal cracks treated as fixed cracks. In addition, potential discrete crack elements were placed mid-section so that a primary crack could be triggered. The crack formation just prior to primary crack formation is shown in Figure 8.12. In the discrete crack treatment of Ingraffea *et al.* (1984), only transverse cracks were modelled because axial symmetric elements were used. Yao and Murray (1993) have given a three-dimensional discrete crack solution that enabled longitudinal cracks to be studied. Ingraffea *et al.* (1984) remesh at each crack growth increment so that the crack could grow according to the condition of local symmetry, whereas Yao and Murray (1993) used a fixed element mesh and orientated the cracks along the element faces that most nearly corresponded to the principal stress trajectory. The cracking in a pull–pull tension specimen (Ingraffea *et al.*, 1984), at the moment the first secondary crack extends to the edge of the concrete confinement is shown in Figure 8.13. Ingraffea *et al.* (1984) give an empirical

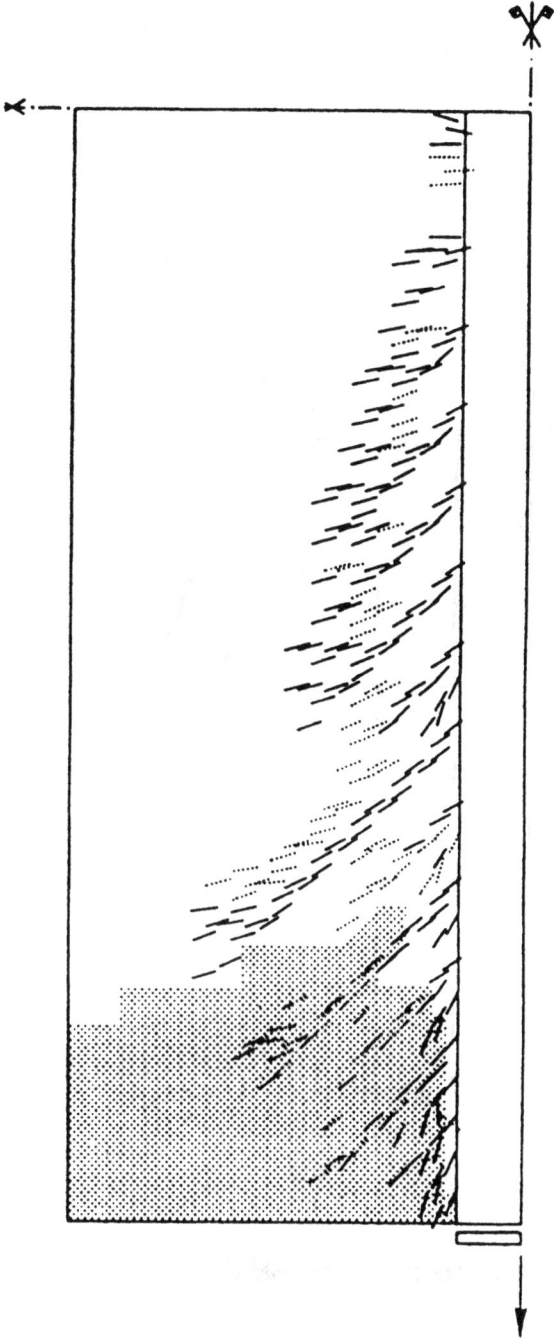

Figure 8.12 Crack formation in the pull–pull tension specimen at impending primary crack formation (- - - - active transverse cracks, · · · · · inactive transverse cracks, longitudinal cracks shaded). (After Rots, 1988.)

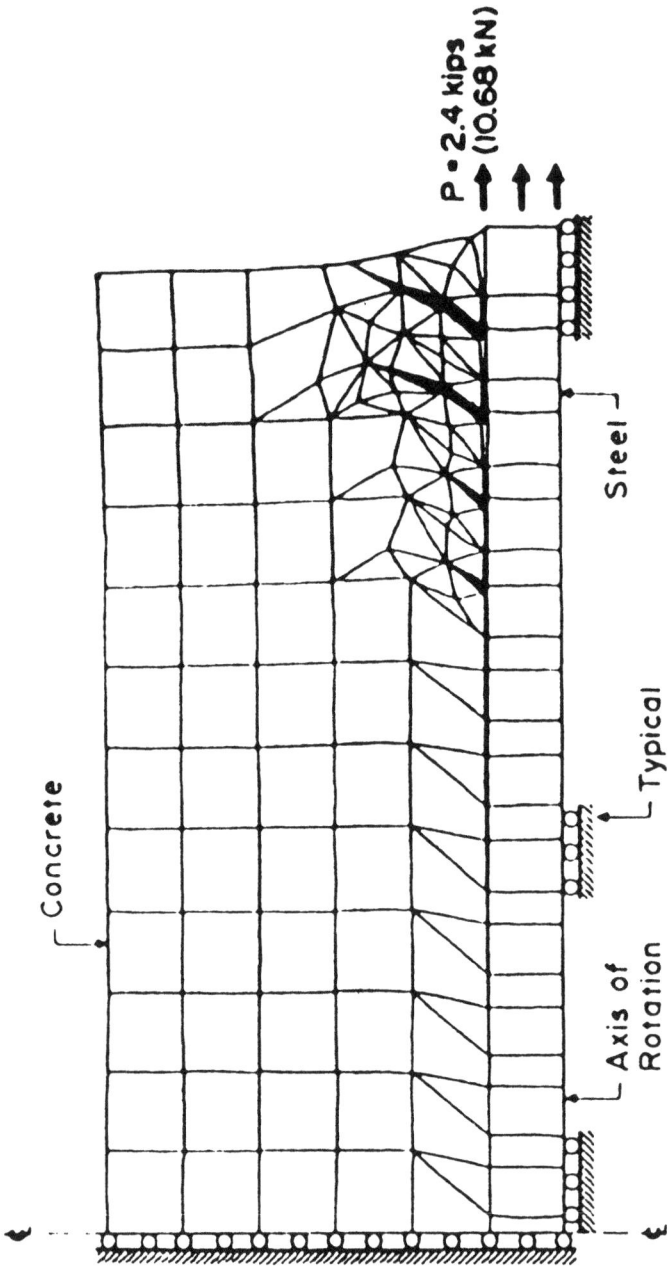

Figure 8.13 Crack formation in the pull–pull tension specimen showing four secondary transverse cracks. (After Ingraffea *et al.*, 1984.)

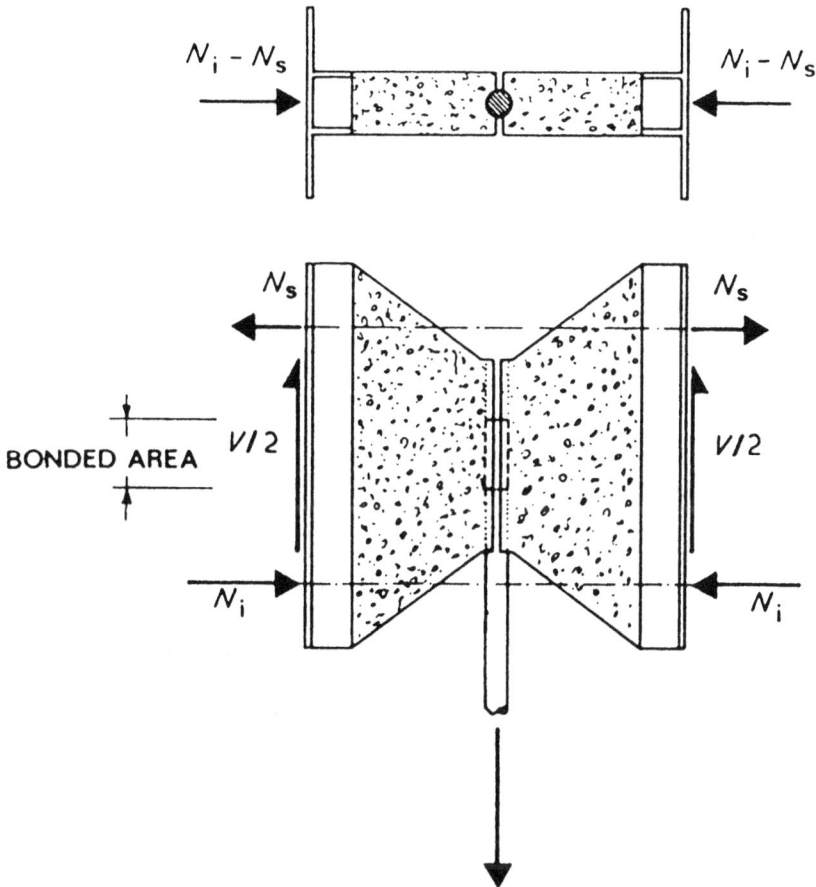

Figure 8.14 Specimen for measuring stage IV bond stress-slip relationship. (After Gambarova *et al.*, 1989a.)

expression for the slip at the end of the specimen that includes the effect of secondary cracking. Gambarova and Giuriani (1985) have criticized the analysis of Ingraffea *et al.* (1984) because they neglected crushing at the reinforcement ribs which becomes the dominant factor for a relatively small slip.

Stage IV of the shear bond-slip relationship has been less studied than the earlier stages, but is important for the proper design of stirrups in beams (Gambarova *et al.*, 1989a). A realistic test method for the study of stage IV reinforcement bond behaviour, shown in Figure 8.14, enables the effect of the confining pressure of transverse reinforcement to be examined (Gambarova *et al.*, 1989a,b). The longitudinal split is cast into the specimen by inserting plexiglass separators into the formwork. Typical bond stress-slip curves for 18 mm diameter round deformed reinforcing bar with crescent-shaped lugs are shown in Figure 8.15. In these tests the confining pressure

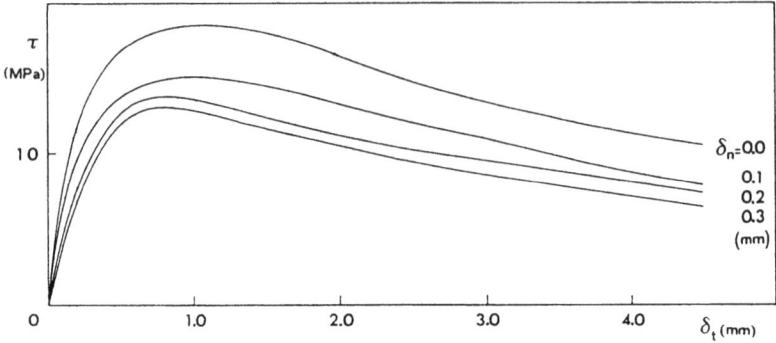

Figure 8.15 Bond stress-slip relationship for stage IV, split opening $\delta_n = 0$, 0.1, 0.2, 0.3 mm. (After Gambarova *et al.*, 1989a.)

was adjusted to allow the longitudinal splits to open by varying amounts. The bond stress-slip relationship can be idealized into a trilinear model (see Figure 8.16) where the first linear portion is stage I, the second linear portion stages II and III, and the final plateau value, stage IV. The drop in shear resistance with bond slip in stage IV need not be modelled since the drop is slight over a considerable slip if there is sufficient confining pressure.

Analytical estimates of the stiffening effect of the concrete have been made using idealized bond stress-slip relationships (Giuriani, 1982; Wu *et al.*, 1991). Wu *et al.* (1991) assumed that stage I was the most important in determining the tensile stiffening, whereas Giuriani (1982) assumed stage II to be more important (see Figure 8.17 for the idealizations). Since in both cases multiple cracking is assumed, Giuriani's (1982) idealization seems preferable. For moderate reinforcement the exponential decay in slip u_s from a crack face is given by

$$u_s = u_{0s} \exp(-\lambda x) \tag{8.14}$$

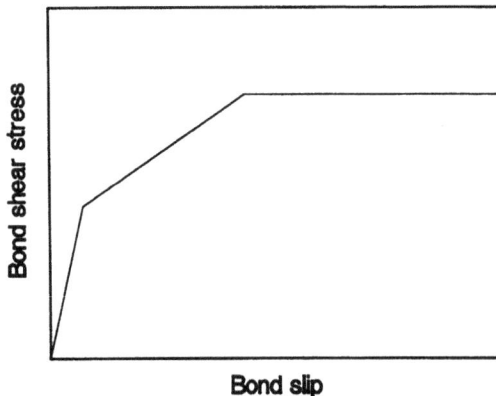

Figure 8.16 Idealized bond stress-slip relationship. (After Gambarova and Giuriani, 1985.)

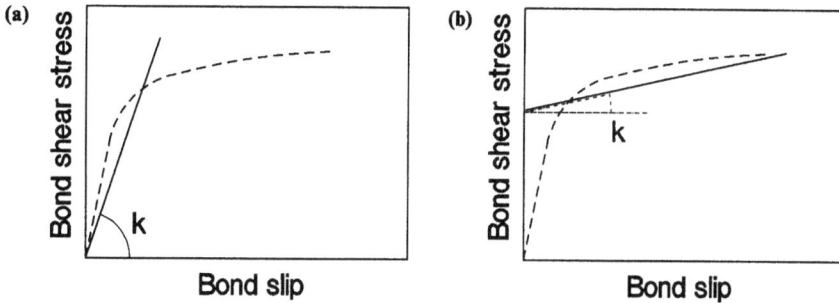

Figure 8.17 Idealized bond stress-slip relationships used to determine tensile stiffening: - - - - - experimental curve, ——— idealization. (a) Wu *et al.*, 1991; (b) Giuriani, 1982.

where u_{0s} is the slip at the crack face and $\lambda^2 = 4k/E_s d$, k being defined in Figure 8.17, E_s is the Young's modulus of steel and d is the diameter of the reinforcement. The transfer length, $1/\lambda$, is clearly much longer under Giuriani's (1982) assumption than that of Wu *et al.* (1991).

8.4.2 Anchorage to concrete

Anchoring elements are used to transmit local loads into reinforced concrete components. The load transmitted can be tension, shear or a combination of the two. Unless the anchorage is near the edge of a member, tensile loads are the most critical and it is anchorages transmitting tensile loads that will be discussed here. The anchors can be headed studs or undercut bolts, where the load transfer is mainly due to mechanical interlock, or expansion or grouted bolts, where the transfer is through the bond between the bolt and the concrete. The possible failure modes of anchors loaded in tension are:

(I) The anchor is pulled out of its hole with little concrete damage. This mode occurs in bonded anchors if the bond strength is too small and the bond stress-slip relationship will be similar to that discussed in section 8.4.1. Expansion anchor bolts will pull out if the expansion force is too small to give the required frictional force.

(II) The anchor is placed too near an edge or too close to each other and the concrete splits (Olsson, 1985).

(III) The bolt or the sleeve fails.

(IV) The full strength of the concrete is realized and a concrete failure cone is developed. If the anchor is too near an edge a partial cone will form at a lower load and if there are multiple anchors too close together a combined cone will form usually at a lower load. If the anchor is deeply embedded to a depth greater than the transfer length given by eqn 8.4, a series of failure cones will develop (Goto *et al.*, 1993).

It is failure mode IV, in which the maximum load is attained, that is of most interest. For a headed anchorage, where all the load is taken on the

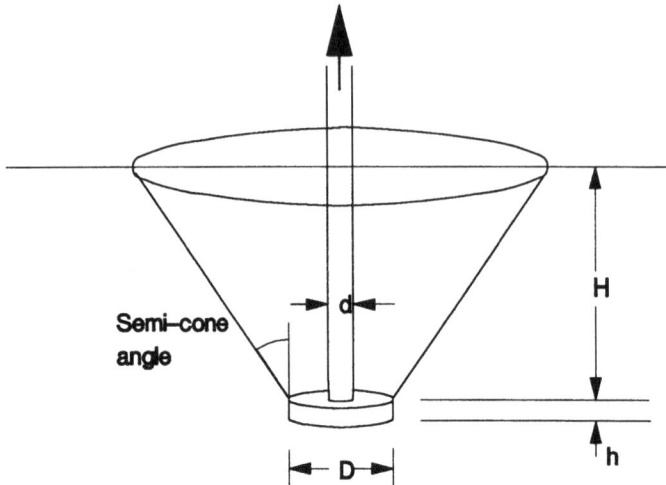

Figure 8.18 Cone fracture from a headed anchorage.

Table 8.2 Dimensions of headed anchor bolts (Eligehausen and Sawade, 1989)

Anchor size	h (mm)	d (mm)	D (mm)	H (mm)	$\bar{H} = H/l_{ch}$
1	10	22	35	120	0.236
2	10	30	45	250	0.492
3	20	50	80	500	0.985

head, failure is by the propagation of a cone fracture (see Figure 8.18). Provided the depth, H, to head diameter, D, is reasonably large, the initial development of a cone fracture is stable even in an ideally elastic-brittle material.[9] The semi-angle generally varies from 45 to 60° and it is assumed to be 45° in ACI 349 (1980). In a series of tests on headed anchor bolts of varying size, Eligehausen and Sawade (1989) found the semi-angle to be 52.5°. For very large embedded lengths LEFM can be used and assuming that the semi-cone angle is 52.5°, Eligehausen and Sawade (1989) have calculated that the cone fracture in an elastic-brittle material will grow stably until $a/L \approx 0.45$. At instability the nominal pull-out stress $(\sigma_N = P/H^2)$ is given by (Eligehausen and Sawade, 1989)

$$\sigma_N = \frac{2.1 f_t}{\sqrt{\bar{H}}} \tag{8.15}$$

where $\bar{H} = H/l_{ch}$ is the non-dimensional depth of the anchor. Eligehausen and Sawade (1989) performed pull-out tests on a series of different size specimens (see Table 8.2). The properties of the concrete used in these tests

[9] Two-dimensional analysis by Ballarini *et al.* (1986, 1987) shows that a stable crack will be initiated if $H/D \geq 2$.

Table 8.3 Properties of concrete (Eligehausen and Sawade, 1989)

Compressive strength, f_c	22.9 MPa
Tensile strength, f_t	1.8 MPa
Young's modulus, E	23.5 GPa
Fracture energy, G_{If}	70 J/m^2
Characteristic length, l_{ch}	508 mm

are shown in Table 8.3, from which the characteristic length can be calculated to be 508 mm, and thus the non-dimensional embedment depths are as shown in Table 8.2. Eligehausen and Sawade (1989) found that LEFM predicted the nominal pull-out strengths reasonably well (see Figure 8.19). The accuracy of this prediction is surprising given the rather small non-dimensional embedment depth. For small embedment depths LEFM will not apply and the pull-out strength would be expected to tend to a finite value as $H \rightarrow 0$. The transition in behaviour from large embedment depths to small is similar to other size effects and could be expected to be covered by Bažant's (1993) size effect law. Assuming that the SEL given by eqn 4.32 applies and for very deep embedments the pull-out strength is given by eqn 8.15, we have taken the results of Eligehausen and Sawade (1989) for the shortest anchor to calculate the SEL

$$\sigma_N = f_t \frac{7.55}{(1 + 12.9\bar{H})^{1/2}} \tag{8.16}$$

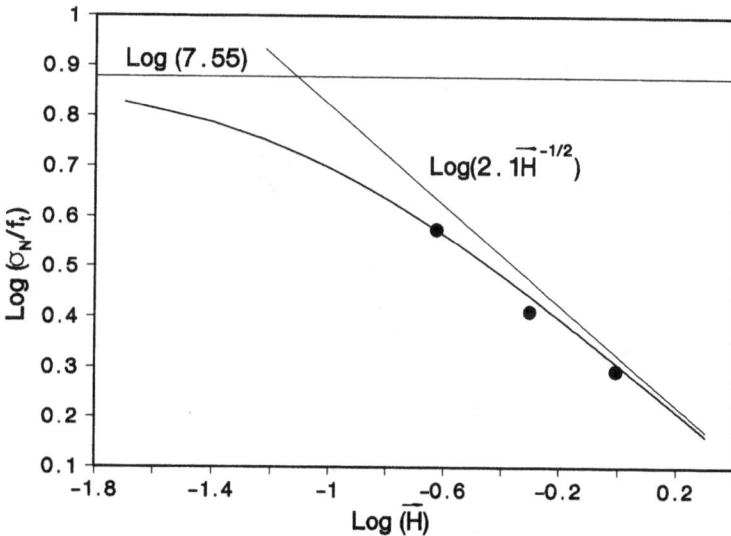

Figure 8.19 Normalized pull-out strength as a function of the non-dimensional embedment depth for headed anchor bolts (● experimental points from Eligehausen and Sawade, 1989; —— eqn 8.17).

The predictions of eqn 8.16 are shown in Figure 8.19. It is difficult to conclude that the SEL is more efficient than simple LEFM because one of the data points was used for the prediction. The SEL gives the limiting nominal pull-out strength as 13.6 MPa as $H \rightarrow 0$, which is equivalent to an average normal stress over the conical surface of 1.31 MPa. This stress value compares well with the tensile strength of 1.8 MPa given for the concrete in Table 8.3.

In an earlier series of tests the pull-out force for expansion bolts whose embedded depths varied from 20 to 150 mm were measured (Eligehausen and Pusill-Wachtsmuth, 1982; Eligehausen, 1987). The semi-angle of the pull-out cone was larger for this series at about 60°. The test results came from concrete whose compressive strengths, f_c, varied from 10 to 50 MPa, but most had a strength close to 20 MPa. The pull-out strengths shown in Figure 8.20 have been adjusted to a strength of 20 MPa by assuming that the strength is proportional to $f_c^{2/3}$. Evaluation of the data by regression analysis by Eligehausen and Pusill-Wachtsmuth (1982) yielded an empirical

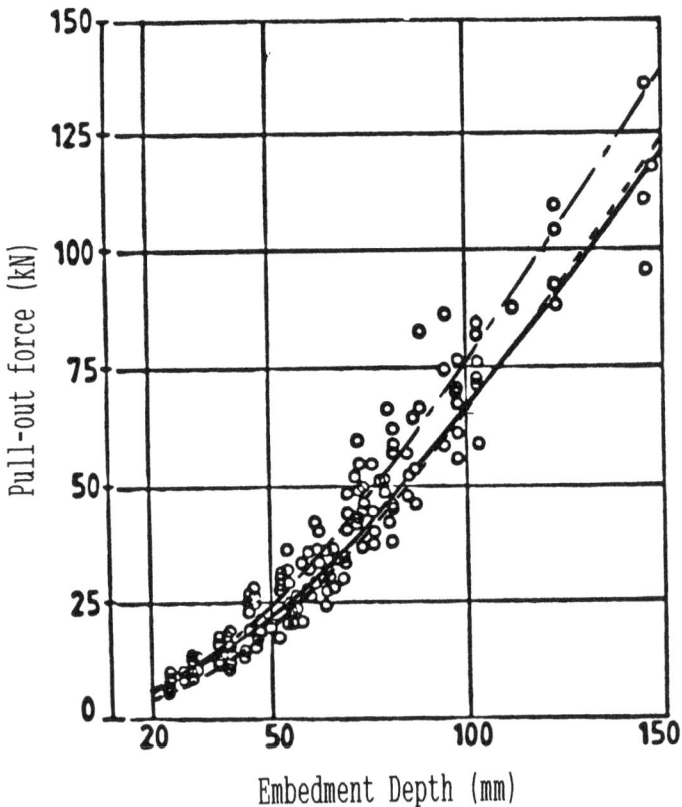

Figure 8.20 Pull-out strength of expansion bolts (- — - modified eqn 8.15; - - - - modified eqn 8.16; —— eqn 8.17). (After Eligehausen and Pusill-Wachtsmuth, 1982.)

equation

$$P = 7.4H^{1.54}f_c^{2/3} \qquad \text{(N, mm, MPa)} \qquad (8.17)$$

that gave a correlation coefficient of 0.97. Although the semi-angle of the pull-out cone is different in the analyses of the later results given above, it is interesting to examine their predictions. To correct for the different compressive strength and the different characteristic length, the pull-out force has been assumed to be proportional to $f_c^{2/3}$ and the characteristic length proportional to $f_c^{-0.3}$, in keeping with eqns 8.2 and 8.4. The LEFM prediction of Eligehausen and Sawade (1989), eqn 8.15, and the SEL derived by us, eqn 8.16, are superimposed on the earlier results of Eligehausen and Pusill-Wachtsmuth (1982) in Figure 8.20. The SEL prediction is extremely close to the empirical relationship (eqn 8.17). The LEFM prediction generally overestimates the pull-out force, but the overestimate is less than 13%. Eligehausen (1987) compares the failure loads measured in his tests on expansion bolts with the values predicted by ACI 349 (1980) and a number of other empirical predictions. ACI 349 (1980), in common with most other predictions, uses a strength of materials approach and assumes that failure occurs when the tensile stress σ_t over the projected area of a pull-out cone with a semi-angle of 45° is given by

$$\sigma_t = 0.33\phi\sqrt{f_c} \qquad \text{(MPa, MPa)} \qquad (8.18)$$

where ϕ is a strength reduction factor. ACI 349 (1980) under-estimates the strength of the expansion anchor bolts shown in Figure 8.20, even using a strength reduction factor of 1, but overestimates the strength of the anchor bolts shown in Figure 8.19, even when a strength reduction factor of 0.85 is used. Such are the dangers of using a strength of materials approach which does not give the correct scaling.

Elfgren et al. (1987) have analysed anchor pull-out using the fictitious crack model of Hillerborg et al. (1976). The most interesting case they analysed was where the crack followed a slightly curved path so that the opening was mode I. The initial direction of the crack was at 73° to the vertical. Unfortunately, only one case was analysed for a bolt embedded to a depth of 150 mm, the tensile strength was assumed to be 3 MPa and the fracture energy 100 J/m². The maximum load predicted from the finite element study was 340 kN. An experimental test on a bolt of similar geometry, but located only 300 mm from an edge, gave a failure load of 206 kN (the tensile strength of the concrete was 2.9 MPa). Using eqns 8.15 and 8.16, the predictions from LEFM and the SEL are 211 kN and 195 kN, respectively. It is surprising that these predictions should be more accurate than the prediction where the FPZ is modelled.

Standard pull-out tests are used to determine whether the in-place strength of concrete has reached a specified level and are reviewed by Carino (1991). The first such test to gain acceptance was the Danish LOK-TEST system

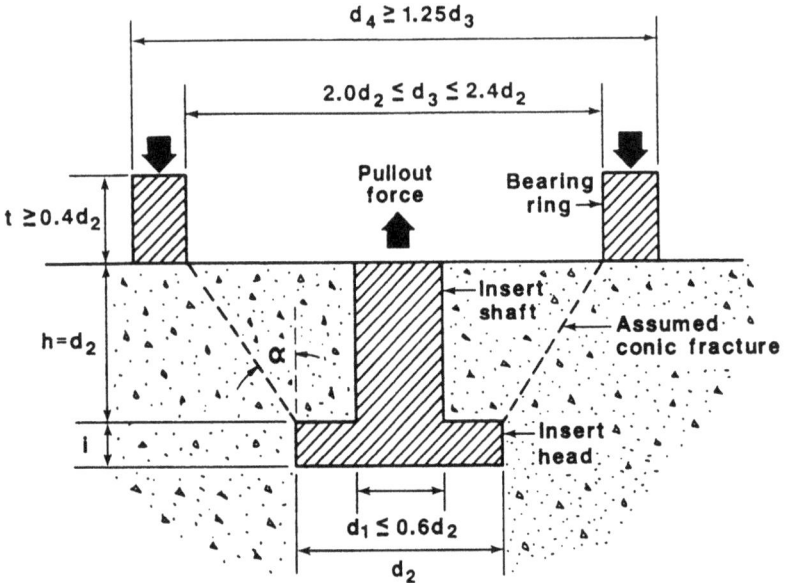

Figure 8.21 The ASTM C900 pull-out test. (© ASTM.)

which was based on the research of Kierkegaard-Hansen (1975). The ASTM standard pull-out test is covered by ASTM C900 (1987). In the standard method a bearing ring is used to react the pull-out force (see Figure 8.21) which to some extent controls the fracture path. Attempts have been made to relate the pull-out strength to the compressive strength of concrete. The correlation relationship originally proposed by the LOK-TEST system for aggregate up to 32 mm was

$$P = 9.48 + 0.829 f_c \qquad \text{(kN, MPa)} \qquad (8.19)$$

However, Bickley (1982) reported considerable differences from eqn 8.19 and proposed different coefficients for differing strength ranges. ACI Committee 228 (1988) recommends establishing an empirical relationship for a range of strengths in the particular concrete being used. Hellier *et al.* (1987) have analysed the pull-out test geometry using the fictitious crack model, assuming that local symmetry is maintained at the crack tip. They found that the primary crack followed an essentially conical path, but with a semi-angle somewhat less than the 35° determined by the test geometry. When the primary crack had grown along a little more than half of the conical surface, a secondary sub-surface crack was initiated under the influence of the reaction ring. This secondary crack grew both towards the free surface and the bolt head. The analysis was terminated shortly after the secondary crack broke through to the free surface, because the value of the maximum compressive stress at this point became sufficient to cause a direct shear failure of the remaining ligament. The same crushing failure mode was predicted in an earlier finite element model using smeared cracks (Ottosen, 1981). The pull-out load when the program was terminated was 22.2 kN which was close to the failure load obtained experimentally.

8.5 Concrete pipes

The failure of pipes would seem to be a fruitful area for the application of fracture mechanics, but there seems to have been no study since that of Gustafsson (1985) which was reported by Hillerborg (1989). In the absence of large defects, the strength of a pipe will tend to its 'strength of materials' value as the size increases. Un-reinforced pipes generally fail in either bending or local crushing. In bending of a thin walled pipe the FPZ will spread from the point of maximum stress around the circumference (see Figure 8.22a). The non-dimensional size will therefore depend upon the ratio of the pipe diameter to the characteristic length, l_{ch}, and the thickness to diameter ratio will be a weak variable. Four hinges must form under local crushing (see Figure 8.22b) as the FPZ spreads from the inside of the pipe to the outside under the local loads and from the outside of the pipe to the inside at the hinges located at the other quarter points. Hence, the non-dimensional size of the pipes under local crushing depends on the ratio of the thickness to characteristic length. Thus, thin walled pipes are non-dimensionally much smaller under local crushing than under bending. Therefore, concrete pipes can be expected to behave in a more ductile fashion under local crushing than under bending.

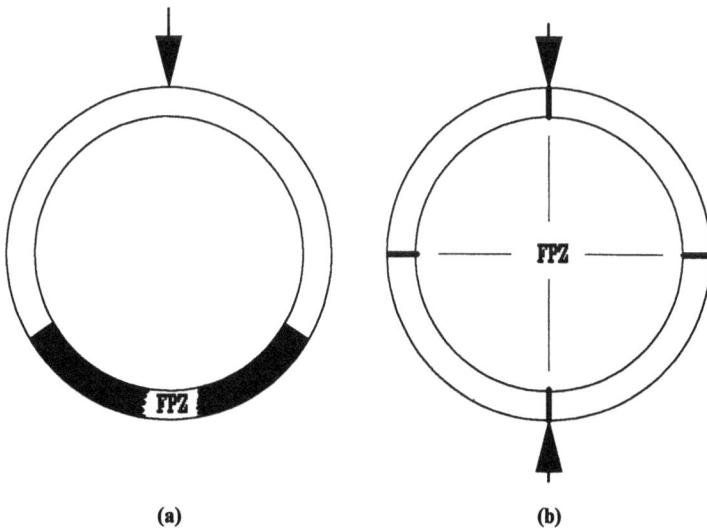

(a)   (b)

Figure 8.22 Failure of un-reinforced concrete pipes (a) in bending and (b) in local crushing.

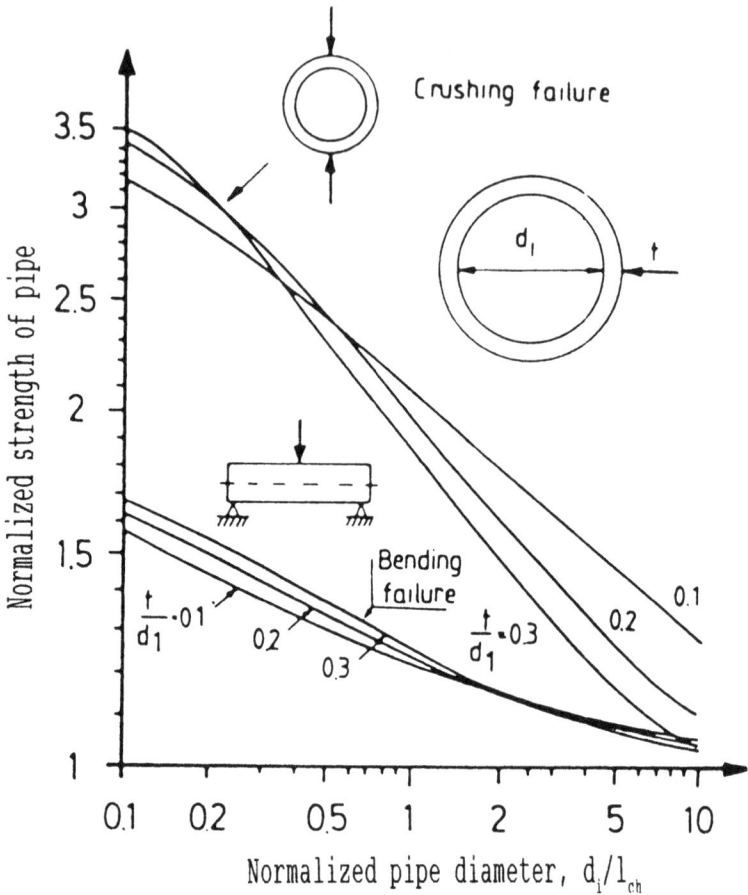

Figure 8.23 Conventional elastic strength of pipes in bending and crushing normalized by the tensile fracture strength of concrete. (After Gustafsson, 1985.)

The ratio of the conventional ultimate elastic strength, σ_u, to the tensile fracture strength, f_t, of the concrete, calculated by Gustafsson (1985) on the basis of the fictitious crack model assuming a linear stress-displacement relationship in the FPZ, is shown in Figure 8.23. The results for both bending and local crushing are presented in terms of the non-dimensional diameter. Not surprisingly then, the 'strength' in local crushing is much higher than the 'strength' in bending. It is reported by Hillerborg (1989) that Scandinavian pipe manufacturers have redesigned their pipes by means of Figure 8.23.

8.6 Summary

The strategy for the design of concrete structures needs to change to take advantage of the advances that have been made in fracture mechanics.

Naturally in an evolving discipline there is no one, universally accepted way of applying fracture mechanics. Some approaches seek to use empirical methods to extend the now well-established LEFM to account for the large FPZ. However, it will greatly impede the development of a rational approach to the design of concrete structures if such empirical approaches are accepted. Empirical methods will be found inadequate to deal with new situations and there will be a never ending need to update the various empirical expressions. Most importantly, such an approach fails to take account adequately of the size effect in concrete structures which is the major contribution of fracture mechanics to concrete design. The design method is not determined by the material properties of concrete alone, size is most important. Small concrete components behave in a ductile fashion and in the limit can be designed by conventional strength of materials. Very large structures with large cracks are brittle and their behaviour can be determined by classic LEFM. However, the ultimate strength of very large structures can only be determined by LEFM if a large crack can grow stably. In an inherently unstable geometry any crack will be small relative to the FPZ at the ultimate load, classic LEFM will not apply, and the FPZ will need to be modelled. The SEL can help in understanding some of the size effects, but only applies if the crack is relatively large at the ultimate load. The empirical approaches using LEFM do not show the true nature of the behaviour of concrete structures and can only be used within the limitations imposed by the data used to obtain them.

In the present chapter it has been shown that if the FPZ is modelled an accurate prediction of the failure modes can be predicted within the limitations imposed by the inherent inhomogeneity of concrete. Some of the advances that have been made in applying statistics to the variations in strength of concrete have been discussed in Chapters 6 and 7. Such work will enable rational factors of safety to be applied to calculations using average values to ensure a safe design.

There are two basic methods of modelling the FPZ: either it can be treated in a discrete fashion with the fictitious crack, or it can be treated as a smeared crack providing the strain-softening relationship is based on a non-local strain. There seems to be no more need to choose one of these two methods of describing the FPZ as 'the Method' than there does between using finite elements or boundary elements in general stress analysis. Each method has its uses and different people will be more confident of using one or the other. Where codes will be called to play their part will be in specifying the material properties of the concrete and, as has been seen in this chapter, the CEB-FIP model code has already gone some way in this regard.

Concrete structures of the future can be more efficiently designed using fracture mechanics, especially if the FPZ is modelled.

References

ACI 349 (1980) Code Requirements for Nuclear Safety Related Concrete Structures.

ACI 228 (1988) *ACI J.*, **85**, 446.

ACI Committee 446 (1989) Fracture of Concrete: Concepts, Models and Determination of Material Properties, Report ACI 446-1R 91.

ACI Committee 446 (1992) Fracture Mechanics: Application to Concrete Structures and Implications with Regard to the Code, Report ACI 446-2R.

Alexander, M.G. (1988) *J. Cem. Concr. Aggregates*, **10**, 9.

Alexander, M.G. and Blight G.E. (1986) In: *Fracture Toughness and Fracture Energy of Concrete* (ed. F.H. Wittmann), p. 323, Elsevier Science Publishers, Amsterdam.

Alexander, M.G., Tait, R.B. and Gill, L.M. (1989) In: *Fracture of Concrete and Rock: Recent Developments* (eds. S.P. Shah, S.E. Swartz and B. Barr), p. 317, Elsevier Applied Science, London.

Alfaiate, J., Pires, E.B. and Martins, J.A.C. (1994) In: *Localized Damage Computer-Aided Assessment and Control III* (eds. M.H. Aliabadi, A. Carpinteri, S. Kalisky and D.J. Cartwright), p. 185, Computational Mechanics Publications, Southampton.

Alvaredo, A.M., Hu, X.Z. and Wittmann, F.H. (1989a) In: *Fracture of Concrete and Rock: Recent Developments* (eds. S.P. Shah, S.E. Swartz and B. Barr), p. 51, Elsevier Applied Science, London.

Alvaredo, A.M., Shah, S.P. and John, R. (1989b) *ASCE J. Eng. Mech.*, **115**, 366.

Andonian, R., Mai, Y.W. and Cotterell, B. (1979) *Int. J. Cem. Composites*, **1**, 151.

Anonymous (1976) *Design of Gravity Dams*, p. 30, US Dept. of the Int., Bureau of Reclamation, Water Resources Technical Publication.

Arrea, M. and Ingraffea, A.R. (1982) *Mixed-Mode Crack Propagation in Mortar and Concrete*, Rpt. 81-13, Dept. Struct. Eng., Cornell Univ., Ithaca, N.Y.

ASTM C900 (1987) *Standard Test Method for Pullout Strength of Hardened Concrete*.

ASTM E399 (1990) *Method of Test for Plane Strain Fracture Toughness of Metallic Materials*.

ASTM E813 (1989) *Standard Test Method for J_{Ic}, a Measure of Fracture Toughness*.

ASTM E1152 (1989) *Standard Test Method for Determining J-R Curves*.

Atkins, A.G. and Mai, Y.W. (1985) *Elastic and Plastic Fracture: Metals, Polymers, Ceramics, Composites, Biological Materials*, Ellis Horwood Ltd., Chichester.

Atkinson, C., Avila, J., Betz, E. and Smelser, R.E. (1972) *J. Mech. Phys. Solids*, **30**, 97.

Aveston, J., Cooper, G.A. and Kelly, A. (1971) In: *Proceedings National Physical Laboratory*, p. 15, IPC Science and Technology Press, London.

Aveston, J. and Kelly, A. (1973) *J. Mater. Sci.*, **8**, 352.

Aveston, J., Mercer, R.A. and Sillwood, J.M. (1974) *Composites—Standards, Testing and Design*, Conference Proceedings, p. 93, National Physical Laboratory, Teddington.

Babut, R. and Brandt, A.M. (1978) In: *Testing and Test Methods of Fibre Cement Composites* (ed. R.N. Swamy), p. 479, The Construction Press, Lancaster.

Bailey, J.E., Chanda, S. and Eden, N.B. (1986) *Fracture Mechanics of Ceramics* (eds. R.C. Bradt *et al.*) Vol. 7, p. 157, Plenum Press, New York.

Baldie. K.D. (1985) PhD Dissertation, University of London.

Baldie, K.D. and Pratt, P.L. (1986) In: *Cement-Based Composites: Strain Rate Effects of Fracture, MRS Symposia Proc.* (eds. S. Mindess and S.P. Shah), Vol. 64, p. 61, Materials Research Society, Pittsburgh.

Ballarini, R., Shah, S.P. and Keer, L.M. (1984) *Eng. Fract. Mech.*, **20**, 433.

Ballarini, R., Keer, L.M. and Shah, S.P. (1986) *Proc. R. Soc. London*, **A404**, 35.

Ballarini, R., Keer, L.M. and Shah, S.P. (1987) *Int. J. Fract.*, **33**, 75.

Ballatore, E., Carpinteri, A., Ferrara, G. and Melchiorri, G. (1990) *Eng. Fract. Mech.*, **35**, 145.

Barenblatt, G.I. (1959) *J. Appl. Math. Mech.*, **23**, 622.
Barenblatt, G.I. (1962) In: *Advances in Applied Mechanics*, **7**, 55, Academic Press, New York.
Barr, B.I.G. and Hasso, E.B.D. (1985) *Mag. Concr. Res.*, **37**, 162.
Bascoul, A., Detriche, C.H., Ollivier, J.P. and Turatsinze, A. (1989a) In: *Fracture of Concrete and Rock: Recent Developments* (eds. S.P. Shah, S.E. Swartz and B. Barr), p. 327, Elsevier Applied Science, London.
Bascoul, A., Ollivier, J.P. and Poushanchi, M. (1989b) *Cem. Concr. Res.*, **19**, 81.
Bažant, Z.P. (1984) *ASCE J. Eng. Mech.*, **110**, 518.
Bažant, Z.P. (1993) *ASCE J. Eng. Mech.*, **119**, 1828.
Bažant, Z.P. (1994) *ASCE J. Eng. Mech.*, **120**, 593.
Bažant, Z.P. and Cao, Z. (1987) *ACI Struct. J.*, **84**, 44.
Bažant, Z.P. and Cedolin, L. (1979) *ASCE J. Eng. Mech.*, **105**, 297.
Bažant, Z.P. and Gambarova, P. (1980) *ASCE J. Struct. Div.*, **106**, 819.
Bažant, Z.P. and Gambarova, P. (1983) *ASCE J. Struct. Eng.*, **110**, 2015.
Bažant, Z.P. and Jirásek, M. (1994) *J. Eng. Mater. Tech.*, **116**, 256.
Bažant, Z.P. and Kazemi, M.T. (1991) *Int. J. Fract.*, **51**, 121.
Bažant, Z.P. and Lin, F.B. 1988 *ASCE J. Struct. Eng.*, **114**, 2493.
Bažant, Z.P. and Oh, B.H. (1983) *Mater. Struct.*, **16**, 155.
Bažant, Z.P. and Ožbolt, J. (1990) *ASCE J. Eng. Mech.*, **116**, 2485.
Bažant, Z.P. and Pfeiffer, P.A. (1987) *ACI Mater. J.*, **84**, 463.
Bažant, Z.P. and Prat, P.C. (1988) *ASCE J. Eng. Mech.*, **114**, 1672.
Bažant, Z.P. and Tsubaki, T. (1980) *ASCE J. Struct. Div.*, **106**, 1947.
Bažant, Z.P. and Xi, Y. (1991) *ASCE J. Eng. Mech.*, **117**, 2623.
Bažant, Z.P., Belytschko, T.B. and Chang, T.P. (1984) *ASCE J. Eng. Mech.*, **110**, 1666.
Bažant, Z.P., Kim, J.K. and Pfeiffer, P.A. (1986) *ASCE J. Struct. Eng.*, **112**, 289.
Bažant, Z.P., Xi, Y. and Reid, S.G. (1991) *ASCE J. Eng. Mech.*, **117**, 2609.
Bažant, Z.P., Ožbolt, J. and Eligehausen, R. (1994) *ASCE J. Struct. Eng.*, **120**, 2377.
Beaudon, J.J. (1986) In: *Fracture Toughness and Fracture Energy of Concrete* (ed. F.H. Wittman), p. 11, Elsevier, Amsterdam.
Begley, J.A. and Landes, J.D. (1972) In: *Fracture Toughness*, ASTM STP 514, p. 1, American Society for Testing and Materials, Philadelphia.
Bentur, A., Diamond, S. and Mindess, J. (1985) *J. Mater. Sci.*, **20**, 3610.
Bergman, B. (1986) *J. Mater. Sci. Letters*, **5**, 611.
Berry, J.P. (1964) In: *Fracture Processes in Polymeric Solids* (ed. B. Rosen), p. 236, Wiley and Sons, New York.
Berthaud, Y. (1989) In: *Fracture of Concrete and Rock* (eds. S.P. Shah and S.E. Swartz), p. 644, Berlin, Springer-Verlag.
Berthelot, J.M. and Robert, J.L. (1987) *J. Acoustic Emission*, **6**, 43.
Bickley, J.A. (1982) *Concr. Int.*, **4**, 44.
Biolzi, L. (1990) *Eng. Fract. Mech.*, **35**, 187.
Biolzi, L., Cangiano, S., Tognon, G. and Carpinteri, A. (1989) *Mater. Struc.*, **22**, 429.
Birchall, J.D., Howard, A.J. and Kendall, K. (1981) *Nature*, **289**, 388.
Birchall, J.D., Howard, A.J. and Kendall, K. (1982) *Proc. Br. Ceram. Soc.*, **32**, 25.
Boca, P., Carpinteri, A. and Valenti, S. (1990) *Eng. Fract. Mech.*, **35**, 159.
Boggs, H.L., Jansen, R.B. and Tarbox, G.S. (1988) In: *Advanced Dam Engineering* (ed. R.B. Jansen), p. 493, Van Nostrand Reinhold, New York.
Bowling, J. and Groves, G.W. (1979) *J. Mater. Sci.*, **14**, 443.
Brandt, A.M. (1980) *Int. J. Cem. Comp.*, **2**, 35.
Broberg, K.B. (1987) *Eng. Fract. Mech.*, **28**, 663.
Brühwiller, E. and Wittmann, F.H. (1989a) *Eng. Fract. Mech.*, **35**, 117.
Brühwiller, E. and Wittmann, F.H. (1989b) *Eng. Fract. Mech.*, **35**, 563.
Bueckner, H.F. *Z. Angew. Math. Mech.*, **50**, 529.
Carino, N.J. (1991) In: *Handbook on Non-Destructive Testing of Concrete* (eds. V.M. Malhotra and N.J. Carino), p. 39, CRC Press, Boca Raton.
Carpinteri, A. (1989a) *ASCE J. Eng. Mech.*, **115**, 1375.
Carpinteri, A. (1989b) *Mater. Struc.*, **22**, 429.
Carpinteri, A. (1990) *Int. J. Fract.*, **44**, 57.

Carpinteri, A. (1991) *Int. J. Fract.*, **51**, 175.
Carpinteri, A., Columbo, G., Ferrara, G. and Guiseppetti, G. (1987) In: *Fracture of Concrete and Rock* (eds. S.P. Shah and S.E. Swartz), p. 131, Springer-Verlag, Berlin.
Carpinteri, A., DiTommaso, A. and Fanelli, M. (1986) In: *Fracture Toughness and Fracture Energy of Concrete* (ed. F.H. Wittmann), p. 117, Elsevier, Amsterdam.
CEB-FIP Model Code 1990, First Predraft 1988, Bulletin d'Information, 190a,b Comité Euro-International du Béton, Lausanne.
Cen, Z. and Mier, G. (1992) *Fatigue Fract. Eng. Mater. Struc.*, **15**, 911.
Cervera, M., Oliver, J., Herrero, E. and Oñate, E. (1990) *Eng. Fract. Mech.*, **35**, 573.
Chappell, J.F. and Ingraffea, A.R. (1981) *A Fracture Mechanics Investigation of the Cracking of Fontana Dam*, Rpt. 81-7, School of Civil and Environmental Engineering, Cornell University, Ithaca.
Chhuy, S., Baron, J. and François, D. (1981) In: *Advances in Fracture Research* (ed. D. François), **4**, 1507.
Chhuy, S. Cannard, G. Robert, J.L. and Acker, P. (1986) In: *Brittle Matrix Composites* (eds. A.M. Brandt and I.H. Marshall), Vol. 1, p. 341, Elsevier Applied Science, London.
Chuang, T.J. and Mai, Y.W. (1989) *Int. J. Solids Struct.*, **25**, 1427.
Chudnovsky, A., Chaoui, K. and Moet, A. (1987) *J. Mater. Sci. Lett.*, **6**, 1033.
Coleman, B.D. (1958) *J. Mech. Phys. Solids*, **7**, 60.
Cook, J. and Gordon, J.E. (1964) *Proc. R. Soc.*, **A282**, 508.
Cornelissen, H.A.W., Hordijk, D.A. and Reinhardt, H.W. (1986) *Heron*, **21**, 45.
Corns, C.F., Tarbox, G.S. and Schrader, E.K. (1988) In: *Advanced Dam Engineering* (ed. R.B. Jansen), p. 466, Van Nostrand Reinhold, New York.
Cotterell, B. (1965) *Int. J. Fract. Mech.*, **1**, 96.
Cotterell, B. (1966) *Int. J. Fract. Mech.*, **2**, 526.
Cotterell, B. (1970) *Int. J. Fract. Mech.*, **6**, 189.
Cotterell, B. and Kamminga, J. (1992) *Mechanics of Pre-Industrial Technology*, Cambridge University Press, Cambridge.
Cotterell, B. and Mai, Y.W. (1987) *J. Mater. Sci.*, **22**, 2734.
Cotterell, B. and Mai, Y.W. (1988a) *Mater. Forum*, **11**, 341.
Cotterell, B. and Mai, Y.W. (1988b) *Adv. Cem. Res.*, **1**, 75.
Cotterell, B. and Mai, Y.W. (1991) In: *Fracture of Engineering Materials and Structures* (eds. S.H. Teoh and K.H. Lee), p. 348. Elsevier Applied Science, London.
Cotterell, B. and Rice, J.R. (1980) *Int. J. Fract.*, **16**, 155.
Cotterell, B., Mai, Y.W. and Foote, R.M.L. (1988) In: *Engineering Applications of New Composites* (eds. S.A. Paipetis and G.C. Papanicolaou), p. 186, Omega Scientific, Wallingford, UK.
Cotterell, B., Paramasivan, P. and Lam, K.Y. (1992) *Mater. Struct.*, **25**, 14.
Cotterell, B., Mai, Y.W. and Lam, K.Y. (1995) *Cem. Concr. Res.*, **25**, 408.
Cox, H.L. (1952) *Br. J. Appl. Phys.*, **3**, 72.
CP110 (1972) *The Structural Use of Concrete, Part 1. Design Materials and Workmanship*, London.
Curtin, W.A. (1991) *J. Am. Ceram. Soc.*, **74**, 2837.
Curtin, W.A. (1993a) *J. Mech. Phys. Solids*, **41**, 35.
Curtin, W.A. (1993b) *J. Mech. Phys. Solids*, **41**, 217.
Daniels, H.E. (1945) *Proc. R. Soc.*, **A186**, 405.
Daniels, H.E. (1989) *Adv. Appl. Prob.*, **21**, 315.
de Borst, R. and Nauta, P. (1985) *Eng. Comp.*, **2**, 35.
de Vekey, R.C. and Majumdar, A.J. (1970) *J. Mater. Sci.*, **5**, 183.
Diamond, S. and Bentur, A. (1985) In: *Application of Fracture Mechanics to Cementitious Composites* (ed. S.P. Shah), p. 87, Martinus Nijhoff, Dordrecht.
Diaz, G. and Kittl, P. (1988) *Int. J. Struct. Mech. Mater. Sci.*, **24**, 209.
Divaker, M.P., Fafitis, A. and Shah, S.P. (1987) *ASCE J. Struct. Eng.*, **113**, 1046
Du, J.J., Kobayashi, A.S. and Hawkins, N.M. (1987) In: *Fracture of Concrete and Rock* (eds. S.P. Shah and S.E. Swartz), p. 199, Springer-Verlag, Berlin,.
Du, J.J., Hawkins, N.M. and Kobayashi, A.S. (1989) In: *Fracture of Concrete and Rock: Recent Developments* (eds. S.P. Shah, S.E. Swartz and B. Barr), p. 297, Elsevier Applied Science, London.
Duan, K., Mai, Y.-W. and Cotterell, B. (1995) *J. Mater. Sci.*, **30**, 1405.
Dugdale, D.S. (1960) *J. Mech. Phys. Solids*, **8**, 100.

Eden, N.B. and Bailey, J.E. (1984a) *J. Mater. Sci.*, **19**, 150.
Eden, N.B. and Bailey, J.E. (1984b) *J. Mater. Sci.*, **19**, 2677.
Eden, N.B. and Bailey, J.E. (1985a) *J. Mater. Sci.*, **20**, 1137.
Eden, N.B. and Bailey, J.E. (1985b) *J. Mater. Sci.*, **20**, 3419.
Eden, N.B. and Bailey, J.E. (1986) In: *Proc. 8th Int. Congress on the Chemistry of Cement*, Vol. 3, p. 382, Rio de Janeiro.
Eden, N.B. and Bailey, J.E. (1988) *Adv. Ceram.*, **22**, 249.
Eden, N.B. and Bailey, J.E. (1989) In *Fracture Mechanics of Concrete Structures: from Theory to Applications* (ed. L. Elfgren), RILEM Report 90-FMA, Chapman and Hall, London.
Elfgren, L. Ohlsson, U. and Gylltoft, K. (1987) In: *Fracture of Concrete and Rock* (ed. S.P. Shah and S.E. Swartz), p. 269, Springer-Verlag, Berlin.
Elices, M., Guinea, G.V. and Planas, J. (1992) *Mater. Struct.*, **25**, 327.
Eligehausen, R. and Pusill-Wachtsmuth, P. (1982) *IABSE Periodica*, No. 1.
Eligehausen, R. and Sawade, G. (1989) In: *Fracture Mechanics of Concrete Structures: from Theory to Applications* (ed. L. Elfgren) p. 281, Chapman and Hall, London.
Erdogan, F. and Sih, G.C. (1963) *J. Basic Eng.*, **85** 507.
Evans, A.G. (1972) *J. Mater. Sci.*, **7**, 1137.
Evans, A.G. and Fuller, E.R. (1974) *Metall. Trans. A*, **5**, 27.
Evans, A.G. and Wiederhorn, S.M. (1974) *J. Mater. Sci.*, **9**, 270.
Ewing, P.D. and Williams, J.G. (1974) *Int. J. Fract.*, **10**, R135.
Ewing, P.D. Swedlow, J.L. and Williams, J.G. (1976) *Int. J. Fract.*, **12**, 85.
Fanelli, M. and Giuseppeti, G. (1990) *Eng. Fract. Mech.*, **35**, 525.
Fardis, M.N. and Buyukzturk, O. (1979) *ASCE J. Eng. Mech. Div.*, **105**, 255.
Fenwick, R.C. and Pauley, T. (1968) *ASCE J. Struct. Div.*, **94**, 2325.
Ferrara, G. and Morabito, P. (1989) In: *Fracture of Concrete and Rock: Recent Developments* (eds. S.P. Shah, S.E. Swartz and B. Barr), p. 377, Elsevier Applied Science, London.
Finnie, I. and Saith, A. (1973) *Int. J. Fract.*, **9**, R484.
Finnie, I. and Weiss, H.D. (1974) *Int. J. Fract.*, **10**, R136.
Foegl, H. and Mang, H.A. (1982) *ASCE J. Struct. Div.*, **108**, 2681.
Foote, R.M.L. (1986) PhD Dissertation, University of Sydney, Sydney, Australia.
Foote, R.M.L., Cotterell, B. and Mai, Y.W. (1980) In: *Advances in Cement Composites* (eds. D.M. Roy, A.J. Majumdar, S.P. Shah and J.A. Manson), p. 135, Materials Research Society, Pittsburg.
Foote, R.M.L., Cotterell, B. and Mai, Y. W. (1986a) In: *Fracture Toughness and Fracture Energy of Concrete* (ed. F.H. Wittmann), p. 91, Elsevier, London.
Foote, R.M.L., Mai, Y.W. and Cotterell, B. (1986b) *J. Mech. Phys. Solids*, **34**, 593.
Foote, R.M.L., Mai, Y.W. and Cotterell, B. (1987) In: *Fibre Reinforced Concrete Properties and Applications*, SP-105 (eds. S.P. Shah and G.B. Batson) Vol. 2, p. 55, American Concrete Institute, Detroit.
Galileo, G. [1638] (1914) *Dialogues Concerning Two New Sciences*, Trans. H. Crew and A. de Salvico, Macmillan, New York.
Gambarova, P.G. and Giuriani, M. (1985) *ASCE J. Struct. Div.*, **111**, 1161.
Gambarova, P.G. and Karakoç, C. (1982) In: *Bond in Concrete* (ed. P. Bartos), p. 82, Applied Science Publishers, London.
Gambarova, P.G., Rosati, G.P. and Zasso, B. (1989a) *Mater. Struct.*, **22**, 35.
Gambarova, P.G., Rosati, G.P. and Zasso, B. (1989b) *Mater. Struct.*, **22**, 347.
Gao, Y.C., Mai, Y.W. and Cotterell, B. (1988) *J. Appl. Maths. Phys.*, **39**, 550.
Gardner, N.J. (1990) *ACI Struct. J.*, **87**, 66.
Gerstle, W., Ingraffea, A.R. and Gergely, P. (1982) In: *Bond in Concrete* (ed. P. Bartos), p. 97, Applied Science Publishers, London.
Gilbert, R.I. and Warner, R.F. (1978) *ASCE J. Struct. Div.*, **104**, 85.
Giuriani, E. (1982) In: *Bond in Concrete* (ed. P. Bartos), p. 107, Applied Science Publishers, London.
Gol'dstein, R.V. and Salaganik, R.L. (1974) *Int. J. Fract.*, **10**, 507.
González-Vidosa, F., Kotsovos, M.D. and Pavlovic, M.N. (1988) *ACI Struct. J.*, **85**, 241.
Gopalaratnam, V.S. and Shah, S.P. (1985) *ACI Mater. J.*, **82**, 310.
Goto, Y. (1971) *J. Am. Concr. Inst.*, **68**, 244.
Goto, Y. and Otsuka, K. (1979) *Technology Report Tohoku University*, **44**, 49.

Goto, Y., Obata, M., Maeno, H. and Kobayashi, Y. (1993) *J. Struct. Eng.*, **119**, 1168.

Griffith, A.A. (1921) *Trans. R. Soc. London, Ser. A*, **221**, 163.

Griffith, A.A. (1925) *Proc. 1st Int. Conf. Appl. Mech.* (eds. C.B. Biezeno and J.M. Burgers) Delft, Technische Boekhandel en Drukkerij, 55.

Guinea, G.V., Planas, J. and Elices, M. (1992) *Mater. Struct.*, **25**, 212.

Guo, Z.H. and Zhang, X.Q. (1987) *ACI Mater. J.*, **84**, 278.

Gurney, C. and Hunt, J. (1967) *Proc. R. Soc. London*, **A299**, 508.

Gurney, C. and Pearson, S. (1948) *Proc. R. Soc. London*, **A192**, 537.

Gustafsson, P.J. (1985) *Fracture Mechanics Studies of Non-Yielding Materials Like Concrete*, Report TVBM-1007, Div. Building Materials, University of Lund, Lund.

Hand, F.R., Pecknold, D.A. and Schnobrich, W.C. (1973) *ASCE J. Struct. Eng.*, **99**, 1491.

Hannant, D.J. (1978) *Fibre Cements and Fibre Concretes*, John Wiley and Sons, Chichester.

Hannant, D.J., Hughes, D.C. and Kelly, A. (1983) *Phil. Trans. R. Soc. London*, **A310**, 175.

Harde, N.A. (1991) *Computer Simulated Crack Propagation in Concrete Beams by Means of Fictitious Crack Method and Boundary Element Method*, Dept. of Building Tech. and Struct. Engn., University of Aalborg, Denmark.

Harris, B. (1980) *Metal Sci.*, **14**, 351.

Harris, B., Varlow, J. and Ellis, C.D. (1972) *Cem. Concr. Res.*, **2**, 447.

Hashida, T., Takahashi, H., Kobayashi, S. and Fukagawa, Y. (1990) *Cem. Concr. Res.*, **20**, 687.

Helfet,T.L. and Harris, B. (1972) *J. Mater. Sci.*, **7**, 494.

Helfinstine, J.D. (1980) *J. Am. Ceram. Soc.*, **60**, 113.

Hellier, A.K., Sansalone, M., Carino, N.J., Stone, W.C. and Ingraffea, A.R. (1987) *Cem. Concr. Aggregates*, **9**, 20.

Hibbert, A.P. and Hannant, D.J. (1982) *Composites*, **13**, 105.

Higgins, D.D. and Bailey, J.E. (1976a) *J. Mater. Sci.*, **11**, 1995.

Higgins, D.D. and Bailey, J.E. (1976b) In: *Proc. Conf. on Hydraulic Cement Paste*, p. 283, E & FN SPON, New York.

Hillerborg, A. (1980) *Int. J. Cem. Comp.*, **2**, 177.

Hillerborg, A. (1983) In: *Fracture Mechanics of Concrete* (ed. F.H. Wittmann), p. 223, Elsevier, Amersterdam.

Hillerborg, A. (1985a) *Mater. Struct.*, **107**, 407.

Hillerborg, A. (1985b) *Mater. Struct.*, **107**, 291.

Hillerborg, A. (1989) In: *Fracture Mechanics of Concrete Structures: from Theory to Applications* (ed. L. Elfgren), p. 314, Chapman and Hall, London.

Hillerborg, A., Modeer, M. and Petersson, P.E. (1976) *Cem. Concr. Res.*, **6**, 773.

Hilsdorf, H.K. and Brameshuber, W. (1985) In: *Application of Fracture Mechanics to Cementitious Materials* (ed. S.P. Shah), p. 361, Martinus Nijhoff, Dordrecht/New York/Lancaster.

Hilsdorf, H.K. and Brameshuber, W. (1991) *Int. J. Fract.*, **51**, 61.

Hoek, E. and Bieniawski, Z.T. (1965) *Int. J. Fract. Mech.*, **1**, 137.

Hordijk, D.A., Reinhardt, H.W. and Cornelissen, H.A.W. (1987) In: *Proc. SEM-RILEM Int. Conf. on Fracture of Concrete and Rocks* (eds. S.P. Shah and S.E. Swartz), p. 138, Springer-Verlag, New York.

Horvath, R. and Petersson (1984) *The Influence of the Size of the Specimen on the Fracture Energy of Concrete*, Rpt. TVBM-5005, Division of Building Materials, University of Lund, Lund.

Hu, X.Z. and Mai, Y.W. (1992a) *J. Mater. Sci.*, **27**, 3502.

Hu, X.Z. and Mai, Y.W. (1992b) *J. Am. Ceram. Soc.*, **75**, 848.

Hu, X.Z. and Wittmann, F.H. (1989) In: *Fracture of Concrete and Rock: Recent Developments* (eds. S.P. Shah, S.E. Swartz and B. Barr), p. 307, Elsevier Applied Science, London.

Hu, X.Z. and Wittmann, F.H. (1990) *J. Mater. Civil Eng.*, **2**, 15.

Hu, X.Z. and Wittmann, F.H. (1991) *Cem. Concr. Res.*, **21**, 1118.

Hu, X.Z. and Wittmann, F.H. (1992a) *Cem. Concr. Res.*, **22**, 559.

Hu, X.Z. and Wittmann, F.H. (1992b) *Mater. Struct.*, **107**, 407.

Hu, X.Z., Cotterell, B. and Mai, Y.W. (1985) *Proc. R. Soc. London*, **A410** 251.

Hu, X.Z., Cotterell, B. and Mai, Y.W. (1986a) In: *Fracture Mechanics of Concrete* (ed. F.H. Wittmann), p. 91, Elsevier, New York.

Hu, X.Z., Cotterell, B. and Mai, Y.W. (1986b) In: *Proc. 3rd Int. Symp. Development of Fibre Reinforced Cement and Concrete* (eds. R.N. Swamy *et al.*), Paper 6.4, Sheffield, Cement & Concrete Association, UK.

Hu, X.Z., Cotterell, B. and Mai, Y.W. (1988a) *Phil. Mag. Lett.*, **57**, 69.
Hu, X.Z., Mai, Y.W. and Cotterell, B. (1988b) *Phil. Mag.*, **58**, 292.
Hu, X.Z., Mai, Y.W. and Cotterell, B. (1989) *J. Mater. Sci.*, **24**, 3118.
Hu, X.Z., Mai, Y.W. and Cotterell, B. (1991) *Phil. Mag.*, **64**, 1265.
Hu, X.Z., Mai, Y.W. and Cotterell, B. (1992) *Phil. Mag.*, **66**, 173.
Hunt, R.A. and McCartney, L.N. (1979) *Int. J. Fract.*, **15**, 365.
Hussain, M.C., Pu, S.L. and Underwood, J. (1973) In: *Fracture Analysis*, ASTM SP560, p. 2.
Hutchinson, J.W. (1983) *J. Appl. Mech.*, **50**, 1042.
Hutchinson, J.W. and Paris P.C. (1977) In: *Elastic-Plastic Fracture*, ASTM STP 668, p. 37, American Society for Testing and Materials, Philadelphia.
Inglis, C.E. (1913) *Trans. Inst. Naval Archit.*, **55**, 219.
Ignacio, C., Prat, P.C. and Bažant, Z.P. (1992) *Int. J. Solids Struct.*, **29**, 1173.
Ingraffea, A.R. (1990) *Eng. Fract. Mech.*
Ingraffea, A.R. and Gerstle, W.H. (1985) In: *Applications of Fracture Mechanics to Cementitious Composites* (ed. S.P. Shah), p. 247, Martinus Nijhoff, Dordrecht/Boston/Lancaster.
Ingraffea, A.R. and Panthaki, M.J. (1985) *US–Japan Seminar on Finite Element Analysis of Reinforced Concrete Structures*, Vol. 1, p. 71, ASCE, New York.
Ingraffea, A.R. and Saouma, V. (1984) In: *Application of Fracture Mechanics to Concrete Structures* (eds. G.C. Sih and A. Di Tommaso), Martinus Nijhoff, Holland.
Ingraffea, A.R., Gerstle, W.H., Gergely, P. and Saouma, M. (1984) *Struct. Eng.*, **110**, 871.
Irwin, G.R. (1948) In: *Fracturing of Metals*, ASM, Cleveland, p. 147.
Irwin, G.R. (1957) *J. Appl. Mech.*, **24**, 361.
Irwin, G.R. (1958) In: *Handbuch der Physik*, Vol. VI (ed. W. Flugge), p. 551, Springer-Verlag, Berlin.
Irwin, G.R. (1960) In: *Structural Mechanics, Proc. 1st Symp. on Naval Structural Mechanics*, p. 557.
Isida, M. (1973) In: *Methods of Analysis and Solution of Crack Problems* (ed. G.C. Sih), p. 56, Noordhoff, Amsterdam.
Jakus, K., Coyne, D.C. and Ritter, J.E. (1978) *J. Mater. Sci..*, **13**, 2071.
Jayatilaka, A. De S. and Trustrum, K. (1977) *J. Mater. Sci.*, **12**, 1426.
Jenq, Y.S. and Shah, S.P. (1985) In: *Application of Fracture Mechanics to Cementitious Composites* (ed. S.P. Shah), p. 319, Martinus Nijhoff, Dordrecht/Boston/Lancaster.
Jenq, Y.S. and Shah, S.P. (1985) *ASCE J. Eng. Mech. Div.*, **111**, 1227.
Jenq, Y.S. and Shah, S.P. (1986a) *ASCE J. Struct. Eng.*, **112**, 19.
Jenq, Y.S. and Shah, S.P. (1986b) In: *Fracture Toughness and Fracture Energy of Concrete* (ed. F.H. Wittmann), p. 499, Elsevier, Amsterdam.
Jenq, Y.S. and Shah, S.P. (1988) *Int. J. Fract.*, **38**, 123.
Jirásek, M. and Bažant, Z.P. (1994) *ASCE J. Eng. Mech.*, **120**, 1521.
John, R. and Shah, S.P. (1986) *J. Cem. Concr. Aggregates*, **8**, 24.
Johnston, C.D. (1982) *Cem. Concr. Aggregates*, **2**, 53.
Kalisky, S. (1989) *Plasticity Theory and Engineering Applications*, Elsevier, Amsterdam.
Karihaloo, B.L. and Nallathambi, P. (1988) *A Notched Beam Test: Mode I Fracture Toughness*, Final Report to RILEM TC89-FTM.
Karihaloo, B.L. and Nallathambi, P. (1989a) *Mater. Struct.*, **22**, 185.
Karihaloo, B.L. and Nallathambi, P. (1989b) *Cem. Concr. Res.*, **19**, 603.
Keer, J.G. (1984) In: *New Reinforced Concretes, Concrete Technology and Design* (ed. R.N. Swamy) Vol. 2, p. 52, Surrey University Press.
Kelly, A. and Macmillan, N.H. (1986) *Strong Solids*, 3rd Edition, Claredon Press, Oxford.
Kendall, K. and Birchall, J.D. (1985) *Mater. Res. Soc. Symp. Proc.*, **42**, 143.
Kendall, K., Howard, A.J. and Birchall, J.D. (1983) *Phil Trans. R. Soc.*, **A310**, 139.
Kendall, K., Alford, N.McN., Tan, S.R. and Birchall, J.D. (1986) *J. Mater. Res.*, **1**, 120.
Kesler, C., Naus, D. and Lott, J. (1972) In: *Proc. Int. Conf. on Mechanical Behaviour of Materials*, Vol. 4, p. 113, Soc. Mater. Sci. Japan, Kyoto.
Kierkegaard-Hansen, P. (1975) *Nordisk Betong*, **3**, 19.
Kim, J.K. and Mai, Y.W. (1991) *Comp. Sci. Techn.*, **41**, 333.
Kim, J.K., Zhou, L.M. and Mai, Y.W. (1993) *J. Mater. Sci.*, **28**, 3923.
Kinnunen, S. and Nylander, H. (1960) Punching of Concrete Slabs Without Reinforcement, *Meddelande* no. 38, Institutionen för Byggnadsttik Kungliga Tekniska Högskolan, Stockholm.

Kormeling, H.A. and Reinhardt, H.W. (1981) *Determination of the Fracture of Normal Concrete and Epoxy Modified Concrete*, RPT 1 5-83-18, Steven Laboratory, Delft University of Technology.

Knab, L.I., Walker, J.N., Clifton, J.R. and Fuller, E.R. (1984) *Cem. Concr. Res.*, **14**, 339.

Knab, L.I., Jennings, H., Walker, J.N., Clifton, J.R. and Grimes, J.W. (1986) In: *Fracture Toughness and Fracture Energy of Concrete* (ed. F.H. Wittmann), p. 241, Elsevier Science Publishers, Amsterdam.

Knehans, R. and Steinbrech, R. (1982) *J. Mater. Sci. Lett.*, **1**, 327.

Knott, J.F. (1973) *Fundamentals of Fracture Mechanics*, Butterworths, London.

Kobayashi, A.S., Hawkins, N.M., Barker, D.B. and Liaw, B.M. (1985) In: *Applications of Fracture Mechanics to Cementitious Composites* (ed. S.P. Shah), p. 25, Martinus Nijhoff Publishers, Dordrecht.

Krafft, J.M., Sullivan, A.M. and Boyle, R.W. (1961) In: *Proc. Symp. Crack Propagation, Cranfield*, p. 8, College of Aeronautics, Cranfield.

Krause, R.F. and Fuller, E.R. (1984) *ASTM STP 855*, 309.

Krenchel, H. (1964) *Fibre Reinforcement*, Akademisk Forlag, Copenhagen.

Lam, K.Y., Cotterell, B. and Phua, S.P. (1991) *J. Amer. Ceram. Soc.*, **74**, 2527.

Landes, J.D. and Begley, J.A. (1974) In: *ASTM STP 560* p. 170, American Society for Testing and Materials, Philadelphia.

Larbi, J.A. (1993) Microstructure of the Interfacial Zone around Aggregate Particles in Concrete, *Heron*, **38**(1), 69.

Lawn, B.R. (1993) *Fracture of Brittle Solids*. Cambridge University Press, Cambridge.

Lawrence, P.J. (1980) *J. Mater. Sci.*, **7**, 351.

Lenain, J.C. and Bunsell, A.R. (1979) *J. Mater. Sci.*, **14**, 321.

Li, S.H., Shah, S.P., Li, Z.J. and Mura, T. (1993) *Int. J. Solids Struct.*, **30**, 1429.

Li, V.C. (1985) In: *Application of Fracture Mechanics to Cementitious Materials* (ed. S.P. Shah), p. 431, Martinus Nijhoff, Dordrecht/Boston/Lancaster.

Li, V.C. and Ward, R. (1988) In: *Proc. Int. Workshops on Fracture Toughness and Fracture Energy: Test Methods for Concrete and Rock* (ed. H. Mihashi), p. 139, Tohoku University, Japan.

Li, V.C., Chan, C.M. and Leung, K.Y. (1987) *Cem. Concr. Res.*, **17**, 441.

Li, V.C., Wang, Y. and Backer, S. (1990) *Composites*, **21**, 132.

Li, V.C., Ward, R. and Hamza, A.M. (1992) In: *Applications of Fracture Mechanics to Reinforced Concrete* (ed. A. Carpinteri), p. 503, Elsevier Applied Science, London.

Liang, R.Y.K. and Li, Y.-N. (1991) *Comp. Mech.*, **7**, 413.

Liaw, B.M., Leang, F.L., Du, J.J., Hawkins, N.M. and Kobayashi, A.S. (1990) *ASCE J. Eng. Mech.*, **116**, 429.

Lim, T.Y., Paramasivam, P. and Lee, S.L. (1987a) *ACI Struct. J.*, **84**, 524.

Lim, T.Y., Paramasivam, P. and Lee, S.L. (1987b) *ACI Mater. J.*, **84**, 286.

Linsbauer, H.N. (1989a) *Eng. Fract. Mech.*, **35**, 541.

Linsbauer, H.N. (1989b) In: *Fracture Mechanics of Concrete Structures: from Theory to Applications* (ed. L. Elfgren), p. 329, Chapman and Hall, London.

Linsbauer, H.N. and Rossmanith, H.P. (1984) *Eng. Fract. Mech.* **19**, 195.

Linsbauer, H.N., Ingraffea, A.R., Rossmanith, H.P. and Wawryzynek, P.A. (1989) *J. Struct. Eng.*, **115**, 1599.

Llorca, J. and Elices, M. (1993) *Eng. Fract. Mech.* **44**, 341.

Luong, M.P. (1986) In: *Fracture Toughness and Fracture Energy of Concrete* (ed. F.H. Wittmann), p. 249, Elsevier Science Publishers, Amsterdam.

McCartney, L.N. (1979) *Int. J. Fract.*, **15**, 477.

Mahajan, R.V. and Ravi-Chandar, K. (1989) *Int. J. Fract.*, **41**, 235.

Mai, Y.W. (1979a) *J. Mater. Sci.*, **14**, 2091.

Mai, Y.W. (1979b) *Int. J. Cem. Composites*, **1**, 151.

Mai, Y.W. (1988) *Mater. Forum*, **11**, 232.

Mai, Y.W. (1992) In: *Applications of Fracture Mechanics to Reinforced Concrete* (ed. A. Carpinteri), p. 201, Elsevier Applied Science, London.

Mai, Y.W. and Gurney, C. (1975) *Phys. Chem. Glasses*, **16**, 70.

Mai, Y.W. and Hakeem, M.I. (1984) *J. Mater. Sci.*, **19**, 501.

Mai, Y.W. and Lawn, B.R. (1987) *J. Amer. Ceram. Soc.*, **70**, 290.

Mai, Y.W., Foote, R.M.L. and Cotterell, B. (1980) *Int. J. Cem. Comp.*, **16**, 155.

Mai, Y.W., Hakeem, M.I. and Cotterell, B. (1983) *J. Mater. Sci.*, **18**, 2156.

Mai, Y.W., Barakat, B., Cotterell, B. and Swain, M.V. (1990) *Phil. Mag.*, **62**, 347.

Mai, Y.W., Hu, X.Z., Duan, K. and Cotterell, B. (1992) In: *Fracture Mechanics of Ceramics* (eds. R.C. Bradt, D.P.H. Hasselman, D. Munz, M. Sakai and V.Ya. Shevchenko), Vol. 10, p. 387, Plenum Press, New York.

Majaumdar, A.J. and Walton, P.L. (1985) In: *Application of Fracture Mechanics to Cementitious Composites* (ed. S.P. Shah), p. 157, Martinus Nijhoff, Dordrecht/Boston/Lancaster.

Maji, A. and Shah, S.P. (1988) In: *Bonding in Cementitious Materials, MRS Symp. Proc.* (eds. S. Mindess and S.P. Shah) Vol. 114, p. 55, Materials Research Society, Pittsburg.

Maji, A., Ouyang, C. and Shah, S.P. (1990) *J. Mater. Res.*, **5**, 206.

Malmberg, B. and Skarendahl, H. (1978) In: *Testing and Test Methods of Fibre Cement Composites* (ed. R.N. Swamy), Proc. RILEM Symp., The Construction Press, Lancaster.

Malvar, L.J. (1992) *ACI Struct. J.*, **89**, 569.

Malvar, L.J. and Warren, G.E. (1988) *Exp. Mech.*, **28**, 266.

Marsden, E.W. (1969) *Greek and Roman Artillery—Historical Development*, Clarendon Press, Oxford.

Mazars, J., Pijaudier-Cabot, G. and Saouridis, C. (1991) *Int. J. Fract.*, **51**, 159.

Melin, S. (1986) *Int. J. Fract.*, **30**, 103.

Melin, S. (1987) *Int. J. Fract.*, **32**, 257.

Melin, S. (1989) *Mater. Struct.*, **22**, 23.

Mindess, S. (1983) In: *Fracture Mechanics of Concrete* (ed. F.H. Wittmann), p. 1, Elsevier, New York.

Mindess, S. (1984) *Cem. Concr. Res.*, **14**, 431.

Mindess, S. (1985) In: *Application of Fracture Mechanics to Cementitious Composites* (ed. S.P. Shah), p. 617, Martinus Nijhoff Publishers, Dordrecht/Boston/Lancaster.

Mindess, S. (1991a) In: *Fracture Mechanics Test Methods for Concrete* (eds. S.P. Shah and A. Carpinteri), p. 231, Chapman and Hall, London.

Mindess, S. (1991b) In: *Toughening Mechanisms in Quasi-Brittle Materials* (ed. S.P. Shah), p. 271, Kluwer Academic Publishers, Dordrecht.

Mindess, S. and Diamond, S. (1982a) *Mater. Struct.*, **15**, 107.

Mindess, S. and Diamond, S. (1982b) *Cem. Concr. Res.*, **12**, 569.

Mindess, S. and Nadeau, J. (1977) *Bull. Am. Ceram. Soc.*, **54**, 478.

Mobasher, B., Ouyang, C.S. and Shah, S.P. (1991) *Int. J. Fract.*, **50**, 199.

Modeer, M. (1979) *A Fracture Mechanics Approach to Failure Analysis of Concrete Materials*, University of Lund, Report TVBM-1001.

Morton, J. (1979) *Mater. Struct.*, **12**, 393.

Murdock, J.W. and Kesler, C.E. (1958) *J. Am. Concr. Inst.*, **55**, 221.

Muskhelishvili, N.I. (1953) *Some Basic Problems of the Mathematical Theory of Elasticity*, Noordhoff, Groningen.

Naik, T.R. (1991) In: *Handbook on Non-Destructive Testing of Concrete* (eds. V.M. Malhotra and N.J. Carino), p. 83, CRC Press, Boca Raton.

Nallathambi, P. and Karihaloo, B.L. (1986) *Eng. Fract. Mech.*, **25**, 315.

Naaman, A.E. and Shah, S.P. (1975) In: *RILEM Symposium Fibre-Reinforced Cement and Concrete*, p. 171, The Construction Press, Lancaster.

Naaman, A.E., Reinhardt, H.W., Fritz, C. and Alwan, J. (1993a) *Mater. Struct.*, **26**, 522.

Naaman, A.E., Reinhardt, H.W., Fritz, C. and Alwan, J. (1993b) In: *Structural Engineering in Natural Hazards Mitigation*, p. 1396, ASCE, New York.

Nadeau, J.S., Mindess, S. and Hay, J.M. (1974) *J. Am. Ceram. Soc.*, **57**, 51.

Nishioka, K., Kamimi, N., Yamakawa, S. and Shirakawa, K. (1975) In: *Fibre Reinforced Cement and Concrete*, p. 425, The Construction Press, Lancaster.

Olsson, P.-A. (1985) *A Fracture Mechanics and Experimental Approach on Anchorage Splitting*, Nordic Concrete Research Publication No. 4, The Nordic Concrete Federation.

Ono, H. and Ohgishi, S. (1989) In: *Fracture Toughness and Fracture Energy* (eds. H. Mihashi, H. Takahashi and F.H. Wittmann), p. 73, Balkema, Rotterdam.

Orowan, E. (1948) *Rept. Prog. Phys.*, **12**, 185.

Ottosen, N.S. (1981) *J. Struct. Div. ASCE*, **107**, 591.

Ouyang, C.S. and Shah, S.P. (1992) *Cem. Concr. Res.*, **22**, 1201.
Ouyang, C.S., Mobasher, B. and Shah, S.P. (1990) *Eng. Fract. Mech.*, **37**, 901.
Pak, A.P. and Trapeznikov (1981) In: *Adv. Fract. Res.* (ed. D. Francois), **4**, 1531.
Palaniswamy, K. and Knauss, W.G. (1978) In: *Mechanics Today* (ed. S. Nermat-Nasser), Vol. 4, p. 87, Pergamon Press, New York.
Paris, P.C. and Sih, G.C.M. (1965) In: *Fracture Toughness Testing and its Application*, ASTM STP 381 (ed. W.F. Brown), p. 30, American Society for Testing and Materials, Philadelphia.
Patterson, W.A. and Chan, H.C. (1975) *Composites*, **6**, 102.
Pauley, T. and Loeber, P.J. (1974) *SP 42*, American Concrete Institute, 1.
Petersson, P.E. (1985) *Crack Growth Development of Fracture Zones in Plain Concrete and Similar Materials*, Rpt. TVBM-1006 Div. Bldg. Mats., Lund Institute of Technology.
Phoenix, S.L. (1993) *Composites Sci. Techn.*, **48**, 65.
Phoenix, S.L. and Raj, R. (1992) *Acta Metall. Mater.*, **40**, 2813.
Piggott, M.R. (1980) *Load Bearing Fibre Composites*, Pergamon Press, Oxford.
Pinchin, D.J. and Tabor, D. (1978) *J. Mater. Sci.*, **13**, 1261.
Planas, J., Elices, M. and Guinea, G.V. (1992) *Mater. Struct.*, **25**, 305.
Poon, C.S. and Groves, G.W. (1987) *J. Mater. Sci.*, **22**, 2148.
Portela, A., Aliabadi, M.H. and Rooke, D.P. (1992) *Int. J. Num. Meth. Eng.*, **33**, 1269.
Radon, J.C., Lever, P.S. and Culver, L.E. (1977) In: *Fracture*, **3**, 113, University of Waterloo Press, Waterloo, Canada.
Raiss, M.E., Dougill, J.W. and Newman, J.B. (1989) In: *Fracture of Concrete and Rock: Recent Developments* (eds. S.P. Shah, S.E. Swartz and B. Barr), p. 243, Elsevier Applied Science, London.
Reinhardt, H.W. (1984) *Heron*, **2**, No. 2.
Reinhardt, H.W. and Hordijk, D.A. (1988) In: *France–US Workshop on Strain Localisation and Size Effect Due to Cracking and Damage*, Cachan, France.
Reinhardt, H.W., Cornelissen, H.A.W. and Hordijk, D.A. (1986) *ASCE J. Struct. Eng.*, **112**, 2462.
Reinhardt, H.W., Cornelissen, H.A.W. and Hordijk, D.A. (1989) In: *Fracture of Concrete and Rock* (eds. S.P. Shah and S.E. Swartz), p. 117, Springer-Verlag, Berlin.
Rescher, O.J. (1990) *Eng. Fract. Mech.*, **35**, 503.
Rice, J.R. (1968) In: *Fracture* (ed. H. Liebowitz) **2**, p. 191, Academic Press, New York.
Rice, J.R. (1972) *Int. J. Solids Struct.*, **8**, 751.
Rice, J.R., Paris, P.G. and Merkle (1973) In: *Progress in Flow, Growth and Fracture Toughness Testing*, ASTM STP 536, p. 231, American Society for Testing and Materials, Philadelphia.
RILEM (1985) *Mater. Struct.*, **18**, 285.
RILEM (1990a) *Mater. Struct.*, **23**, 457.
RILEM (1990b) *Mater. Struct.*, **23**, 461.
Ringot, E., Ollivier, J.P. and Maso, J.C. (1987) *Cem. Conc. Res.*, **17**, 411.
Roelfstra P.E. and Wittmann, F.H. (1986) In: *Fracture Toughness and Fracture Energy* (ed. F.H. Wittmann), p. 163, Elsevier Applied Science, London.
Rokugo, K., Iwasa, M., Seko, S. and Koyangi, W. (1989) In: *Fracture of Concrete and Rock* (eds. S.P. Shah, S.E. Swartz and B. Barr), p. 513, Elsevier Applied Science, London.
Roger, S.A., Brooks, S.A., Sinclair, W., Groves, G.W. and Double, D.D. (1985) *J. Mater Sci*, **20**, 2853.
Rooke, P.P. and Cartwright, D.V. (1976) *Compendium of Stress Intensity Factors*, H.M. Stationery Office, London.
Rots, J.G. (1985) *Bond-Slip Simulations using Smeared Cracks and/or Interface Elements*, Research RPT., 85-01, Struct. Mech., Dept. Civil Eng., Delft University of Technology.
Rots, J.G. (1986) In: *Fracture Toughness and Fracture Energy of Concrete* (ed. F.H. Wittmann), p. 137, Elsevier Applied Science, London.
Rots, J.G. (1988) *Computational Modelling of Concrete Fracture*, Dissertation, Delft Univ. of Technology.
Rots, J.G. (1991) *Int. J. Fract.*, **51**, 45.
Rots, J.G. and de Borst, R. (1987) *ASCE J. Eng. Mech.*, **113**, 1739.
Rubinstein, A.A. (1991) *Int. J. Fract.*, **47**, 291.
Saito, M. (1984) *Int. J. Cem. Comp. Lightweight Concr.*, **6**, 143.

Saito, M. and Imai, S. (1983) *Am. Concr. Inst. J.*, **67**, 431.

Sakamoto, M. and Indrawan, B. (1995) *Preliminary Report of the January 17, 1995 Great Hanshin Earthquake*, Kajima Corporation, Kabori Research Complex, Inc., Tokyo.

Salih, A.L. and Aliabadi, M.H. (1994) In: *Localized Damage Computer-Aided Assessment and Control III* (eds. M.H. Aliabadi, A. Carpinteri, S. Kalisky and D.J. Cartwright), p. 185, Computational Mechanics Publications, Southampton.

Saouma, V.E., Ayari, M.L. and Boggs, H. (1987) In: *Fracture of Concrete and Rock* (eds. S.P. Shah and S.E. Swartz), p. 311, Springer-Verlag, Berlin.

Saouma, V.E., Broz, J.J. and Boggs, H. (1991) *J. Mater. Civil Eng.*, **3**, 219.

Schneider, U. and Diederichs, U. (1983) In: *Fracture Mechanics of Concrete* (ed. F.H. Wittmann), p. 207, Elsevier Science Publishers, Amsterdam.

Shah, S.P. (1988) *Mater. Struct.*, **21**, 145.

Shah, S.P. and McGarry, F.J. (1971) *ASCE J. Eng. Mech. Div.*, **97**, 1663.

Sih, G.C. (1973a) *Eng. Fract. Mech.*, **5**, 365.

Sih, G.C. (1973b) *A Special Theory of Crack Propagation in Mechanics of Fracture, Methods, Analysis and Solutions of Crack Problems*, Noordhoff, Groningen.

Sih, G.C. (1974) *Int. J. Fract.*, **10**, 305.

Sinclair, W. and Groves, G.W. (1985) *J. Mater. Sci.*, **20**, 2846.

Slate F.O. and Hover, K.C. (1984) In: *Fracture Mechanics of Concrete: Material Characterisation and Testing* (eds. A. Carpinteri and A.R. Ingraffea), p. 137, Martinus Nijhoff Publishers, The Hague.

Smith, E. (1994) *Mech. Mater.*, **17**, 369.

Stang, H. and Shah, S.P. (1986) *J. Mater. Sci.*, **21**, 21.

Stefanou, G.D. (1993) *Eng. Fract. Mech.*, **44**, 137.

Strange, P.C. and Bryant, A.H. (1979) *ASCE J. Eng. Mech Div.*, **105**, 337.

Streit, R. and Finnie, I. (1980) *Exp. Mech.*, **20**, 17.

Struble, L., Stuzman, P. and Fuller, E.R. (1989) *J. Am. Ceram. Soc.*, **72**, 2295.

Suidan, M. and Schnobrich, W.C. (1973) *ASCE J. Struct. Eng.*, **99**, 2109.

Sumi, Y., Nemat-Nasser, S. and Keer, L.M. (1985) *Eng. Fract. Mech.*, **22**, 759.

Sutcu, M. (1989) *Acta Metall.*, **37**, 651.

Swamy, R.N. and Hussin, M.W. (1989) In: *Fibre Reinforced Cements and Concretes* (eds. R.N. Swamy and B. Barr), p. 90, Elsevier Applied Science, London.

Swanson, P.L., Fairbanks, C.J., Lawn, B.R., Mai, Y.W. and Hockey, B.J. (1987) *J. Am. Ceram. Soc.*, **70**, 279.

Swartz, S.E. and Go, C.G. (1984) *Exp. Mech.*, **24**, 129.

Swartz, S.E and Refai, T.M.E. (1987) In: *Fracture of Concrete and Rock* (eds. S.P. Shah and S.E. Swartz), p. 242, Springer-Verlag, Berlin.

Swartz, S.E. and Taha, N.M. (1990) *Eng. Fract. Mech.*, **35**, 137.

Swartz, S.E., Lu, L.W. and Tang, L.D. (1988) *Mater. Struct.*, **21**, 33.

Swenson, D.V. and Ingraffea, A.R. (1991) *Int. J. Fract.*, **10**, 73.

Swift, D.G. and Smith, R.B.L. (1978) In: *Testing and Test Methods of Fibre Cement Composites* (ed. R.N. Swamy), p. 463, The Construction Press, Lancaster.

Tada, H., Paris P.C. and Irwin, G.R. (1973) *The Stress Analysis of Cracks Handbook*, Del Research Corp., Hellertown.

Tait. R.B. (1984) PhD Thesis, University of Cape Town.

Tait, R.B. and Garrett, G.G. (1986) *Cem. Concr. Res.*, **16**, 143.

Tait, R.B., Diamond, S., Askers, S.A.S. and Mindess, S. (1990) In: *Micromechanics of Failure of Quasi-Brittle Materials* (eds. S.P. Shah, S.E. Swartz and M.L. Wang), p. 52, Elsevier Applied Science, London.

Takaku, A. and Arridge, R.G.C. (1973) *J. Phys. D: Appl. Phys.*, **6**, 2038.

Tarng, K.-M., Chern, J.-C. and Chen, H.-W. (1991) *J. Chinese Inst. Eng.*, **14**, 173.

Thouless, M.D. and Evans, A.G. (1988) *Acta Metall.* **36**, 517.

Trustrum, K. and Jayatilaka, A. De S. (1983) *J. Mater. Sci.*, **18**, 2765.

Turner, C.E. (1973) *Mater. Sci. Eng.*, **11**, 275.

Turner, C.E. (1979) In: *Post Yield Fracture Mechanics* (ed. D.G.H. Latzko) Applied Science Publishers, Barking, UK.

Van Mier, J.G.M. (1989) In: *Micromechanics of Failure of Quasi-Brittle Materials* (eds. S.P. Shah, S.E. Swartz and M.L. Wang), p. 33, Elsevier Applied Science Publishers, London.

Van Mier, J.G.M. (1991) In: *Fracture Processes in Concrete, Rock and Ceramics* (eds. J.G.M. van Mier, J.G. Rots and A. Bakker), Vol. 1, p. 27, E&FN Spon, London.

Visalvanich, K. and Naaman, A.E. (1983) *J. Am. Concr. Inst.*, **80**, 128.

Ward, R.J. and Li, V.C. (1989) In: *Fracture of Concrete and Rock* (eds. S.P. Shah. S.E. Swartz and B. Barr), p. 645, Elsevier Applied Science, London.

Wawrzynek, P.A. and Ingraffea, A.R. (1987a) *Theor. and Appl. Fract. Mech.*, **8**, 137.

Wawrzynek, P.A. and Ingraffea, A.R. (1987b) *Engn. Comput.*, **3**, 13.

Wells, A.A. (1961) In: *Proc. Symp. Crack Propagation, Cranfield*, **1**, p. 210, College of Aeronautics, Cranfield.

Wells, A.A. (1963) *Brit. Weld. J.*, **11**, 35.

Wells, J.K. and Beaumont, P.W.R. (1985) *J. Mater. Sci.*, **20**, 97.

Weibull, W. (1939) *Ingen. Vetenskaps. Akad. Hand.*, No. 151.

Weibull, W. (1951) *J. Appl. Mech.*, **18**, 293.

Wecharatana, M. and Shah, S.P. (1982) *J. Struct. Div. ASCE*, **108**, 1400.

Wecharatana, M. and Shah, S.P. (1983) *Cem. Concr. Res.*, **13**, 819.

Weidmann, G.E. and Williams, J.G. (1975) *Polymer*, **16**, 921.

Weiderhorn, S.M., Freiman, S.W., Fuller, E.R. and Simmons, C.J. (1982) *J. Mater. Sci.*, **17**, 3460.

Williams, L.S. (1956) *Trans. Br. Ceram. Soc.*, **55**, 287.

Williams, M.L. (1957) *J. Appl. Mech.*, **24**, 109.

Williams, J.G. and Ewing, P.D. (1972) *Int. J. Fract. Mech.*, **8**, 441.

Wilson, W.K. (1966) In: *Plane Strain Fracture Toughness Testing of High Strength Metallic Materials*, ASTM STP 410, p. 75, American Society for Testing and Materials, Philadelphia.

Wiss, Janney Elstner Associates, Inc. and Musuer Rutledge Consulting Engineers (1987) Final Report: Collapse of the Thruway Bridge at Schoharie Creek, New York State Thruway Authority, Albany, New York.

Wittmann, F.H. (1983a) In: *Fracture Mechanics of Concrete* (ed. F.H. Wittmann), p. 43, Elsevier, New York.

Wittmann, F.H. (1983b) *Concr. J., Japan Concr. Inst.*, **21**, 19.

Wittman, F.H. (1985) In: *Application of Fracture Mechanics to Cementitious Composites* (ed. S.P. Shah), p. 593, Martinus Nijhoof Publishers, Dordrecht.

Wittmann, F.H. and Hu, X.Z. (1991) *Int. J. Fract.*, **51**, 19.

Wittmann, F.H., Roelfstra, P.E., Mihahashi, H., Huang, Y.Y., Zhang X.H. and Noniwa, N. (1987) *Mater. Strut.*, **20**, 103.

Wittmann, F.H., Rokugo, K., Bruhwiler, E., Mihashi, H. and Simonin, P. 1988 *Mater. Struct.*, **21**, 21.

Wu, C.H. (1978) *J. Appl. Mechs.*, **45**, 553.

Wu, X.R. and Carlsson, A.J. (1991) *Weight Functions and Stress Intensity Factor Solution*, Pergamon Press, Oxford.

Wu, Z., Yoshikawa, H. and Tanabe, T. (1991) *ASCE J. Struct. Eng.*, **117**, 715.

Yao, B. and Murray, D.W. (1993) *ASCE J. Struct. Eng.*, **119**, 2813.

Yu, Y.Z. (1981) *J. Water Conservancy*, No. 6 (in Chinese).

Yu, Y.Z. (1989) In: *Fracture Mechanics of Concrete Structures, from Theory to Applications* (ed. L. Elfgren), p. 355, Chapman and Hall, London.

Yuzugullu, O. and Schnobrich, W.C. (1973) *ACI J.*, **70**, 1973.

Zaitsev, Y. (1983) In: *Fracture Mechanics of Concrete* (ed. F.H. Wittmann), p. 251, Elsevier, New York.

Zaitsev, Y.B. and Wittmann, F.H. (1981) *Mater. Struct*, **14**, 365.

Zech, B. and Wittmann, F.H. (1977) In: *Trans. 4th. Int. Conf. on Structural Mechanics in Reactor Technology* (eds. T.A. Jaeger and B.A. Boley), Vol. H, J1/11, p. 1, European Communities, Brussels.

Zhou, L.M., Kim, J.K. and Mai, Y.W. (1992) *J. Mater. Sci.*, **27**, 3155.

Ziegeldorf, S. (1983) In: *Fracture Mechanics of Concrete* (ed. F.H. Wittmann), p. 371, Elsevier, New York.

Index

anchorages 273–279

Barenblatt's hypotheses 11–12, 14, 19, 30, 34–35, 168, 174
bond slip *see* reinforcement bond slip
borehole fracture test 263–265
boundary element method 131
break-off (BO) test 260–262
bridge plinth 257–260
brittle/ductile behaviour 14–15
brittleness number 143, 147

Castigliano's theorem 19–20
characteristic length 4–5, 12, 120, 122, 144, 168, 177, 250, 256, 279
compliance 5–9, 73–76, 80–84, 101
Cook–Gordon mechanism 66–67
crack band model 116–117, 124–128, 150–152, 265–266, 268
crack extension force 2, 5, 7–8, 20, 30–32
crack growth resistance 29–32, 37–38, 59–60, 147–148, 167–173, 176–180, 188–189, 223
crack opening displacement (COD) 16–20, 34, 163
crack opening mode 10–11
crack path direction 21–29, 149, 151, 201, 256, 260, 268, 278
crack path stability 28–29
crack profile 133, 135–136, 164
crack tip opening displacement (CTOD) 13, 16–20, 121, 132–135, 138, 151, 166, 179, 249

damage 43, 126
dams 251–257
discrete crack model *see* fictitious crack
ductile/brittle behaviour *see* brittle/ductile behaviour
Dugdale model 16–19

effective crack length *see* equivalent crack
engineers' theory of bending applied to
cementitious materials 116, 117–118
fibre reinforced composites 153–160
equivalent crack 116, 118–124

failure mechanisms
cementitious materials 51–63
fibre reinforced composites 63–69
interfaces in fibre reinforced composites 66–69

fibres
efficiency factors 105
critical length 63–66, 105
critical volume fraction 105–106
pull-out 30, 106–110, 111
reinforcing 40
snubbing 106
fibre bridging zone (FBZ) 44, 50–51, 69–70, 71–72, 97–101, 160, 164–166, 168–169, 171, 173–176, 179
fibre-reinforced composites
Type I 44–47, 110–111, 153, 212–216
Type II 47–50, 111–113, 153–180, 216–222
toughness 113–114, 157–158
fictitious crack 16–17, 19, 116–117, 119, 128–138, 150–152, 173–177, 208, 259, 268–270, 277, 280
finite element method 8, 125–130, 135, 160, 256, 259
fracture energy 3, 12, 18, 30, 34, 38, 43, 61–63, 78, 86–93, 113, 129–130, 132, 135, 138, 142–144, 149–152, 166, 179, 204, 207, 249, 256
fracture mode *see* crack opening mode
fracture path *see* crack path direction; crack path stability
fracture process zone (FPZ) 3–4, 8, 11, 12, 15, 17–18, 21–24, 30, 32–34, 37, 38, 41–44, 48, 50, 54, 69, 71–78, 92, 93, 97–98, 116, 119–121, 125–126, 145, 148–150, 152, 160, 163, 173–177, 179, 207–210, 251, 252, 259, 281
fracture stability 8–9, 145–147
fracture toughness 12, 13–15, 59–60, 112, 121, 145, 162–163, 166, 174, 263–265
fracture work *see* fracture energy

Griffith fracture *see* linear elastic fracture mechanics

in situ tests 260–265

J-integral 33–38, 78–79, 261–262

K-superposition 10–11, 131–138, 144, 150, 152, 161–177, 179

Levy's rule 252–254
linear elastic fracture mechanics (LEFM) 1–3, 9–16, 38, 138, 141, 148–150, 152, 249, 251, 254, 260, 265, 274–277, 281

LOK test 278

machine compliance 8, 9
macro-defect-free cement paste (MDF) 51, 60–63, 69
maximum energy release rate criterion 24
maximum strain energy density criterion 25
maximum tangential strain criterion 24
maximum tangential stress (MTS) criterion 24–26
mean strength 182
median strength 182
microstructure 39–41
mixed mode fracture 21–29, 148–152
monolithic structure see dams; bridge plinth
multi-cutting technique 73–76
multiple cracking 45, 47, 110–111, 213

non-linear fracture mechanics (NLFM) 32–38
non-local strain 94, 126–127
notch effect 145–146

Pareto distribution 188, 194, 198, 200, 216
pipes 279–280
plastic work 3
plastic zone 15
punching failure 265–266

reinforcement bond slip 267–273

shear retention factor 151
size effect 1, 5–6, 14, 17, 19–20, 27–28, 71, 90, 139–145, 177–179, 206–210, 259, 265–266, 275–277
slow crack growth 224–226
smeared crack concept see crack band model
snap back 118, 147
softening modulus 125
specific fracture energy see fracture energy
statistical strength
  bundles 211–212
  computer simulation 200–206, 218–222
  effect of crack growth resistance 188–189, 223
  expected number of cracks 197
  fibre-reinforced composites 211–222
  materials with crack growth resistance 188–189
  single phase material 184–189
  small specimens 186–187
  two-phase materials 190–206
strain softening
  bilinear 86, 119–120, 250

compliance methods 80–84
direct tension measurement 78
fibre reinforced composites see stress–strain curves
J-integral method 78–79
linear 84
mixed mode 93–97
Mode I 4, 40–44, 77–87, 125–126, 138, 250
piece-wise 86, 138
power law 84
stress–displacement curves
  cementitious materials see strain softening
  fibre reinforced composites see stress–strain curves
stress intensity factor 9–13, 119, 121–124, 132, 139, 163–164, 167–168, 188, 191, 201–204, 208, 216, 219, 224–226
stress locking 151
stress–strain curves
  cementitious materials see strain softening
  fibre reinforced composites
    Mode I 44–50, 101–104
    Type I 110–111
    Type II 111–113
surface energy 2–3

T-stress 20–21, 28, 29
tensile strength 4–5, 17, 90, 119, 138, 177, 249, 263, 277, 280
three point bend specimen 14
tied-crack model 54–55
time-dependent crack growth
  creep strain 233–234
  cyclic fatigue 224, 226, 228–229, 230–231, 244–248
  dynamic fatigue (constant stress rate) 224, 226, 227–228, 230, 239, 242–244
  single crack theory 226–229
  static fatigue (constant stress) 224, 226, 227, 230, 237, 240–241
  statistical theory
    single phase 229–235
    two phases 235–243
toughness index see fibre-reinforced composites, toughness

Weibull
  analysis 181–184, 186, 223
  modulus 183, 187, 188, 189, 190, 207, 212, 219
weight function 13

Young's modulus 2, 73, 102, 105, 153, 177, 249

For Product Safety Concerns and Information please contact our EU
representative GPSR@taylorandfrancis.com
Taylor & Francis Verlag GmbH, Kaufingerstraße 24, 80331 München, Germany